T4-AFY-595

SCIENCE TECHNOLOGY AND SOCIETY IN THE TIME OF ALFRED NOBEL

Other Titles of Interest

COGGIN, P. A.
Technology and Man

DANZIN, A.
Science and the Second Renaissance of Europe

ENCEL, S. & RONAYNE, J.
Science, Technology and the Public Policy

FELD, B. T.
A Voice Crying in the Wilderness:
Essays on the Problems of Science and World Affairs

GOLDSMITH, M.
Science, Technology and Global Problems:
Issues of Development. Towards a New Role for
Science and Technology

GVISHIANI, J.
Science, Technology and Global Problems

RADHAKRISHNA, S.
Science, Technology and Global Problems:
Views from the Developing World

SEGERSTEDT, T.
Ethics for Science Policy

STANDKE, K. H.
Science, Technology and Society:
Needs, Challenges and Limitations

STARR & RITTERBUSH
Science, Technology and the Human Prospect

THOMAS, D. B.
Science, Technology and Global Problems:
Integration of Science and Technology with Development

UNITED NATIONS OFFICE FOR SCIENCE AND TECHNOLOGY
Science, Technology and Global Problems

URQUIDI, V. L.
Science, Technology and Global Problems:
Science and Technology in Development Planning

Nobel Symposium (52nd: 1981).

SCIENCE TECHNOLOGY AND SOCIETY IN THE TIME OF ALFRED NOBEL

NOBEL SYMPOSIUM 52
HELD AT
BJÖRKBORN, KARLSKOGA, SWEDEN
17–22 AUGUST 1981

EDITORS
CARL GUSTAF BERNHARD
ELISABETH CRAWFORD
PER SÖRBOM

TECHNICAL EDITOR
ELISABETH HESELTINE

PUBLISHED FOR
THE NOBEL FOUNDATION
BY
PERGAMON PRESS

OXFORD · NEW YORK · TORONTO · SYDNEY · PARIS · FRANKFURT

U.K.
Pergamon Press Ltd., Headington Hill Hall,
Oxford OX3 0BW, England
U.S.A.
Pergamon Press Inc., Maxwell House, Fairview Park,
Elmsford, New York 10523, U.S.A.
CANADA
Pergamon Press Canada Ltd., Suite 104,
150 Consumers Rd., Willowdale, Ontario M2J 1P9, Canada
AUSTRALIA
Pergamon Press (Aust.) Pty. Ltd., P.O. Box 544,
Potts Point, N.S.W. 2011, Australia
FRANCE
Pergamon Press SARL, 24 rue des Ecoles,
75240 Paris, Cedex 05, France
FEDERAL REPUBLIC OF GERMANY
Pergamon Press GmbH, 6242 Kronberg-Taunus,
Hammerweg 6, Federal Republic of Germany

First edition 1982

Library of Congress Cataloging in Publication Data

Main entry under title:

Science, technology, and society in the time of Alfred Nobel.

Includes index.
1. Science—History—Congresses.
2. Technology—History—Congresses.
3. Science—Social aspects—History—Congresses
4. Technology—Social aspects—History—Congresses. 5. Nobel, Alfred Bernhard, 1833-1896. I. Bernhard, Carl Gustaf. II. Crawford, Elisabeth T. III. Sörbom, Per, 1940-
Q124.6.S385 1982 509'.034 82-11254

British Library Cataloguing in Publication Data

Science, technology and society in the time of Alfred Nobel.
1. Science—History—Congresses
I. Bernhard, Carl Gustaf
509 Q125

ISBN 0-08-027939-2

Printed in Great Britain by A. Wheaton & Co. Ltd., Exeter

FOREWORD

This volume comprises papers presented at Nobel Symposium No. 52, on 'Science, technology and society in the time of Alfred Nobel', which took place at Björkborn near Karlskoga from 17 to 22 August 1981. Also included are comments on the papers in each of five sessions prepared by invited discussants.[1]

The Symposium was held to mark the eightieth anniversary of the first awarding of the Nobel prizes in 1901. In choosing the theme for the Symposium, the organizers sought to gain insight into three different aspects of Alfred Nobel's work and time.

First, Alfred Nobel (1833–1896) lived during a period when science and technology, separately and concurrently, were instrumental in reshaping the daily life, work and thinking of large populations: hence, the intrinsic interest in examining the interfaces of science and technology as well as their impact upon society during a time which many have seen as the source of our present-day scientific and technological world.

Second, Alfred Nobel partook of the scientific and technological enterprise of his time through his invention of explosives. Knowledge about the scientific and technological strivings and about society in the time of Alfred Nobel is important to gain a perspective on his life's work and to understand its successes and failures.

Third, Alfred Nobel used his huge fortune to create the Nobel prizes, thus leaving a legacy which extended from his own time into the present. Since the prizes served to highlight and to legitimize developments in the disciplines concerned (physics, chemistry and physiology or medicine) during the late nineteenth century, they naturally enter into discussions of those developments. This is illustrated by several of the papers in Sections II and III of this volume. However, it also seemed fitting that a symposium that marked the first awarding of the prizes should examine the early history of the Nobel prize institution. This matter was discussed at a special session: its genesis, purpose and results are presented at the end of the present volume.

In this volume, the 'time' of Alfred Nobel is interpreted liberally and is not restricted to his lifetime. This is intentional; it results from the choice of participants, who have both general knowledge about science and technology in the late nineteenth and early twentieth centuries, and, more importantly, can treat specialized subjects that highlight the interdependence of science,

[1] Information about the organization and programme of the Symposium, as well as a list of participants, are given at the end of this volume.

technology and society. The time period covered in the papers extends roughly from the 1860s to the outbreak of the First World War. Depending on the object of his study and the concepts used, however, authors situate their work at different points within the general period. This volume contains many examples of such a 'conceptual', as opposed to a chronological, periodization; the study of 'industrialization', for instance, necessitates looking back at the early part of the period, or even before, whereas the authors who focus on its consequences – e.g. changes in managerial structure or in the engineering profession – naturally place their work at the end of the overall period.

Other authors have restricted themselves to specific time periods which, in retrospect, appear to be the most interesting to historians of the overall period. These have been characterized variously: for example, 'la belle époque', 'fin-de-siècle' and 'the turn of the century'. The latter term not only has great utility, in that it can be stretched to cover several decades, but it is also a pertinent one in the present context, since the turn of the century constituted the high point of such new specialities as physical chemistry and biochemistry. Although 'fin-de-siècle' covers the same time period, in popular usage it denotes the mood of world-weariness – probably best described by the term *Kultur-pessimismus* – that some writers felt was characteristic of the closing of the past century. The term is also carried over to physics and is used to characterize a set of attitudes toward scientific knowledge – 'descriptionism' being chief among them – which emerge from the writings of physicists who were taking stock of their discipline at the closing of the century.

To the historian of science and technology in the 'time' of Alfred Nobel, it seems proper and natural to restrict the analysis of scientific and technological ideas, theories and innovations to the specific time period under study. Whenever necessary, however, he goes back in time in order to trace their origins. For the working scientist who examines the history of his own discipline, it may seem equally natural to extend the time perspective forward, and to base his evaluations of ideas, theories and innovations, at least in part, on what that science foretold of the present-day state of knowledge and its applications. In this volume, both perspectives are represented, although the latter is a minority one.

This confrontation of viewpoints was also intended by the organizers: The description of the Symposium that accompanied the letters of invitation was as follows: 'By bringing together some thirty-five scholars, who have all been engaged in studies of different aspects of science and technology in the late nineteenth and early twentieth centuries, we hope, not so much to arrive at a synthesis of the current state of knowledge about these matters (this being perhaps overly ambitious) but to facilitate an exchange of viewpoints across disciplinary boundaries.' As this volume shows, the hope for the latter was amply fulfilled. Although the papers do not add up to a synthesis of science, technology and society in the time of Alfred Nobel, they nevertheless point up the richness of problems for research – as well as of approaches and materials to treat them – that this period offers the historian of science and technology.

Stockholm and Paris, November 1981

CARL GUSTAF BERNHARD
ELISABETH CRAWFORD
PER SÖRBOM

CONTENTS

IV The interchange of technology and society: 1860–1914

V Specialized aspects of scientific and technological change

Round table: The use of archival materials concerning the Nobel prizes in science and medicine for research in the history of science: Problems and methods

Appendices

CHAIRMAN'S INTRODUCTION

Today, the Nobel prize is one of the few international awards known by name to a large part of the non-scientific public, and it is probably the only prize that almost every active scientist knows about. Considerably fewer people know about the man behind the prizes, and very few have any idea of his connection with Björkborn, Karlskoga, i.e. the place where the Nobel Foundation, in August 1981, arranged an international Nobel Symposium on 'Science, Technology and Society in the Time of Alfred Nobel'.

When Alfred Nobel's will was to be executed, the question arose as to whether France, Italy or Sweden should be regarded as the home of this lonely cosmopolitan bachelor, known as the 'most wealthy European vagabond' – born in Sweden, educated in Russia and with residences in Paris, San Remo and Björkborn.

For several reasons – apart from the legal aspects – Björkborn came to be regarded as his domicile, although it did not come into his hands until two years before his death in San Remo, Italy, in 1896. It is noteworthy that he had his beautiful and expensive Russian horses – bought in St Petersburg – moved to Björkborn from Paris where he used 'se promener en voiture en le Bois de Boulogne'. In between his frequent international and intercontinental business trips, Nobel might be found in the winter-garden of his elegant home on the Avenue de Malakoff, entertaining a select group of international guests, in – if necessary – five languages.

This summer happens to represent exactly 100 years since Nobel built himself a well-equipped laboratory in Sevran, near Paris, which offered facilities for his varied experimental ventures into chemistry, physics and even human physiology. The new laboratory was far better suited for those purposes than the laboratories that had mainly been used for the development of explosives. In general, most of Nobel's inventions served to further technical developments, and the applications of his inventions resulted in the deposition of 355 patents.

His dangerous experiments on explosives were – in the beginning – carried out with an almost unbelievable single-minded optimism and lack of theoretical knowledge, and they took many lives, including that of his youngest brother. These experiments were performed in various parts of the world, the first being done in Stockholm in 1863 in cooperation with his father, Immanuel Nobel. They led him from the practical application of nitroglycerine, including an appropriate ignition device, via several intermediate steps to the invention of dynamite or 'security explosive', patented in 1867.

Although Nobel subsequently directed his interest to the question of security, the severe accidents that occurred in various countries brought forth protests and temporary prohibitions. In addition, the marketing of his products gave rise to endless patent disputes. Powerful persistence and an inventive fervour carried him through these adversities. A new era had begun, with marvellous possibilities for opening mines, clearing ground for cities and building roads, tunnels and channels on the ground and under the water – but also for military applications. Indeed, Nobel's inventions in explosives led to a worldwide industry – one of the very first multinational enterprises – and he became a rich man. It had been a great adventure from the first experiments (for which he and his father were honoured by the Royal Swedish Academy of Sciences in 1868) to the work in the laboratory in Bofors. It was built by Nobel in 1894 for the development of, among other things, artificial silk and rubber and for studies that would serve industry in Bofors, which is involved in modern defence technology.

However, these thirty years of adventure must have left many mental scars, which, together with the pathos and even tragedy of his relations with the opposite sex, complicated the life of this sensitive and lonely man – an inventor of explosives who nourished literary ambitions and had a deep interest in the peace movement. Was his large donation – his last gift to society – to be regarded as a sign of penitence or as the erection of a monument for eternity? We shall never know.

In his emotional life, the pendulum was always swinging between optimism and deep pessimism; he often tried to hide his sensitivity behind ironic or even cynical remarks. A complicated man, whom no thoughtful person can dismiss simplistically as just another inventor; but who rather deserves our sympathetic understanding. He was a man of great charm and winning ways, although – it seems to me – not altogether an appealing personality. But who is?

We are all children of our time. Although Nobel had outstanding ability and the power to rise high above the norm of contemporary human achievement, he could not help being influenced by the habits, taboos and strivings of his time, 'la belle époque', which, to us, in our world, does not seem altogether attractive either. Knowledge about the scientific and technological strivings and the society of his time would be helpful in better understanding Alfred Nobel; that was the topic of the Symposium, arranged eighty years after the first Nobel prizes were awarded.

During the preparation of this Symposium, the Chairman of the Organizing Committee, Professor Sten Lindroth, passed away. We feel deeply the sad loss of a most dear friend, a remarkable person and an outstanding scholar. Knowing his enthusiasm and strong sense of commitment to the success of this Symposium, I felt my own responsibility in taking over the Chairmanship to be all the greater. It is in this capacity that I thank all those who have contributed to the Symposium and to this volume.

The Symposium was made possible by the gracious support of The Bank of Sweden Tercentenary Foundation, the Nobel Foundation and the Bofors Company.

CARL GUSTAF BERNHARD
The Royal Swedish Academy of Sciences, Stockholm, November 1981

I

Changing conditions of scientific work in the late nineteenth and early twentieth centuries

DIVISION OF LABOUR AND THE COMMON GOOD: THE INTERNATIONAL ASSOCIATION OF ACADEMIES, 1899-1914*

BRIGITTE SCHROEDER-GUDEHUS

Institut d'histoire et de sociopolitique des sciences, Université de Montréal, Montréal, Province of Quebec, Canada

'Ours is an age of organisation.' Speaking to the National Academy of Sciences in Washington at its Semi-Centennial Anniversary in 1913, the Secretary of the Royal Society addressed himself to the problems and prospects of international cooperation in research.[1] In Arthur Schuster's view, scientific research had entered upon a new era, in which the collaborative progress of knowledge could no longer be entrusted exclusively to the informal interaction between individual scientists nor to the uncoordinated efforts of a constantly increasing number of independent international associations.

During the preceding decades, international scientific collaboration had changed considerably. The expansion of international activities had been spectacular, but the changes had been even more significant regarding their form and structure. The handful of international scientific associations that existed in the middle of the nineteenth century had grown to more than five hundred in 1913. Whereas only one or two scientific congresses had been held per year in the 1850s, by the turn of the century their number had reached an annual average of thirty.[2] In many cases, cooperation had emerged in response to needs arising from scientific and technical progress, and much of this cooperation was undertaken at the initiative and upon the responsibility of governments.[3] At the same time, there was no indication that this prodigious development of organized collaboration had caused any decline in the traditional patterns of scholarly relations, such as personal contacts and correspondence, exchange of publications, study and research abroad, etc.

International scientific collaboration had thus developed into a network of spectacular breadth and diversity. It is difficult to imagine how this phenomenon could be completely recorded, let alone systematically evaluated. The absolute importance of international collaboration for the advancement of scientific knowledge is hardly ever called into question, although -- to the best of our knowledge -- this importance has not been systematically evaluated for specific fields or specific periods.[4] Interesting as it may be, it is not the only problem

*Translated from the French by Alan Duff.

that comes to mind when one considers the impressive expansion of international activities in the late nineteenth century.

It does indeed seem rather peculiar at first sight that this phenomenon of massive 'internationalization' should have occurred at a period of increasing international tension and the gradual exacerbation of the great power rivalries. Public opinion was deeply involved in these rivalries which, in some instances, had become important elements of collective self-awareness and of a national sense of purpose. One should not, of course, presume that leading scholars in universities or research institutions – the social group with which this paper is concerned – were necessarily unanimous on specific matters of foreign policy. It is, however, fairly obvious that few, if any, would have taken exception to the principle that the nation's power and prestige should be given highest priority. It was a matter of self-awareness, social recognition, and – with science policy still in its infancy – of professional interest, when scientific communities in all countries were endeavouring to see their contribution to the fatherland's strength and glory properly recognized by the government and the public.[5]

As the so-called 'science-based' industries began to prove their importance in the struggle for world markets, the argument that scientific pre-eminence was crucial to great power status became all the more convincing. The Franco-German War had demonstrated the value of advanced technical equipment, and learned opinion in France was quick to point out that the defeat was largely due to the lack of attention – and money – given to scientific and technical instruction. Scientific capabilities emerged as a military asset. No less a figure than Bismarck, speaking in 1894, declared that the chemists and technologists showed greater merit in helping to maintain peace than did any pacifistic predisposition on the part of governments. For, in his opinion, governments were merely hesitating to unleash a war until they could be assured of having control over the latest discoveries.[6] It is true that a close link between research and industrial or military implementation existed in only a few scientific and technical disciplines, but no reluctance was shown in making claims for the great utility of the products of all branches of organized knowledge, whether those products were economic or cultural, already in evidence or still far off. Service to the nation provided the basis for social legitimation, and the State consequently found itself enjoined to finance research for the sake of the national interest.

It was inevitable, then, that such declarations assume nationalistic overtones – one of the favourite arguments being that without adequate backing for research the country was in danger of being outstripped by the scientific advances of other nations. Allusions to international competition were freely couched in belligerent terms: 'confrontation' was, of course, peaceful, but it was nonetheless necessary to 'fight'; to an English observer, the high level of training in Germany was a 'precision weapon';[7] the World Exhibition of 1900 was 'the *Sedan* of French industry';[8] and the Germans, quite unperturbedly, listened to the claim that their country's greatness rested upon two pillars – the army and science.[9]

How was it possible to reconcile this attachment of science to national interest with the commitment to internationalism of science and the scientists' day-to-day experience of common purpose and common standards? There is no doubt that the norms of science had an international constitution. If any proof were needed, it is certainly provided by the institution of the Nobel prizes: the very possibility of creating the award depended entirely on the worldwide recognition and acceptance of the criteria the awarding body would bring to bear on its decisions. Ought there not to have been at least a potential contradiction between the patriotic commitment involved in pursuing the national interest and the fervent enthusiasm for international collaboration attested to by the growing number of meetings, associations and projects and the speeches that accompanied this collaboration? The dilemma of the scientist-citizen has been the subject of historical studies, the contents or conclusions of which we cannot undertake to summarize here.[10] It is no mere chance, however, that such studies should all be concerned with periods following the month of August 1914. The First World War had succeeded in shattering the belief that science, by its universality, kept its practitioners 'out of the fray'. In fact, in all countries engaged in the War, scientists participated enthusiastically in the war effort; those who could not expect their competence to be of immediate economic or military utility did not need to disarm, but threw themselves at the front of the psychological war. The impassioned eloquence of the possessors of higher knowledge gave rise to a literature of which the scientific collectivity has little reason to be proud.[11]

The War by no means wiped out of the recollection of the warmongering professors the memory of international cooperation. The proliferation of congresses and associations was recalled at times with nostalgia, at times with bitterness or even scorn. Should one then adopt the outlook of those who were witness to the events? Is it sufficient to study the rapid expansion of international scientific collaboration before the War to discover those of its elements that would enable us to assess the authenticity of the internationalism by which this growth seemed to be inspired? Would not this enable us to gauge the extent of the dilemma in which those who advocated internationalism found themselves?

To adopt this approach would, in my opinion, mean following the wrong path. Today's habit of viewing questions of international scientific collaboration, explicitly or implicitly, against a background of the norms of universality and communality incurs the risk of distorting our instruments of analysis. It does not, in fact, appear that the majority of those living at the time viewed the problems of international scientific collaboration either exclusively or even essentially in terms of alignment along nationalist or internationalist positions. The question most widely discussed at the time was quite different, lying closer to concern with the production of knowledge than to reflection on world order: this question was raised by the transformation that scientific activity had undergone. The advances in knowledge and in social organization had caused what had been for generations an integrated branch of knowledge to be split up into myriad disciplines and specializations. Scientific activity, which had formerly been a vocation open only to a minority, was now in the process of becoming a source of employment for a great number of people, who were required to have neither brilliance of mind nor the power of synthesis but merely competence in a clearly restricted domain. Whether or not one deplored this fragmentation or resolutely accepted its consequences, research had ceased to be viewed by many as an essentially individual occupation. Henceforth, it was to obey the principles of the division of labour and to become an organized activity, a *Grossbetrieb* (a grand enterprise). In the light of these discussions, the various forms of international scientific collaboration appear primarily as prolongations at the international level of organizational requirements that had initially appeared at the local level. One is tempted to say that the international dimension of the collaboration thus organized – far from being the principal issue at stake in the discussion – was no more than an accessory feature, although, as we shall see, it need only be placed in a broader intellectual and political context for things to appear less simple. The purpose of this paper, however, is simply to put forward a new perspective. It is an attempt to probe the dynamics of international collaboration without isolating it from the intellectual, political and social texture of scientific life in a given period.

A Formal and Organized Corporate Undertaking: The International Association of Academies

In order to keep the discussion within manageable limits, I shall concentrate on one instance of organized international collaboration, the International Association of Academies (IAA). Arbitrary as it may appear, the choice is justified, since it would be difficult – short of examining a large number of organizations – to find a more appropriate example. The national academies (or their equivalents) were, of course, not altogether similar in their mandate, structure, composition, style and – last but not least – their relation to the state.[12] Yet they were comparable. They were expected – and accustomed – to adopt a position with respect to the workings of science, research and higher education by regarding those branches both from a general point of view and in their relations with society as conceived by the academy members. As a result, the academies have left to posterity considerable quantities of source material and documents.

The decision to study an association of academies could certainly be questioned, on the grounds that at the end of the nineteenth century academies were generally rather removed from actual research work, so that they could scarcely be equated with 'science' or with 'scientists' as such. That is true. However, scholars were appointed to membership in academies in recognition of a scientific achievement or a social position, or both. In that respect – the differences in prestige notwithstanding – academies can be considered to reflect important aspects of scientific thought and practice at a given time.

The creation of the IAA in 1899 was both an expression of the tendency towards international organization and a reaction against certain aspects of that tendency; and, viewed from either angle, it was a significant event.[13] In the area of cooperation in the basic sciences and in the humanities, it was the first 'umbrella' organization of a non-governmental type. The founding members included the most important academies on the European continent, as well as the Royal Society, and the National Academy of Sciences in Washington; subsequently, other academies joined as well. By the eve of the First World War, the IAA had twenty-two members.[14] The structures of the organization were extremely flexible; the Council met annually, the General Assembly every fourth year. The IAA's official mandate was very general: to stimulate international scientific cooperation, to encourage contacts among scientists of different countries and – through appropriate discussion and coordination – to foresee and hence forestall 'regrettable collisions', i.e. the overlapping of work. In 1914 the scale of its activities was unimpressive. Having no national juridical status, the IAA was not entitled to receive donations and therefore possessed neither the authority nor the funds necessary to bring effective influence to bear in determining the direction of research. Later, particularly after the outbreak of the War, it was believed that a further reason for the failure of the IAA was to be found in the lack of cooperative spirit among its members: the French attacking the Germans for their disproportionate over-representation,[15] and the Germans reproaching the French for their jealous nationalism. The IAA did not survive the War.[16]

The frenzied search that later ensued to unearth from the lacklustre history of the IAA arguments to justify the exclusion of the central powers from the International Research Council has caused important features of this experience to become obscured. What in fact was involved was a first attempt to centralize the coordination of international scientific activities, i.e. to organize those activities internationally in keeping with the 'great international scientific enterprise' (*internationaler wissenschaftlicher Grossbetrieb*).[17] It is there that the main interest of the Association lies; and it is there, too, that the chief motives of the founding members must be sought. There was, of course, no lack of internationalist rhetoric in the history of the IAA. Even before its official foundation, and, later, from one general assembly to another, the idea of a science without boundaries was exalted in official speeches. It was rare for anyone to mention the IAA without at the same time recalling the impressive succession of great minds who had foreseen – indeed ardently desired – the emergence of a federation of academies: Francis Bacon, Gottfried Wilhelm Leibniz, Auguste Comte, Alexander von Humboldt. . . .[18]

Nonetheless, international scientific collaboration was mainly presented as attending to more immediate needs, such as those of reaching agreement over nomenclature and units of measurement; coordinating observations made simultaneously at different places using identical instruments and methods; reducing expenditures through joint participation in large-scale projects;[19] and, finally, cushioning the impact of an avalanche of scientific information and making possible its rational utilization. It was readily recognized within the academies how greatly this development in international scientific collaboration derived from the changes that had occurred in its material condition, how much it had benefited from the improved efficiency of printing and publishing techniques and – last but not least – how greatly it had profited from the more substantial funds that had ultimately become available to certain academies.[20]

The academies had always maintained relations with each other, and there was no question of minimizing this secular tradition. There seems, however, to have been considerable agreement to the effect that those traditional relations represented no more than a semblance of cooperation, with, for instance, the nomination of foreign associate members being considered a mark of high esteem rather than an engagement to collaborate. As a French Academician observed regretfully in 1879: 'Apart from these reciprocal gestures of politeness, each remains at home.'[21] The IAA's objective, therefore, was to replace such largely ritualistic international relations with a '*formal and organised* concertation of the scientific endeavours of all nations'[22] [my italics]. It was by anchoring those relations to a permanent structure of cooperation that the founders of the IAA hoped to establish the preconditions indispensable to the implementation of large-scale projects.

The projects that the IAA initiated, sponsored or endeavoured to supervise were of an astonishing diversity, but none of them was completed before the War put an end to the Association.[23] One should not be surprised to find scarcely any signs of impatience at the slow progress of these projects. The partners in this cooperation were realists, and the implementation of large-scale projects was not their sole aim; they had other aims as well, although the statutes – with customary discretion – make scarcely any reference to them.

One of their objectives was of a financial nature, i.e. to discover potential sources of financial backing and to support those who approached such backers by bringing the prestige of the Association to tip the scales in their favour. The financial background to those projects presented an indeed remarkable mosaic. Thus, for instance, the production of an encyclopaedia of Islam was financed not only by contributions from member academies involved in the project but also by other academies, such as that of Algiers, and by the German Society of Oriental Studies, the Senate of the Town of Hamburg, the Dutch colonial administration, and the Italian government. The Association, or its subcommissions, did not hesitate to make direct approaches to political authorities; it was thus, for instance, that they succeeded in obtaining from several Indian princes the sum of £800 for the edition of the Mahâbhârata. . . .[24]

The Centralized Coordination of International Activities

There can be no doubt, then, that the prestige of the Association was exploited in the quest for financial backing.[25] The founders' calculations were not, however, confined to the purely financial aspects of support. In this context, by seeking financial resources, the academies were raising the entire problem of their position within the national scientific communities. In a memorandum of 1892, i.e. during the preparatory period of the IAA, a member of the Viennese Academy presented three main arguments in favour of inter-academy collaboration: such collaboration would make it possible to avoid parallel endeavours; it would enable financial resources to be pooled; and it would also make it possible to tap public funds more profitably. In elaborating the latter point he said:

> 'International coordination of research has become imperative, and the fact that the Academies have not yet succeeded in reaching agreement on this issue has already led to the creation of independent organizations – in the form of scientific congresses or unions, for instance. These bodies are constantly undertaking increasingly numerous projects and thus appropriating an ever-growing amount of the public funds which the governments might more readily have apportioned to their Academies. Have not the Academies been specially created to assist the State in all that it would like to see undertaken for the encouragement and advancement of scientific research? The more financial backing the State accords to extra-academic circles and projects, the less money will be left to fund the Academies.'[26]

The Association emerged, then, as an endeavour by certain established scientific élites to reinforce – by claiming an internationally privileged role – an authority that was being undermined at home by the growth and differentiation of national scientific communities, the creation of extra-university research institutes, and particularly by the development of learned societies.[27] All of this indicates how necessary it is to interpret the creation and functioning of the IAA not only as an outcome of the intrinsic internationality of scientific activities but also as a reflection of the tensions present in the very heart of the different national scientific communities. The situation did indeed vary from one country to another, and it would be far too sweeping to claim that all learned societies came into being and developed in competition with the academies. But such societies were closer to the evolution of disciplines, more directly involved with economic and social life, and hence readier to adapt to its demands and to seize the opportunities presented.[28] Since they were more representative of scientific life than the academies, those societies succeeded in taking their place beside the academies and the universities in discussions between science and the state. And the resultant unease of the academies, particularly in the Germanic world, is reflected in documents: in official declarations and ceremonial speeches, great insistence is placed on the mission of the academies, which alone were capable of conceiving and directing large-scale research projects; likewise, alone were capable of preserving the unity of all the sciences, and of countering the harmful fragmentation of disciplines and specialities. All of this, in the opinion of the academies, should have been sufficient reason to merit public respect and governmental support.[29]

The upsurge of international activity, and particularly the impressive increase in the number of congresses, associations and other international bodies, was due far more to the initiative of the learned societies than to that of the academies.[30] Thus, by becoming an international body, the academies could entertain the hope of taking the lead, of retaliating, and – by rebound effect – of reassuming authority at the national level. Consequently, the pressing proposal to establish centralized coordination of international congresses and associations through the intermediary of the IAA became incorporated into the strategy of redefining the relations of power within the scientific communities. It was only logical, then, that the academies should dream of obtaining from governments the right of scrutiny over decisions made by the public authorities to whom the learned societies and other scientific bodies regularly turned in search of moral or material support. Throughout the history of the IAA, ideas similar to those expressed by Friedrich von Hartel in 1892[31] continued to be put forward. In 1899 Joseph Lister, then President of the Royal Society, stressed the advantages of a mechanism 'by name of which suggestions made for international cooperation in scientific inquiries could be thoroughly discussed by the leading men of science from a purely scientific point of view, before definite proposals are made with a view to official action by the governments of the countries concerned'.[32]

The IAA was capable of providing such a mechanism, and certain persons, such as George Ellery Hale, went so far as to wish to see it exercise functions of control. While inveighing bitterly against the American Academy for its indifference towards the IAA, he spoke with unremitting enthusiasm of the IAA's powerful influence and, in 1913, of the imminent measures 'which would bring the Association into a still more commanding position in international science'.[33]

Of course, not all the members of the IAA entertained such powerful ambitions for the Association. In fact, a draft resolution by the French academies to the effect that any project for international cooperation should be obliged to pass through the IAA ran into opposition from the Academy of Berlin, which claimed that such a ruling would place the IAA in a monopolistic position – and that was a move of which this Academy did not approve.[34] The spirit of collaboration, therefore, did have its limits. And those limits began precisely at the point at which collaboration interfered too much with freedom of movement. One may, however, judge the seriousness of purpose behind the French proposal in the light of what was to happen after the War. Although the idea of a central coordinating authority was not adopted before 1914, it nevertheless triumphed in 1919 with the creation of the International Research Council, in which the French Academy exerted a decisive influence.[35]

The IAA – and in fact the entire network of international scientific organizations at the time – did not emerge, then, as a natural outgrowth engendered by the internationality of scientific knowledge; nor was it, simply, a response to the needs created by the spectacular growth, differentiation and increased rapidity in the transmission of knowledge or the improvement of means of communication; it was also a response to the requirements of socio-professional strategies – both those employed within the scientific communities and those directed towards the governments and the public. Documents and literature abound in references of this kind. By comparison, the conflict between patriotism and internationalism – the potential dilemma of the scientist-citizen of which I spoke earlier – is relatively more difficult to document. The vision of a fraternity of science, rising above national boundaries and offering an inspiring example to politicians, is relatively common, but such references are found mainly in ceremonial speeches, obituaries[36] or in the writings of committed pacifists. Is one to say that the development of international cooperation and of a formal, organized association was regarded by scientists at the time as being beyond any political consideration? Undoubtedly it was not, even though the connections with politics remained well above the problems of personal conscience. It was by considering itself an *organization* that international scientific collaboration acquired its political dimension.

The Era of Organization

Interdisciplinary differentiation, division of labour, and the transformation of research into a collective, organized activity – such were the topics of discussion among men of science, and indeed of all disciplines, at the turn of the century.[37] This transformation was not greeted with equal enthusiasm by all members of the scientific body. Notwithstanding, even those who deplored the fragmentation of science into disciplines and specialities could not gainsay the need for mass-scale observation and hence for the introduction of division of labour; even those who maintained that the individual alone was capable of creative effort were forced to admit the importance of concerted effort;[38] and even those who feared that concerted effort would lead to an inundation of the creative spirit by mindless factual information had little more to contribute to the discussion than more or less outspoken admonitions.[39]

The introduction of the principle of the division of labour in scientific operations was not limited to the natural sciences but was extended to the humanities. Indeed, scholars in the humanities had actually played a decisive role in the foundation of the IAA, which, at the outset, sponsored the great lexicographic projects and encyclopediae. One should not be surprised by the fact that differences of opinion on the subject of organization did not follow disciplinary lines. The proponents of the divison of labour claimed that scientific activity should be opened up to considerations of efficiency and rational organization. The formulation of these statements seems to indicate that they were as much inspired by the concern for seeing these claims put into practice as they were by the concern to give them prominence in the public image of science. The idea stressed constantly is that by viewing its organization and internal functioning from the point of view of efficiency and productivity, science is in tune with the mainstream of modern civilization, participating in this way in the progress of the whole nation and offering its own specific contribution to the common good. In accepting the differentiation of functions, hence the division of labour, as an internal principle of organization, science is giving proof of its willingness – and ability – to act together with the dynamic elements of modern society.

The archives of the academies themselves – particularly the Academy of Berlin – afford sound substantiation for these remarks. The subject of the organization of research was widely broached in speeches and discussions; endorsement of the principle of the division of labour was often couched in arguments for efficiency and social utility; and, here and there – beneath the somewhat overemphatic assurance of the language – one may detect concern to prevent the isolation of intellectuals from the national community. In fact, by stressing the need to coordinate the preparatory work for large-scale projects and to ensure long-term continuity, the academies sought to put in the limelight their organizational abilities, i.e. the services that they alone could offer to science and society.[40] This intent often led to a remarkably frank stressing of the social nature of scientific undertakings: 'Science, too, has its social problem', said the secretary of the philosophy/history section of the Academy of Berlin, in 1890.

> 'Just like the *Grosstaat* [big State] and *Grossindustrie* [big industry], *Grosswissenschaft* [big science] has shown itself to be an indispensable element of our cultural development, of which the Academies are – or should be – the legitimate guardians. . . . *Grosswissenschaft* needs circulating capital as much as *Grossindustrie.* If this capital is lacking, the Academy is reduced to a mere decoration, and we can hardly complain if the public . . . considers us superfluous.'[41]

This passage is interesting because it shows how the problem of short-term utility is discreetly sidestepped. Furthermore, the fact that those remarks were addressed by an ancient historian (Theodor Mommsen) to a church historian (Adolf Harnack) underlines the political profitability of the analogy with industry. The popularity of references to industry, or to the 'factory',[42] remains astonishing even if one takes into consideration the interest that could be aroused at the time by the scientific organization of industrial labour. In Germany, even more than elsewhere, this popularity was great, no doubt because the country's economic performance had been decisive to the formation and advance of the Empire. It was, then, an important feature of collective experience and a cause for pride. It is no surprise that the researcher as an individual should have counted for little in this conception of scientific advance. Provided that the system could succeed in getting everyone to work to the best of his ability and specialist skill, the vision of a research *Betrieb* on the industrial model seemed acceptable. The favourable response – and obvious success – that the idea of the rational organization of scientific activities met with in Germany was to provide an inexhaustible source of inspiration for the denunciatory literature of the War years.[43]

However, as is shown by the words of Arthur Schuster quoted at the beginning of this paper, enthusiasm for the rational organization of scientific activity was far from being confined to the Germans. Within this context, overriding national allegiances emerged as a condition for the effectiveness of the international division of labour:

> '. . . the number of skilled observers living at any one time is not great, and they are spread over many lands,'

wrote Sir Michael Foster in 1884 (choosing as illustration to his remarks the observation of an eclipse of the sun).

> 'The problems of the future must be faced by the best men, and why should not these men work together? Why should not the best men be selected – now an Italian, now a German, now a Frenchman – because they are best to do the work for which they are best fitted? It is only in this way that we can get the best work done in the future.'[44]

The English physiologist was convinced that only an organization, and more specifically an international organization, could enable the sciences to advance; this was a conviction he drew from the very teachings of his own discipline:

> 'One of the most salient features of animals is a division of parts whereby each part does its best to fulfil the work required of it. On the other hand, all the parts of the body are so united that every part works for the common good. Just as in the body politic there are laws and unwritten customs which regulate the actions of the members, so also with the workers in science. Differentiation has proceeded to a great degree amongst scientific workers; each inquirer has now to limit his inquiries not only to one science, but to one part of that science, and there is no doubt that in the future division of labour will have to proceed still further.
>
> 'So much for division; but what about integration? . . . human wit may well devise some tie that will bind all the workers of the world together by one indissoluble knot. What is wanted in science is organisation. . . .'[44]

It is rare to find the principles of the physiological division of labour, functional adaptation and the harmonious coordination of functions used with such directness to demonstrate the advantages of international scientific cooperation;

yet it is equally rare not to find this same reasoning cropping up whenever the issue of research organization is raised. The corollary to the division of labour – i.e. the harmonious integration of partial contributions – had a chosen place in the reasoning of the academies. Was not their mission not only to organize scientific work but also to use their unique ability to thwart centrifugal tendencies?[45] The more the academies continued to remind their governments that their task consisted both in organizing research – that is, in distributing the various tasks – and in safeguarding the coherence of the branches of knowledge, the more the IAA became represented as the ideal organization to meet the requirements of coordination and integration whenever the task lay beyond the means of a single academy.

Conclusions

Scientific internationalism – the 'ideology' that was supposed to enable men of science to benefit from the advantages of international coordination, collaboration and validation, yet to remain irreproachable patriots[46] – advanced all the more easily in that the two poles of potential tension were separated by a seemingly neutral no-man's land: common enthusiasm for the rational organization of scientific work.

One need only look closely at the texts, then, to realize that scientific internationalism was neither conceived nor regarded as a force to offer opposition to constraints that might one day or another bear upon the individual at the political level. Naturally, in speculating on the future of international collaboration, ready expressions of confidence were heard with regard to the ability of science to keep open the channels of communication among the nations if ever the day should come when the ways of diplomacy were blocked. The reference to diplomacy is noteworthy. Men of science did not hesitate to perceive the possibility of a divergence between the postulates of science and the imperatives of their nation's diplomacy, i.e. the *technique* of conducting foreign relations as compared with the substantive issues at stake. It cannot be merely a matter of style or wording. Except in the writings of outspoken pacifists, there are scarcely any statements to be found about the bridges of science holding up in the event of conflicts involving national interest.

In addressing the First General Assembly of the IAA in Paris in 1900, the president of the Institut de France described the mission of men of science as being 'to replace the struggles of war by those of work'. However, so little faith did he have in this mission that, in the same breath, he proposed an alternative one, which was to console humanity in the interim by offering it a glimpse of the splendours of a distant future. . . .[47] In 1906, in a commemorative speech, the Rector of the University of Berlin (who was himself an active proponent of international collaboration through the IAA) explained what was to his mind the motive force behind this collaboration: 'to gain an *international* training in the *national* interest.'[48] '*Par la science, pour la patrie*' (through science, for the country) was the catchword of the French Association for Scientific Advancement. One could not find a better definition of pre-War scientific internationalism as it was generally understood.

On the practical side, there are abundant examples that show how, for the greater good of the undertaking, the participants succeeded in setting aside their concern for national prestige. There were other instances in which national vanities — the desire to contribute to the glory of the country — led either to failure of projects or to delay in their realization. Although political considerations emerged only discreetly on official occasions, they were more freely aired in the less cushioned atmosphere of internal discussions. As illustration, one example — culled from the history of the Berlin Academy — must suffice: the projected trilingual edition of the *Encyclopedia of Islam* did not meet with unanimous approval at the Academy. The pros and cons were weighed against, amongst other considerations, the demands of colonial policy. While some found in this policy — and particularly in Germany's Middle-East policy — arguments in favour of a financial contribution by the Academy, others countered these by pointing to the small proportion of Muslims in the German colonies. The latter saw little interest in financing a project from which the chief benefit would be drawn by England, France, Russia and Holland. . . .[49] The example is extreme, perhaps, but nonetheless instructive. One may, of course, wonder whether colonial policy really constituted a prime preoccupation of the participants in the debate, or whether colonial considerations merely served as a handy argument in support of institutional interests. But this is of scant importance; what is significant is the fact that considerations of this nature should have had tactical value.

It would be mistaken, then, to believe that the relation between the exigences of politics and the principles of science was seen as a threat in which the demands of politics disturbed the peace of mind of those who wished to abide by the principles of science. Far from it: in the world of politics, men of science seemed to find elements of moral support and social recognition. The extension of collaborative relations beyond national borders certainly responded to scientific needs. But how fond people were at the time of stressing that in this respect science was joining in the general development of interstate relations and that the IAA thus corresponded to the world situation.[50] On the eve of the foundation of the Association, one of the secretaries of the Berlin Academy endeavoured to re-situate the event in a historical perspective. He recognized that the nineteenth century had fulfilled its mission, which was the integration of civilized nations. Then he turned to the dream of peace among nations and foresaw another mission for the twentieth century, that of the international association:

'Just as the States have responded to the call of the Tsar and are at this moment gathered in The Hague, so too the German Academies will be meeting this autumn with representatives of the most distinguished foreign academies with the purpose of founding an association.'[51]

The references to international arbitration and to the development of humanity by successive stages of integration are interesting, for they are reflections of political doctrines that were deeply influenced by the theories of evolution.

In the passage quoted above, the foundation of the IAA and the efforts to establish international arbitration are linked with postulates of social evolution: not only to the laws of the struggle for survival, but also to the imperatives of 'biological solidarity'.[52] The IAA and international arbitration are perceived as two expressions of one tendency – the tendency towards organization not only in the name of increased efficiency, but also in the name of functional harmony. Organization is viewed as a mechanism that permits the averaging (in the Spencerian sense) of divergent interests and their cooperation for the common good. An international association of the civilized countries' leading scientific bodies seemed perfectly in keeping, then, not only with current theory of international peace and world order, but also with political practice. One might ask, of course, whether the reference by the secretary of the Berlin Academy to The Hague Conference was an expression of deep conviction or merely a gesture calculated to reassure the established powers and to intimidate possible opponents. The words were no doubt chosen with circumspection. All the same it was not the apolitical brotherhood of men of science, but rather the harmony with the great political choices of the State and the demands of modernization that added the stamp of legitimacy to organized international cooperation in science.

Notes

1. Schuster, A. (1913) Address. In: *The Semi-Centennial Anniversary of the National Academy of Sciences, 1863–1913,* Washington DC, National Academy of Sciences, pp. 19–36.
2. See, for example, Union des Associations Internationales (1957) *Les 1.978 organisations internationales fondées depuis le Congrès de Vienne,* Brussels; Union des Associations Internationales (1960) *Les Congrès internationaux de 1681 à 1899,* Brussels.
3. Obvious examples can be found in fields such as geophysics and geodesics, meteorology, seismology, epidemiology, etc.
4. Only recently, case studies have explored more specifically the interplay between international collaboration and developments in a particular field [see, for example, Cawood, J. (1977) Terrestrial magnetism and the development of international collaboration in the early nineteenth century. *Ann. Sci.,* **34,** 551–587; Devorkine, D. H. (1981) A sense of community in astrophysics: adopting a system of spectral classification. *Isis,* 72 (261), 29–49].
5. The fact that government funding depended to a large extent on approval by the public was clearly recognized. In an article describing to a general readership the work of the Prussian Academy of Science, the authors declared: 'It seems merely equitable that [the Academy] should be accountable not only to the government . . . but also to the people whose taxes contribute to maintain the State's strength not only in the military

domain, but also on the intellectual level (*in geistiger Beziehung*).' [Diels, H. & Waldeyer, W. (1908) Die wissenschaftliche Arbeit der Königlich-Preussischen Akademie der Wissenschaften im Jahre 1907. *Int. Wochenschr. Wiss. Kunst. Tech.*, col. 328.]

6. Address to a delegation of members of the Reichstag (National-Liberal Party), 19 April 1894. In: *Fürst Bismarcks Reden* (s.d.), Vol. 13, Leipzig, Reclam (*Im Ruhestand*), p. 93.
7. Quoted from Haines, G. (1969) *Essays on German Influence on English Education and Science, 1850–1917,* Hamden, Conn., p. 142.
8. Gerault, G. (1902) *Les Expositions universelles envisagées du point de vue de leur résultats économiques*, Paris; quoted from Poidevin, R. (1969) *Les Relations économiques et financières entre la France et l'Allemagne de 1898 à 1914,* Paris, Armand Colin, pp. 348–349.
9. Memorandum of 21 November 1909. In: *Fünfzig Jahre Kaiser-Wilhelm-Gesellschaft . . . 1911–1961* (1961), Vol. 1, Göttingen, pp. 80–94. See also note 5.
10. Forman, P. [(1973) Scientific internationalism and the Weimar physicists: the ideology and its manipulation in Germany after World War I. *Isis,* **64** (222), 151–180] explains the bases of scientific internationalism, the tension between collaboration and competition, and the role played by this ideology at the level of representation. The ideology thus contributes towards rendering (scientific) internationalism and patriotism compatible. Schroeder-Gudehus, B. [(1978) *Les Scientifiques et la paix. La Communauté scientifique internationale au cours des années vingt,* Montréal] examines primarily the political implications of scientific internationalism. See also Kevles, D. J. (1968) George Ellery Hale, the First World War and the advancement of science in America. *Isis,* **59**, 427–437; Kevles, D. J. (1970) Into hostile political camps. *Isis,* **62**, 47–60; Gruber, C. S. (1975) *Mars and Minerva,* Baton Rouge, Louisiana State University Press. Paul, H. [(1972) *The Sorcerer's Apprentice: The French Scientists' Image of German Science 1840–1919,* Gainesville, Fla., University of Florida Press] devotes the major part of his study to the nineteenth century, stressing the tactical use made by men of science of international scientific competition in seeking backing from public funds. Most of the studies dealing with international scientific collaboration in the nineteenth century touch only in passing on the problem of national rivalries [see, for example, Crosland, M. (1978) *Aspects of international scientific collaboration and organization before 1900.* In: Forbes, E. G., ed., *Human Implications of Scientific Advance,* Edinburgh, Edinburgh University Press, pp. 114–125].
11. See, for example, Petit, G. & Laudet, M., eds (1916) *Les Allemands et la science,* Paris, Alcan; Kellermann, H. (1915) *Der Krieg der Geister,* Weimar, Dresden, Heimat und Welt. See also Schroeder-Gudehus (note 10), pp. 63–130.
12. Indeed, not all the academies were organized in the same way: only the four German Academies and the Academy of Vienna had two sections (mathematics/physics and philosophy/history); three French Academies belonged to the IAA: l'Académie des Sciences, L'Académie des Sciences Morales et Politiques, and l'Académie des Inscriptions et Belles-Lettres; the Academy of St Petersburg had three sections (natural sciences, social sciences and literature); the British Academy was founded in 1904 in order to ensure that the British humanities and social sciences were represented at the IAA.
13. The plan of forming a Federation of European Academies had been discussed since the early 1890s; the formation of a cartel of German academies in 1893 (with the exception of the Academy of Berlin, which did not join until 1906), and the contacts between the Royal Society, the cartel, and the Academy of Berlin over the subject of the International Catalogue of Scientific Literature gave impetus to the project of establishing the Association, which was founded on 9–10 October 1899 in Wiesbaden. See His, W. (1902) *Zur Vorgeschichte des deutschen Kartells und der internationalen Association der Akademien. Berichte über die Verhandlungen der Königlich-Sächsischen Gesellschaft der Wissenschaften zu Leipzig, Math.-Phys. Klasse,* Sonderheft.
14. The founding members were the Academies of Vienna, Göttingen, Leipzig, Munich, Paris, Rome (Lincei), St Petersburg, Washington, Berlin, and the Royal Society of London. These were subsequently joined by the Academies of Amsterdam, Brussels, Budapest, Christiania, Tokyo, Copenhagen, Madrid, Stockholm, l'Académie des Inscrip-

tions et Belles-Lettres and l'Académie des Sciences Morales et Politiques de l'Institut de France, the British Academy, and the Société Helvétique des Sciences Naturelles.
As regards the history of the IAA, my information is based essentially on the archives of the Royal Society, the National Academy of Sciences, and the Académie des Sciences, as well as on the following documents and papers: the minutes of the assemblies of the IAA for the years 1900, 1901, 1904, 1910 and 1913; His (note 13); Schuster (note 1), and his commentaries published in *Nature* (**74**, 258; **83**, 370–372; **91**, 322–323); Darboux, J. (1901) *Journal des savants*, January, pp. 5–23.

15. Four, not including the Academy of Vienna.
16. For details see Kevles (note 10).
17. *Sitzungsberichte der mathematisch-physikalischen Klasse der Königlichen Akademie der Wissenschaften, Munich, 16 November 1901*, Vol. 31, p. 420.
18. See, for example, Bouillier, F. (1879) *L'Institut et les Académies de Province*, Paris, Hachette, pp. 358–364.
19. Schuster (note 1), p. 22.
20. Waldeyer, W. (1902) Meeting of 3 July. *Sitzungsberichte der Königlich-Preussischen Akademie der Wissenschaften zu Berlin*, p. 788; Suess, E. (1916) *Erinnerungen*, Leipzig, Hirzel, pp. 49–180.
21. Bouillier (note 18), p. 360; Grau, C. (1975) *Die Berliner Akademie der Wissenschaften in der Zeit des Imperialismus*, Berlin, Akademie-Verlag, p. 201.
22. von Auwers, A. (1900) Meeting of 28 June. *Sitzungsberichte* (Berlin), p. 659.
23. An edition of the works of Leibniz, study of the brain, *Corpus medicorum antiquorum*, annual physico-chemical tables, Mahâbhârata, measurement of a meridian, Septante, organization of the loan of manuscripts, astrographic chart. See also Grau (note 21), pp. 102, 201.
24. His (note 13), pp. 14–15. It was not, in fact, only for the sake of *international* projects that the academies made use of the prestige of the IAA. The American astronomer George E. Hale endeavoured to use the Association to convince potential financial backers of the need to install the National Academy of Sciences in premises fitting to its elevated function, if only in order to provide a suitably dignified setting for one of the Assemblies of the IAA, an honour which would no doubt be accorded to the Academy in the foreseeable future (Letter from Hale to Elihu Root, 29 December 1913; archives of the National Academy of Sciences, *IAA*, 1899–1914).
25. Endorsement of projects by the IAA was given considerable publicity in scientific journals. It certainly helped to attract public and private funds to the Marey Institute, created in Paris for the standardization of instruments for physiological measurements [see, for example, *Rev. gén. Sci.* (1902), p. 497]. One can also assume that the IAA's refusal, in 1907, to pronounce on the question of an international language dealt a heavy blow to that movement.
26. His (note 13), pp. 14–15 (quoting Friedrich von Hartel).
27. Cf. Harnack, A. (1900) *Geschichte der Königliche-Preussischen Akademie der Wissenschaften zu Berlin*, Berlin, Reichsdruckerei, pp. 1002–1003 (quoting a speech by Theodor Mommsen of 2 July 1874).
28. On interdisciplinary differentiation, see von Gizycki, R. (1976) *Prozesse wissenschaftlicher Differenzierung*, Berlin, Duncker & Humblot.
29. Mommsen, T. (1890) Meeting of 3 July. *Sitzungsberichte* (Berlin), p. 792; von Auwers, A. (1900) Meeting of 28 June. *Sitzungsberichte* (Berlin), p. 669; von Waldeyer, W. (1904) Meeting of 4 July. *Sitzungsberichte* (Berlin), p. 623; Diels, H. (1904) Die Akademien der Wissenschaften. *Woche*, **4** (40–53), 2303–2305; von Zittel (1900) Meeting of 16 November. *Sitzungsberichte* (Munich), p. 422.
30. Other institutions were involved in the bibliographical field. The IAA-sponsored *International Catalogue of Scientific Literature* was not the only project aimed at creating an effective instrument of reference. We might mention in addition the Institut international de bibliographie des associations internationales and the attempt to rationalize scientific information undertaken by Wilhelm Ostwald, *Die Brücke* [Ostwald, W. (1933) *Lebenslinien*, Vol. 3, pp. 287–310; Lewandrowski, P. (1979) *Der Kampf Wilhelm Ostwalds zum*

Informations problem in der wissenschaftlichen Forschung. In: *Internationales Symposium anlässlich des 125. Geburtstages von Wilhelm Ostwald,* Berlin, Akademie-Verlag, pp. 149–161].

31. See note 26.
32. Lister to the President of the National Academy of Sciences, 14 April 1899 (Archives of the National Academy of Sciences, *IAA,* 1889–1914).
33. Hale to Elihu Root, 29 December 1913 (Archives of the National Academy of Sciences, *IAA,* 1899–1914).
34. Quoted from Grau (note 21), p. 101.
35. See Schroeder-Gudehus (note 10), pp. 101–160.
36. See, for example, the obituary of Alphonse Milne-Edwards by R. Blanchard (1900) *Rev. gén. Sci.,* pp. 662–666.
37. One of the most enthusiastic supporters of these ideas was Wilhelm Ostwald: 'As the entire intellectual endeavour of mankind is to be understood as an organism, the principle of the division of labour becomes the basic pattern of progress' (Ostwald, note 30, p. 294).
38. See, for example, Hermann Diels, secretary of the philosophical-historical section of the Prussian Academy, who was a strong supporter of international collaboration in general and of the IAA in particular. He made it clear, however, that the collective work of academies was necessarily limited to undertakings, the methods and objectives of which were well established beforehand. In his view, the results of such endeavours were generally not scholarship in its highest sense (*Wissenschaft der höchsten Potenz*) but instruments to facilitate and to secure further progress – 'logarithmic tables for the higher achievements of science'. [*Die Organisation der Wissenschaft.* In: Hinneberg, P., ed. (1912) *Die allgemeinen Grundlagen der Kultur der Gegenwart,* Leipzig, p. 668.]
39. For instance, Roethe, G. (1913) Meeting of 26 June. *Sitzungsberichte* (Berlin), pp. 587, 589.
40. See, for example, von Auwers, W. (1900) Meeting of 28 June. *Sitzungsberichte* (Berlin), p. 668; Harnack (note 27), p. 1043.
41. Mommsen, T. (1890) Meeting of 3 July. *Sitzungsberichte* (Berlin), pp. 792–793.
42. Hartmann, L. M. (1908) *Theodor Mommsen,* Gotha, pp. 94–96; Schwarzschild, K. (1910) Die grossen Sternwarten der Vereinigten Staaten. *Int. Wochenschr. Wiss. kunst. Tech.,* col. 1531–1544; see also Harnack, A. (1905) Vom Grossbetrieb der Wissenschaft. *Preuss. Jahrb.* **119**, 28 January, pp. 193–201.
    On welcoming Adolf Harnack into the Academy of Berlin, Theodor Mommsen expressed his appreciation of the new member's qualities of leadership: scientific work is necessarily a collective effort, but it nevertheless needs to be directed by a leader ('nicht von Einem geleistet, aber von Einem geleitet'). Meeting of 3 July 1890, *Sitzungsberichte* (Berlin), p. 792.
    I might mention in passing that this concept of *Grosswissenschaft* was not restricted to the Germans nor to the period of the turn of the century. Some sixty years later, Frédéric Joliot-Curie was dreaming of improving the efficiency of the laboratory by organizing it according to the principles of an industrial concern: 'There should be one man for each type of machine, one man to run a particle accelerator, one who runs a Wilson cloud chamber, one who runs photographic counters, and so on – and one man in the middle who tells the others to perform various experiments . . .' [quoted from Weart, S. (1979) *Scientists in Power,* Cambridge, Mass., Harvard University Press, p. 50].
43. See Petit & Leudet (note 11); Kellermann (note 11); and literature cited in Schroeder-Gudehus (note 10), pp. 63–97.
44. Foster, M. (1894) The International Medical Congress. *Nature,* **49**, 563; see also, for another even more emphatic expression of enthusiasm, Sarton, G. (1938) L'histoire des sciences et l'organisation internationale [1913]. *Isis,* **29**, 311–325.
45. Diels, H. (1899) Meeting of 29 June. *Sitzungsberichte* (Berlin), p. 594; Harnack (note 27), pp. 981, 1043; see also the appraisal of the integrating function of the plenary sessions of the Academy in Grau (note 21), pp. 9–10.
46. Forman (note 10), pp. 153–156.

47. Comte de Franqueville (1901) *Discours prononcé à l'occasion de la première Assemblée générale de l'Association internationale des Académies*, Paris, Imprimerie nationale, p. 8.
48. Diels, H. (1906) *International Aufgaben der Universität*, Berlin, Universitäts-Druckerei, p. 37.
49. Grau (note 21), p. 205.
50. Waldeyer, W. (1902) Meeting of 3 July. *Sitzungsberichte* (Berlin), p. 788.
51. Diels, H. (1899) Meeting of 29 June. *Sitzungsberichte* (Berlin), pp. 594, 603.
52. These found their political expression, for example, in Léon Bourgeois' 'solidarisme' (see his *Essai d'une philosophie de la solidarité*, Paris, 1902, or *L'Idée de solidarité et ses conséquences sociales*, Paris, 1902).

SOCIAL RESPONSIBILITY IN VICTORIAN SCIENCE

JOHN ZIMAN

H. H. Wills Physics Laboratory, Royal Fort, Bristol, UK

An Authentic Source

We tend to regard the contemporary *problématique* of science, technology and society as unique. Yet it has its historical antecedents and origins. Can we trace these back into the times of Alfred Nobel? Was there any consciousness, then, of science as an instrumental factor in society, as a means of achieving preconceived ends? What did the scientists themselves think they were doing, beyond advancing knowledge? What was important to them, outside their laboratories, lecture halls and book-lined studies? Questions such as these cannot be answered satisfactorily without reference to primary sources.

As a student of science in present-day society, with concerns mainly for the future, I have always regretted a personal lack of scholarly acquaintance with the evidences of the past. The invitation, at short notice, to work up some topic of this kind was accepted, therefore, only on the understanding that this could be no more than a preliminary and unsophisticated exploration of a theme that really calls for serious, professional treatment. Shortage of time is also my excuse for not seeking out and citing the secondary historical literature relevant to this topic.

The obvious primary source lay immediately at hand. In the stockroom of our physics department library were all the old bound volumes of one of the world's most famous and influential journals – *Nature*. This 'weekly illustrated journal of science' was first published on 7 November 1869, and appeared thereafter at the rate of two volumes a year. Since each volume amounts to many hundreds of pages of double-column print, there was clearly more than enough material for my purpose. Indeed, I could only hope to sample this mine of information at random, leafing my way through every tenth volume, from volume 1 to volume 41, thus covering the period from about 1870 to about 1890 in five-year intervals.

This is an inadequate 'data base' on which to construct an edifice of generalizations. It is also significantly biased by editorial policy: *Nature* was not a passive medium for scientific information and opinion. From the first, it proclaimed an intention 'to urge the claims of Science to a more general recognition in Education and in Daily Life'. It is difficult to guess how far such intentions were from the centre of scientific opinion of the day. It was also, at that time, written almost entirely for British subscribers – at Home and in the Empire – and treated all 'American and Continental' news as distinctively foreign.

But the 20 000 or so pages from which this ten percent sample was drawn are uniform in style and attitude. One immediately recognizes the high-minded tone of the late Victorian era of industrial progress, geographical expansion and cultural self-confidence. These voices may not be entirely representative of the world of science at the time of Alfred Nobel, but they are authentic of their day. They express a coherent view, as seen from a central point of vantage, and their words still echo in the thoughts of their successors, a century later.

A Retrospective Definition of a Theme

'Social responsibility in science' is an ill-defined notion, which has become a catch-phrase. But it is a convenient term for a cluster of issues, attitudes and concerns about the relationship between science and society at large. It stands for an outward orientation, away from the production and accumulation of knowledge towards the deliberate use (and abuse) of scientific method, or specific scientific techniques, to meet acknowledged human needs. It shifts the focus of attention away from the 'internal' activities of the scientific community, from actual research discoveries, and from their essentially unpredictable consequences, towards the 'external' social context of science.

'Social responsibility' thus implies an attitude by scientists towards societal problems and needs. Even in our own day, when no one doubts the effectiveness of scientific research as a conscious instrument of social purpose and power, this attitude is not paramount in the everyday thoughts of the working scientist. The sheer technical labour of experiment, observation, theorizing, communication, etc., keeps broader questions of ends and means at a distance. Even now, for example, *Nature* devotes only a few pages of each issue to social and political affairs, before getting down to the real meat of the latest discoveries in molecular biology or cosmology, or plate tectonics.

Nevertheless, around the periphery of the strictly academic domain, there is now an immensely active region of discussion about 'science, technology and society'. Topics relevant to this theme are expounded in books, reported in weekly magazines, taught about in lectures, and gravely considered by government committees. We recognize them under such headings as 'science policy', 'technology assessment', 'the economics of research', 'science and the media' or 'rights and responsibilities of scientists', not to mention the wider social and political problems of war, energy, the environment, population, economic development and so on. These are the characteristic items in the agenda of the modern 'socially responsible scientist'. This is the theme, thus defined retrospectively, to be explored in the writings of the past.

What I have done, then, is to skim through those five old volumes of *Nature*, looking for articles, editorials, letters, or brief news items that might now be considered relevant to the theme of social responsibility in science. All communications referring solely to the *content* of science – claims of new observations, theoretical controversies, general accounts of recent discoveries, and so on – were ignored, along with interminable reports of the proceedings of various learned societies, British, Imperial and foreign. A large number of biographical memoirs had to be passed over, together with various items of 'internal' news, such as appointments, promotions and prizes awarded to individual scientists. Some of this discarded material is, of course, fascinating in its own right, and might yield relevant items to more thorough scrutiny; but the whole theme is too coarse-grained to pass through a finer sieve of definition and selection. Occasional items may also have been missed among the book reviews, which were not looked at carefully.

Perhaps we should greet this selected material in the same spirit as, say, a naturalist living at that time in New Zealand, receiving news of science in England through this single channel, six months in arrears, and sending back his own occasional reports of the meetings of the local philosophical society, out there in the colonies. It was not so much a matter of being minutely informed about events which one could not influence, as of keeping in touch with current attitudes and opinions which one felt one should be sharing.

A Significant Concern

We tend to think of natural scientists in the late nineteenth century as a highly academic community disconnected from social or technological issues, and concerned, almost exclusively, with 'internal' questions, whether cognitive or communal. It was surprising, then, to find that my notes referred to a total of about 300 pages out of the 3074 pages of the five-volume sample. Admittedly, *Nature* was trying to speak to a wider public than it does today, and sought to fulfil some of the functions that are now undertaken by other general science magazines such as *New Scientist*, or by more specialized periodicals such as, say, the *School Science Review*. But even if questions of social relevance were not central to the interests of British scientists at that time, they were certainly not insignificantly marginal. The theme of social responsibility seems to have deeper and more solid roots in Victorian science than is often supposed.

In assessing the thrust of this concern, we must, of course, make due allowance for the immense change that has taken place both 'within' science and 'outside' it. The enormous growth of scientific knowledge and practical technique scarcely needs emphasis. For example, in the 1880s, the world itself was still being explored: the scientific literature is full of reports of geographical, anthropological and botanical discoveries, with all their speculative economic and political implications. The weight of many factors in the interaction between science and society has inevitably been altered by the growth in the scale and scope of research; for example, the cost of supporting science in those days was far below the threshold of national economic concern. Nineteenth-century society itself was not only very different politically and economically; it also had a very different cultural style. For example, it was almost romantically optimistic about its future. This comes out clearly in *Nature*'s 'advertisement' of editorial policy. The intention was to make a major feature of

> 'articles written by men eminent in Science on subjects connected with the various points of contact of Natural knowledge with practical affairs, the public health and material progress; and on the advancement of Science and its educational and civilizing functions' (1870; 1, 323).

Nevertheless, in spite of far-reaching historical changes of content, scope and social context, the initial impression is of continuity of attitudes and themes rather than of a glimpse of an entirely unfamiliar world. This impression derives to some extent from the pleasure of recognizing such hardy perennials as expressions of dissatisfaction with the constitution of London University (1884; 31, 145) or from a splendid debate at Oxford on vivisection (1885; 31, 453); but it survives more serious analysis, as one recognizes the primitive ancestors of most of our present obsessions. One can already detect a general notion of socially responsible science, although the components are combined in somewhat different proportions and oriented towards different immediate goals.

But this general impression of thematic continuity may not stand up to more critical analysis. Let us take it simply as a tentative hypothesis, to be tested by more detailed consideration of particular topics.

Science Education

The broadest interface between nineteenth-century science and nineteenth-century society was through formal education. *Nature* proclaimed its intention to publish

> 'records of all efforts made for the encouragement of Natural knowledge in our Colleges and Schools, and notices of aids to Science teaching' (1870; 1, 34).

About one-third of the 300 pages noted in my sample survey come under this heading.

That interest goes beyond the internal affairs of a community, whose members were often (though not exclusively) employed as teachers in universities, colleges and schools. I do not count the many specific announcements of academic appointments such as:

> 'Mr J. J. Thomson has been elected to fill the post of Cavendish Professor of Experimental Physics in the University of Cambridge, in succession to

> Lord Rayleigh. A numerously signed requisition to Sir Wm. Thomson [Lord Kelvin] to become a candidate was declined' (1884; 31, 179);

or incidental chit-chat, such as:

> 'Cambridge was *en fête* on Monday. Peterhouse, the oldest collegiate institution in the University, was celebrating the six-hundredth anniversary of its foundation. It was stated at the dinner that one-third of the present Fellows were Fellows of the Royal Society' (1884; 31, 179).

It is zeal for science education as a general social influence that is so striking.

This mission was not without support from the outside: in 1875 (11, 241), for example, no less a political eminence than the Marquis of Salisbury was reported as speaking publicly in favour of scientific education. But the general impression is of an enlightened minority group, well persuaded that

> 'the only principle on which a satisfactory course of education can be constructed is, that it is essential for the well-being of every man and woman that he and she should start in life with a well-trained mind and a fair knowledge of the principles and the main facts of everyday life' (1874; 11, 21),

fighting to change a situation where

> 'In most cases where science has been admitted into our schools it has been only on sufferance as a kind of interloper for which any odd corner is good enough' (1874; 11, 21).

At the highest level, this implied an attempt to breach the entrenched positions of the classics and mathematics in the universities and socially prestigious secondary schools. The tone of this attack was not very confident, relying upon apologetic arguments such as that science was not so difficult as classics – hence suitable for weaker pupils (1869; 1, 18) – and could be developed without damaging traditional standards of classical scholarship (1869; 1, 25). The 1870s, in fact, saw the establishment of experimental science as a formal course of study at Cambridge, although there are reports of the unwillingness of the Cambridge colleges to support teaching posts in science (1870; 1, 586), of the weakness of the new Natural Sciences Tripos in comparison with the old-established Mathematical Tripos (1875; 11, 132), and various debates over new regulations and curricular structure (1879–1880; 21, 26, 86, 125). Even at the end of the period, in 1890, there is bitter complaint (41, 25, 265) that 'science men' are practically excluded from fair competition in the examinations for entry into the Indian Civil Service, the most esteemed branch of the Imperial bureaucracy. We observe, indeed, the initial phases of the 'Two Cultures' dichotomy, which still keeps the majority of our mandarins untouched by formal scientific education from the age of sixteen.

They apparently managed these matters better elsewhere, especially in

Germany. At least that was one of the most important arguments for more science education in Britain. The comparisons are odious: the New Statutes (of 1879) 'will scarcely place Cambridge on a level, so far as teaching power goes with a second rate German university' (21, 125). According to H. E. Roscoe (1869; 1, 157), the German university system, although state-supported, was intellectually free, held in high esteem, non-sectarian, cheap for the student, well staffed, scientifically thorough, connected with industry, and generally without defects. An account of the new University of Strasbourg (1885; 31, 557) – admittedly cultural window dressing for the German annexation of Alsace-Lorraine after the Franco-Prussian War – is as admiring and envious of enlightened prodigality as, say, a European description of the Bell Telephone Laboratories in the 1950s, or a present-day American report on an automated Japanese car factory. French education is not set up as an ideal model, but there are numerous favourable references to the much larger scale of expenditure on technical education than in England (1879; 21, 139).

These comparisons apply not only to higher education in science: they also apply to technical training and to science in general education:

> '. . . plenty of time remains in German schools for the teaching of science, which forms so important a part of education throughout that country, and which gives the German a starting point in life so very much superior to that which the average Englishman has, even when educated at our public schools and universities' (1874; 11, 21).

Indeed, the scientific community, through such notably 'responsible' scientists as T. H. Huxley, was obviously much involved in the movement to establish a system of compulsory primary education in Britain. Every effort was to be taken to incorporate elementary science in the standard curriculum of mass education – as Henry Armstrong put it:

> '. . . the advantages to be derived from even the most elementary acquaintance with what may be termed the science of daily life are so manifold that, if once understood by the public, the claims of science to a place in the ordinary school course must meet with universal recognition' (1884; 31, 19).

There was also the challenge of founding new institutions for technical education, where scientific subjects would be taught at an intermediate level for the skilled craftsmen and professional experts needed in advanced industry. This was a grave weakness of Britain compared with its commercial competitors, especially Germany, with its new system of polytechnic institutions (1870; 1, 475), and France, with a long tradition of state-supported *hautes écoles* (1879; 21, 139).

In its twentieth-anniversary number, *Nature* had good cause to celebrate the progress that had been achieved in science education:

> 'Twenty years ago England was in the birth-throes of a national system of primary instruction. This year has seen the State recognition of the necessity of a secondary and essentially a scientific system of education and the Technical Instruction Act marks an era in the scientific annals of the nation. . . . A race is thus springing up which has sufficient knowledge of science to enforce due recognition of its importance and public opinion can now, far more than in the past, be relied on to support its demands' (1889; 51, 1).

Unfortunately, this momentum for reform was not maintained: a century later, we still regret and suffer from the continuing weakness and inadequacy of scientific and technical education in Britain, by comparison with what is achieved in other countries.

But I do not think we should blame our Victorian scientific forebears for these deficiencies. They backed up their rhetoric with both realistic and inventive detail. The administrative threads of such uptight institutions as the Universities of Cambridge (1874; 11, 125) and London (1885; 31, 145, 159, 352) are carefully unknotted for reform. Thoughtful plans for a whole system of 'polytechnics' in London are propounded (1890; 41, 242, 481). The implications of various clauses of the 'Education Code' of the Technical Instruction Act are laboriously spelled out (1890; 41, 385, 505). The accumulated wealth of the Guilds of the City of London is fastened on as an appropriate endowment for the new College of Science to train science teachers (1880; 21, 221).

This involvement in educational reform is not (as so often nowadays) limited to framing general objectives, demanding more financial resources and proposing new institutional arrangements. Henry Armstrong explained in detail his new method of teaching chemistry, right down to the description of an experiment to determine the composition of air (1884; 31, 19). This was followed by a lively correspondence from other science teachers, including the familiar complaint that pedagogic innovations have to be abandoned because:

> 'Science masters are [entirely] at the mercy of examiners both of University examiners, periodically examining a school, and of examiners for open scholarships' (1884; 31, 28).

Armstrong was not alone in advocating a new approach to science teaching. Even today, we emphasize the value of experiments in science education, perhaps echoing C. K. Wead, an American physics professor, who wrote that

> '. . . discovery, with its necessary companions, self-reliance, independent thought, shrewdness of judgement – the very qualities which make a successful man of the world – are all developed by experimental science instead of the too frequent opposite effect which makes anxious business fathers dread too much schooling for the sons who will have to follow them' (1885; 31, 578).

But laboratory teaching was expensive. W. J. Harrison's 'New Method for the Teaching of Science in Public Elementary Schools' in Birmingham was to take demonstration apparatus from school to school in a handcart. As a result: 'Among the boys the half-timers then muster strongly, often getting leave to come in for that lesson only, and sitting with bare arms and rolled-up aprons, just as they run from their work' (1884; 31, 175). And although we sympathize with C. W. Quin's insistence on 'the absolute necessity of beginning scientific training at a very early age', we are still trying to prove his assertion that 'children of 8 or 9 are [not] too young for systematic science teaching' (1869; 1, 209).

The modern scientist tends to regard science teaching as a specialized profession, for which he has little direct concern. One cannot imagine that modern readers of *Nature* or the *New Scientist* would be interested in general accounts of educational problems, such as the 'Report of the Commissioners of Education in the United States for the year 1882–83', drawing attention to the persistence of conditions such as:

> 'The variation in different States of the expenditure on education . . . is still exemplified in the fact that Massachusetts pays fifteen times the amount per head that Alabama pays.'

and:

> 'We have no need to enlarge again here upon the United States difficulty; the education of the negro. The burning question of course is, Who is to pay for it?' (1885; 31, 435).

The educational role of science was evidently felt with greater responsibility in the late nineteenth century than it is today.

Popularization of Science

In Victorian times, there does not seem to have been a clear demarcation between formal instruction and public lecturing about science. The first volume of *Nature* is enthusiastic about 'Lectures for Women' (1870; 1, 488) and, for working men, to 'Help Them Get Things Right' (1869; 1, 71). Those who did the 'arduous work of teaching', free, at the Working Men's College (1870; 1, 511) were among the élite of the science community. In 1875, for example, Professor Roscoe lectured publicly in Glasgow to 3000 people (at thruppence per head) on 'The History of the Chemical Elements' (11, 233). On the other hand, some of the attempts to popularize science were as inept and infuriating as those that can be found in the modern mass media. M. Lichtenstein wrote to the Editor about attending one of the well-endowed Gresham Lectures – given, it seems, by a classical scholar with no knowledge of science:

> 'Choking with Indignation, I left the building, never having heard, in all

my life, either in sermon or lecture, so many false statements uttered in the space of half an hour' (1874; **11**, 28).

Nevertheless, such references to the popularization of science are quite brief. Apart from an account of steps towards the foundation of the science museums at South Kensington (1890; 41, 409), I could find no further reference to this subject in later volumes. It is true that *Nature* itself was not a specialized scientific periodical. Its editorial policy statement distinguished between 'Those portions of the Paper more especially devoted to the discussion of matters interesting to the public at large' and those portions 'more especially interesting to Scientific men'. But that public (like modern subscribers to the *Scientific American*, say) is evidently drawn from a narrow stratum of educated, technically oriented people, already far better informed about Science than the mass of ordinary people. I wonder whether the scientific community of Nobel's time was any more concerned about the mass popularization of science than scientists are today. This sample even suggests a decline in that particular manifestation of social responsibility from the 1870s to the 1890s; but the evidence is not sufficient to support any definite conclusion on this point.

Scientific Technology

Nowadays, 'science' is almost synonymous with 'technology'. At least, one should say, they are symbiotic. The most abstrusely academic research is connected with practical affairs by a few overlapping links of experimentation, inventive application, industrial development and commercial manufacture. The modern scientist is expected to exercise social responsibility very largely through the mechanisms that give direction to the immense social forces of scientific technology.

Our Victorian forebears were certainly as interested in technological innovation as we are. About seventy-five pages of my sample were about recent or prospective advances in various fields of industry and engineering, both civil and military. This was the new age of electricity, celebrated, for example, in a whole series of articles on the 'Progress of the Telegraph' (1875; **11**, 392, 450). The facilities of the new Merchants' Telephone Exchange in New York are described enthusiastically:

> 'If having established the connection the *employé* is obliged to withdraw his telephone, the communication between Edward and John is secret. If while these two are in conversation No 42, James, wishes to correspond with John, for example, the *employé* may join in the conversation of the two interlocutors just like a servant announcing a visitor. If required, conversation may be established between the three subscribers' (1880; **21**, 495).

It is reported that Victoria Station is being lit with sixty electric lights driven from a twenty-horsepower motor (1879; 21, 162); and there is an interesting discussion of the relative merits of various schemes of gas or electric lighting for Paris (1880; 21, 282). In the same volume, there are several brief reports from America about Edison's new carbon-filament lamp (1879; 21, 187, 215), including the information that the shares of his company had recently gone up from $20 to $3500 (1879; 21, 261). This was evidently the trigger for an intensely sarcastic editorial (1880; 21, 341):

> 'What is then the nature of the inventions thus heralded before the world? Regarded quietly, and without prejudice, from a scientific standpoint, what is the value of the discoveries which can thus play havoc on the Stock Exchange?'

But the attack goes beyond 'the reckless and amusing statements made by newspaper correspondents and interviewers', to assertions that 'Mr Edison is thirty-five years behind the time in his invention', and regrets that 'Mr Edison does not devote some time to learn what has been already done in this field'. If only it were possible, then or now, to

> 'Let public opinion insist that the inventor should be allowed to pursue his way unhampered by the officious interference of the unprincipled speculators whom his soul abhors, or by the irrepressible unscientific reporter who is only one degree less reprehensible for the part he plays.'

The attractions of the superstar technological project were also evident. The recently opened St Gotthard Tunnel called up the same sentiments ('The greatest work hitherto attempted by man', 1880; 21, 581) as a present-day description of the flight test of the space shuttle. An admirable series of articles (1869, 1870; 1, 160, 303, 631) analysed various proposals for 'The Projected Channel Railways', rather as one might nowadays analyse schemes for collecting solar power by stationary earth satellites. Unfortunately, history has not yet vindicated their confident judgement:

> 'That a permanent railway across the English Channel will be built, we doubt not: we are equally confident that Messrs Bateman and Revy's scheme of a sunken iron tube on the sea-bed is a practical solution of the problem. No less an authority than the Emperor Napoleon III, after mature consideration of the scheme, wrote to say: "C'est le seul réalisable," and as the design is one that belongs essentially to England, His Majesty's opinion acquires enhanced value and importance' (1870; 1, 631).

Science and War

What is surprising, however, is the obvious fascination of scientists with military technology. We tend to assume that the liaison between science and war (with all the social responsibilities thereby incurred by scientists) was kept fairly quiet until, say, the First World War, and did not become a matter of great public concern until after Hiroshima. But the articles in *Nature* go far beyond portentous frivolities, such as the suggestion that

> '. . . germs of small-pox and similar malignant diseases [should be collected] in cotton or other dust-substances and [loaded into] shells. We should then hear of an enemy dislodged from his position by a volley of typhus or a few rounds of Asiatic cholera' (1870; 1, 562),

or mere news items such as that observation balloons were to be used by the Expeditionary Force going to the Soudan (1885; 31, 368). There are detailed technical reports on current military innovations, such as the use of torpedoes in the American Civil War (1870; 1, 656) or Hiram Maxim's latest invention:

> 'A gun which loads and fires itself is certainly a novelty, and presents many interesting features and possibilities to anyone who takes an interest in implements of warfare' (1885; 31, 414).

The series of articles by Sir Frederick Abel, FRS (of Her Majesty's Chemical Department at Woolwich Arsenal) on 'Smokeless Explosives' (1890; 41, 328, 352) are of special interest. They make clear the great tactical advantages of smokeless powder in land and sea battles and report the considerable efforts of the major European powers to seize these advantages for themselves. Secrecy is an essential factor:

> 'As in the case of *mélinite*, the fabulously destructive effects of which were much vaunted at about the same time, the secret of the precise nature of the smokeless powder [used in the new Lebel rifle] was so well preserved by the French authorities, that surmises could only be made on the subject even by those most conversant with these matters.'

Meanwhile, of course

> 'In Germany, the subject of smokeless powder for small arms and artillery was being steadily pursued in secret, and a small-arm powder giving excellent results in regard to ballistic properties and uniformity, was elaborated at the Rottweil powder-works. . . .'

In these circumstances, it is scarcely surprising that Alfred Nobel applied his inventive skill with blasting agents to the production of a smokeless powder. We may be sure that others would soon have taken the same path to the far more efficient and destructive weaponry that this invention obviously made possible. Incidentally, Abel took pains to combat the widespread fallacy that smokeless powder was also *noiseless*. We, too, on occasion find it

> '. . . somewhat difficult to conceive that, in these comparatively enlightened days — an acquaintance with the first principles of physical science having for many years past constituted a preliminary condition of admission to

> the training establishments of the future warrior – the physical impossibility of such fairy tales which appear to be considered necessary in France for the delusion of the ordinary public, would not at once have been obvious.'

The positive concern of late Victorian science with military technology is further evidenced by the controversy in the pages of *Nature* over the design of battleships. In the half-century since Trafalgar, naval architecture and gunnery had changed out of all recognition – but the heavily armoured iron ships had never been tested in battle against the high-explosive shells of their opposite numbers. Some of the phrases of the editorial of Thursday, 26 February 1885 (31, 381–384) could be brought forward a hundred years and applied to strategic nuclear weapons. A Mr Barnaby, the Director of Naval Construction, had at one time computed the 'relative efficiency' of warships by an algebraic formula compounded of factors such as 'weight of armour per ton of ship's measurement', 'the height of battery port-sills above load water-line', etc., and then later asserted the entirely different principle that 'the fairest available measure of [fighting] power' is the 'displacement or total weight' of the ship. The editorial points out the complete inconsistency of these two principles, and demolishes all such simple numerical indicators.

> 'The fighting power of a ship is . . . composed of several diverse and independent elements; and there is nothing approaching to a consensus of professional opinion as to the relative importance of these elements. To assume that they all vary together with the ship's dimensions, or with her weight in tons, is in the highest degree delusive and absurd. . . .
> '. . . It would be extremely difficult to devise any simple standard by which the popular mind may be fairly impressed with the relative powers of our own and foreign navies; while for purposes of exact comparison or of technical discussion no such standard could be regarded as absolute.'

One can approve the scientific rationality of such contributions to public affairs – but does it match to a higher ideal of social responsibility? It was the Prime Minister, Lord Salisbury, addressing the Institution of Electrical Engineers, on the subject of the electric telegraph – 'that small discovery, worked out by a few distinguished men in their laboratories upon experiments of an apparently trivial character, on matter and instruments not in the first instance, of a very recondite description . . . ' – who drew attention to its most significant social consequence:

> 'I would ask you to think of what is the most conspicuous feature in the politics of our time, the one which occupies the thoughts of every Statesman, and which places the whole future of the whole civilized world in a condition of doubt and question. It is the existence of those gigantic armies held in leash by the various Governments of the world, whose tremendous power may be a guarantee for the happiness of mankind and the main-

tenance of civilization, but who, on the other hand, hold in their hands powers of destruction which are almost equal to the task of levelling civilization to the ground' (1889; 41, 21).

Scientists and Inventors

There is no doubt that nineteenth-century scientists were generally interested in technological progress beyond its purely technical aspects. They took upon themselves much of the credit for the transformation of industry – for example, the progress made in the previous forty years in the 'construction of self-acting machinery' that would surely make living cheaper (1870; 1, 432). The scientific engineers of the Institute of Electrical Engineers would have applauded Lord Salisbury when he looked far into the future:

> 'If it ever does happen that in the house of the artisan you can turn on power as now you can turn on gas – and there is nothing in the essence of the problem, nothing in the facts of the science, as we know them, that should prevent such a consummation taking place – if ever that distribution of power should be so organized, you will then see men and women able to pursue in their own homes many of the industries which now require their aggregation at the factory. You may, above all, see women and children pursue these industries without that disruption of families which is one of the most unhappy results of the present requirements of industry' (1889; 41, 21).

They were not unaware of the social consequences of industrial change. There is a report that beet sugar is ruining the cane-sugar industry of Cuba (1875; 11, 314). There is mention of the introduction of labour-saving machinery, by which its seems 'the Swiss toolmakers annihilated the English watch toolmakers some years ago' (1880; 21, 397). This comes in a fascinating article by Silvanus Thompson on the reports made by a number of skilled artisans who were sent over to Paris for the Exhibition of 1878 – most of whom came away 'fully alive to all the advantages which accrue to an industry from the extension of labour-saving appliances, and from the dissemination of higher technical knowledge', and presumably confident that craftsmen displaced by such appliances would soon find alternative employment. Industrial prosperity was an important national goal, to which, for example, reform of the patent system would undoubtedly contribute (1874; 11, 141).

Nevertheless, it is not without significance that a review of the first twenty years of *Nature* made practically no direct reference to industrial or technological developments (1889; 41, 1). It was all very well for George Gore to argue that 'employment for workmen in this country can be increased' by 'encouragement of experimental scientific research' (1869; 1, 623). There was still a long and irreversible chain of cause and effect from a scientific discovery, through its application as an invention, to the final manufacture that provides employment; and the scientific community did not take any responsibility for what happened further down that chain. This aloofness is all too evident in a comment on Edison:

> 'We are doing no injustice to Mr Edison's splendid genius when we say that it is to the character of the inventor not to that of the scientific thinker, that he aspires' (1880; 31, 341).

Except, perhaps, in relation to military technology, there seems to be no notion of a systematic activity of 'research and development' that would bring science into close interaction with technical problems of direct social significance.

Science and the Environment

Scientists nowadays are urged to take responsible attitudes towards the environment. Every agenda for social action or education on science, technology and society lays great emphasis on issues such as environmental pollution, ecological balance, conservation of resources, industrial health and safety, and so on. These issues are taken up so earnestly that it would seem that no one had ever thought of them before.

This is a grave misconception. The scientists of a hundred years ago were as much concerned as we are to make the world a cleaner, healthier, safer place, where mankind and all other forms of life could prosper. The evidence is quite clear throughout my sample. The thirty-five or so pages on these issues are unequivocally on the same side as our angels: there are no apologetics for dirt, disease, poverty, waste or exploitation of labour.

A long report on scientific aspects of an 'International Health Exhibition' organized by the Society of Arts (1884; 31, 138–143) illustrates the wide range of issues in which scientists were closely involved in social action. For example, there is a call for more stringent legislation, similar in character to that at present in operation in the United States and France, on the adulteration of food and drugs, on the grounds that

> 'These are questions largely affecting the health of the whole nation, and especially affecting the welfare of the poor, who suffer most by the substitution of worthless, inferior or adulterated articles in the fabrication of apparently cheap, but often very dear because worthless, articles of food.'

There were tests on the cooking efficiency and relative smokiness of kitchen stoves, conducted by members of the 'Smoke Abatement Institution'. A new code of public sanitation was proposed. The author of the report, Mr Ernest Hart, had

> 'for several years as Chairman of the National Health Society, . . . occupied

myself with collecting the facts and figures which demonstrate the urgent necessity of improved legislation for the safeguarding of the sanitary construction of our houses, and the improved education and registration of those builders and plumbers to whom we intrust that construction'.

Since *Nature* is not a medical journal, there is little about human diseases as such, but a strong case is made for the foundation of an Institute of Public Health to foster higher education and research in sanitary science.

Water pollution was clearly a major concern. In the very first volume of *Nature* (1870; 1, 578), there was a polemic on 'The Abuse of Water', which could scarcely have been improved on, scientifically, humanely, economically or rhetorically, for the next hundred years. It was followed up with positive proposals, such as General Scott's paper on 'Suggestions for Dealing with the Sewage of London' (1874; 21, 133). It would be unfair to blame the scientists and engineers for the failure to clean up the Thames and other rivers until a much later day.

The risks of accidental explosions were as salient then as now. F. A. Abel makes no bones about the causes of bursting steam boilers: 'Very few explosions in 1873 have been due to the neglect of the attendants, but by far the greater number to that of the boiler owners or the makers,' and calls for statutory inspection (1875; 11, 436). However, he blames poor working practices, often abetted by the miners, for the frightful colliery explosions that were then so frequent. He returned to the same theme in 1885 (31, 469, 493, 518), with a series of articles on explosions caused by the vapours of non-explosive liquids such as paint thinners. Improvements in railway safety are also commended (1874; 11, 173; 1884; 31, 84), and there is a tough article on ensuring stability in the design of ships (1880; 21, 485).

The biological environment is seen mainly through the eyes of the farmer or fisherman — the dangers of the spread of plant diseases such as phylloxera and of pests such as the Colorado beetle and apple blood louse from the New World (1875; 11, 394, 1880; 21, 356) are set off against the remarkable successes of Louis Pasteur against threats to the wine, beer, silkworm, cattle and poultry industries (1884; 3, 138–143). Presumably, the Royal Agricultural Society really deserved the scorn poured on its efforts to foster research on potato disease (1874; 11, 67, 109). The socio-economic origins of scientific ecology are evident in the work of the United States Fish Commission, founded in 1871, when the Congress 'has its attention directed to the alarming decrease in its east coast food fishes' (1884; 31, 128; 1885; 31, 294). These articles emphasize that, although 'The Fish Commission's work in its original conception was really the solution of practical economic problems, and it has in the main adhered to this idea', it had also been permitted to deviate 'from the rather uninteresting study of the shallow fishing-grounds to the rich field of [basic] deep-sea research'. We, on the other hand, might now value the fact that 'the natural history work of the Fish Commission has, of necessity, been mainly of a systematic character, dealing with species and their distribution more than with problems of anatomy, embryology and history'. It is noteworthy, nevertheless, that there is a continuing demand for similar practical scientific support for British Fisheries (1870; 1, 243; 1890; 41, 497).

In some ways, the most surprising items in the whole of my sample were concerned with energy. For all his academic and social eminence, Lord Rayleigh shows no disdain for practical needs when he devotes a major part of a lecture on 'The Dissipation of Energy' to 'The prevention of unnecessary dissipation [which] is the guide to economy of fuel in industrial operations' (1875; 11, 454). There is a gift for the sociology of knowledge in Balfour Stewart's metaphorical interpretation of the energy concept:

> '. . . in the social world we have what may justly be termed two kinds of energy, namely:
> 1. Actual or personal energy [i.e. force of character, persistence, etc.].
> 2. Energy derived from position [i.e. social standing, family fortune, etc.]' (1870; 1, 647).

But Sydney Lupton on 'The Coal Question' (1885; 31, 242) is uncanny. He estimates the amount of accessible coal in Great Britain, extrapolates exponentially from past consumption trends, and calculates that the total supply will be exhausted by 'about AD 1990'. Ignoring oil, he accepts that tides, winds and waterfalls could be exploited, but points out that since we have no monopoly on these sources 'we should compete with our neighbours rather at a disadvantage'. The cost of importing coal would be economically insupportable. Soon

> 'by the scarcity of our coal our pre-eminence in cheapness of manufactures becomes a thing of the past, the means of paying for imported food will gradually cease, and the pressure of population, together with the increased cost of the necessities of life, by emigration, by an increased death-rate, and by a reduced birth-rate, will change the England of to-day into a country like the England of 1780.'

Even his remedies give one a sense of *déjà vu*:

> 'After discussing and rejecting the expediency of limiting or taxing the output or export of coal, on the ground that any such measure would impose a serious burden upon our manufactures and commerce. . . .'

he supports the traditional monetarist policy of Stanley Jevons – make a serious effort to pay off the National Debt! I wonder whether the readers of *Nature* were relieved by an editorial by George Gore, a month later (1885; 31, 357), insisting that all could be well if only governments and people understood that

> 'The practical value of new scientific knowledge as a source of wealth and progress is incomparably greater than that of all the coal deposits, petroleum springs, and gold-fields of the earth.'

Science and the State

Social responsibility *in* science cannot really be disentangled from social responsibility *for* science. Modern science is so totally dependent upon state support that we regard government science policy as one of the principal means by which science is made to respond to social needs. Scientists themselves are held to bear great social responsibility through the influence they are able to exert on such policy.

A century ago, that particular moral burden could scarcely be envisaged. In Victorian Britain, both the scientific community and the central government machine were much smaller, much weaker and much more aloof from one another than they are today. They shared very few common institutions. In all those pages of *Nature*, there are few specific references to government research organizations as such. If an informal survey of government expenditure on science is to be trusted (1870; 1, 589), these consisted of little more than several astronomical observatories, botanical gardens and museums, together with the Ordnance Survey, the Hydrographic Department of the Admiralty, and a few other small sections attached to military establishments such as Woolwich Arsenal. There was simply no government science worth having a policy *for*.

In this respect – as was frequently remarked at the time – Britain was exceptional among advanced industrial nations. There are occasional admiring references to countries such as France (e.g. 1870; 1, 587), where science was much more obviously incorporated within the powerful bureaucratic apparatus of the State. Surprisingly enough, the United States Federal Government is also seen to be much more involved in science than is the British Government, perhaps through the necessity of creating a coherent scientific organization to survey the vast continental territories opening up in the West. There is a familiar sound to the report that Congress had at last created the United States Geological Survey, to put an end to 'the chronic feuds to which so many independent United States Government Surveys with rival objects and officers gave rise' (1880; 21, 197).

There seems nowhere to have been a really strong connection between science and politics. In an age of fierce national rivalry, the fact that each nation was very proud of its 'own' scientists and their achievements for the benefit of their countrymen was not turned into an operational instrument. There are no reports of constraints on travel or of communication between scientists. And yet there is a decided absence of positive international collaboration on scientific projects – witness the seventy stations set up across the Pacific by astronomers from six countries to observe the transit of Venus (1874; 11, 102), without any overall coordination of effort. The important conference in Washington that finally agreed on Greenwich as the Prime Meridian and standard for universal time was *inter*national, not in any sense *trans*national (1884; 31, 7, 82): the delegates, 'in some cases scientific men, in others the ambassadors accredited to the United States, were instructed by their respective Governments specifically for the settlement of [these] questions' – and, in the end, France abstained. We may observe the same clashes of national pride in settling a contemporary version of the same issue – the harmonization of the dates of 'summer time' in the European air travel region!

But the scientists themselves were not indifferent to the vast resources that could be tapped from the State – presumably without significant loss of independence in research. In its very first volume, *Nature* gave full editorial support to the report of the British Association recommending an inquiry into two questions:

> ' I. Does there exist in the United Kingdom of Great Britain and Ireland sufficient provision for the vigorous prosecution of Physical Research?
> 'II. If not, what further provision is needed? and what measures should be taken to secure it?' (1869; 1, 127).

It was confidently argued that a Royal Commission would show such deficiencies in the organization and magnitude of government support for science that a great reform of the system would be inevitable:

> '. . . one of the important results of the analysis will be bringing to light the scattered character of our scientific efforts: almost every department of the State having charge of some scientific institution – the Admiralty of one, the War Office of another, the Board of Trade of a third, and so on, a dispersion which is absolutely prohibitive of harmonious systems of progressive improvement, of efficient superintendence, of economy in expenditure, and of definite responsibility. . . .
> 'We may hope as another most important result, that a central administration of scientific affairs will be shown to be necessary. In all other civilized countries a Minister of State is charged with this duty. It seems absolutely impossible to organise or maintain in an efficient state anything like a harmonious scientific system, without a dominant authority presiding over the whole' (1870; 1, 589).

As was pointed out at a meeting at the Society of Arts,

> 'Mr Chadwick, on one occasion, desiring to ask in the House of Commons a question regarding some scientific matter, found that it affected four different departments, and should therefore elicit a quadruple reply, the horrors of which he evaded by most informally putting the question to the Premier himself' (1870; 1, 575).

Alas, this campaign can scarcely have been successful – for the present situation could be described in just the same terms. There has not yet been what the mid-Victorian reformers thought to be 'probable' – i.e. 'a total re-arrangement of the internal organisation and the official distribution of our scientific institutions, with a view to concentrated superintendence and responsibility' – perhaps for the very reason that this 'will also involve a revision of scientific staffs and salaries, with all the attendant questions of patronage, promotions, distinctions, privileges and pensions' (1870; 1, 589). The whole subject is scarcely mentioned at all in later years. Even the material fruits of this agitation were not prolific. Looking back over the first twenty years of its existence, *Nature*'s report is cool (1889; 41, 1):

> 'The support afforded by the Governments of Western Europe to scientific investigation has been markedly increased within the period which we survey. France has largely extended her subsidies to scientific research, whilst Germany has made use of a large part of her increased Imperial revenue to improve the arrangements for similar objects in her Universities. The British Government has shown a decided inclination in the same direction: the grant to the Royal Society for the promotion of scientific research has been increased from £1000 to £4000 a year; whilst subsidies have been voted for the Marine laboratory at Plymouth, to the Committee on Solar Physics, to the Meteorological Council, and quite recently to the University Colleges throughout the country. . . .'

Science as a Profession

Much more information than can be gleaned from those old volumes of *Nature* would be needed to enter deeper into this important historical issue. It does seem, however, that the demand for better 'provision' for research was not primarily directed towards finding the funds for major pieces of research apparatus, or to setting up larger, more sophisticated research institutions. There were complaints, of course, at the way in which funds for 'large appliances' such as telescopes have to come from private benefactors rather than from public funds (1869; 1, 263: 1870; 1, 316, 375, 409: 1874; 11, 62: 1879; 21, 19, 47: etc.). This was not yet the age of Big Science, with all that that can imply for science, for scientists and for society. The real call was for better endowment of research as a profession, with many more scientists being supported directly in paid employment.

The scientific community of the day was not so small, after all. Leone Levi estimated that there were about 120 learned societies, with an aggregate of 60 000 members. Allowing for overlapping membership,

> 'we arrive at the interesting fact that there are in the United Kingdom, 45 000 men representing the scientific world, or in the proportion of fifteen in every ten thousand of the entire population; the "upper ten thousand" of the aristocracy of learning being thus three times as many as the "upper ten thousand" of the aristocracy of wealth' (1869; 1, 99).

But, despite the value of scientific research to industry and employment,

> 'there is in this country no recognized payment for the labours of scientific discovery, and no provision for the support of men who investigate science. . . . In consequence of the peculiar nature of the occupation, its hopelessness as a source of emolument to the investigator, the great skill and extreme self-denial required, and frequently danger incurred in its pursuit, and the consequent great difficulty of achieving success in it, *scarcely one person in one million of the population of England is exclusively devoted to it*, although a much greater proportion occupy a small amount of their time in its advancement' (1870; 1, 624).

Why was this deficiency not remedied? On what grounds had journalists previously regarded 'the proposal to endow scientific research as a visionary and wild scheme', so that it was still necessary to argue very cautiously 'to show that it is practicable, by means of a judicious application of precarious salaries, to train up a class of scientific investigators, and that it is a safe investment to give endowments to young men before they have reached eminence in their studies?' (1874; 11, 1).

Nature was evidently ahead of public opinion, even among scientists, on this point. Two letters to the Editor in 1870 suggest some of the objections that must have been widely held. A. R. Wallace (1, 288) argued forcefully that

> 'the State has no moral right to apply funds raised by the taxation of all its members to any purpose which is not directly available for the benefit of all',

hence,

> 'if our men of science want more complete laboratories, or finer telescopes, or more expensive apparatus of any kind, who but our scientific associations and the large and wealthy class now interested in science should supply the want.'

Thus,

> 'if we once admit the right of the Government to support institutions for the benefit of any class of students or amateurs, however large and respectable, we adopt a principle which will enable us to offer but a feeble resistance to the claims of less and less extensive interests whenever they happen to become fashion.'

This letter was soundly rebuffed, of course, in an editorial in the same issue (1, 279); but recent political events in Britain and the United States show that it expressed a viewpoint that can still find substantial favour in the public mind.

There must also have been many older scientists who would associate themselves with the 'garrulous old man' who signed himself '*in sicco*' (1, 431) and asked, 'In what way can Government most beneficially interfere with the spontaneous energy of original scientific labourers', and then spelled out all the theoretical objections to paying research workers either at piece rates 'by results' or beforehand for work they were expected to do. It is an elegant bit of rhetoric, not without applicability to the modern scientific profession. Certainly, conditions in our own academic profession are not so far from those envisaged in the draft scheme, where

> 'no candidate is to establish his claim to a permanent endowment until he has previously served an apprenticeship of some ten years, during which he must furnish continual proofs of his aptitude and diligence, and will receive regular payment by results amounting to a continuous salary if his work is satisfactory' (1874; 11, 2).

In other words, science at the time of Alfred Nobel was to be regarded primarily as a *vocation*, from which the scientist was to be rewarded as much in personal satisfaction or public acclaim as in material gain. To pay for research was to corrupt it, not only bringing it down from the transcendental plane of truth for its own sake, but also reducing it to 'hack work' whose intrinsic quality was then suspect. So strong was this sentiment, even outside science, that *Nature* had to combat a proposal that scientists would happily examine patents for free, even though lawyers, of course, would expect to be paid fees for their part in the same work (1874; 11, 191).

Scientists in Society

Not being directly dependent on the production of research results to make a living, the nineteenth-century British scientist had all the freedom he needed to exercise social responsibility. But this spirit of enthusiastic amateurism disconnected science from other social realities and gave to research a superficial air of dilettantism which could easily slip into a frivolous indifference to the social consequences of what was being sought or accidentally found. It has to be admitted that the world of science, a century ago, was far more self-sufficient, far more inward looking and parochial than it can afford to be today.

Yet the scientific community was by no means an isolated, disregarded, powerless sectarian group of 'outsiders'. *Nature* speaks in the confident voice of the professional upper-middle class, well aware of its superiority to the masses, in education, culture and material well-being, and not compelled to kowtow to the aristocracy or the plutocracy. The guest of honour at its ceremonial occasions might well be the Prime Minister of the day, and the leading scientific notables evidently had personal access to the centres of political and economic power. Although the *formal* links between science and the government were limited, there were *informal* connections – especially through higher education – through which scientific opinion could have its influence within the Establishment. This opinion was fragmented and practically unconcerned with the major political and social issues of the time. It could easily be doped into complacency about the ills of poverty and social degradation, or of militarism and imperialism, by a mild dose of social Darwinism (1869; 1, 183). But it was not crudely reactionary, nor vulgarly arrogant, nor shrill with demoralized radicalism.

For all the frustrations that they suffered in their civilizing mission, the mid-Victorian scientists were quite confident that this mission was on its way to success. In science they had the *practical* means of improving society:

> 'It is an important national question, "By what means can employment for workmen in this country be increased?" My reply is "By encouragement of experimental scientific research".'

They also had the *spiritual* means of achieving progress (1870; 1, 623). In their enlightenment, they could ask a leading politician to 'set himself to rescue politics from its present degraded position as a mere theatre for party strife, and to elevate it into something like a science of national life and progress' (1880; 21, 295), thus indicating their very real feeling that the scientific attitude is the ultimate in social responsibility.

Has this attitude changed at all, in the scientific community, in the past century? Might not a modern editorial in *Nature* go on in the same vein as:

> 'There are one or two eminent men of science in parliament, but no one of either party ever seems to think of looking at any measure or any line of conduct apart from party bias, and solely as a matter for scientific consideration. . . . And we should advise those of our public men who are really desirous to discover the science of statesmanship, and to guide their public conduct by its principles, to leave the method of agitation alone for a period, and take to calm but rigid scientific research in their own department, and we are sure the results will surprise even themselves.'

A vain hope, alas, now as then!

The little historical investigation that is reported here is too weakly documented to prove any unusual hypothesis about the past. The whole question of how scientists in the time of Alfred Nobel saw themselves as social actors begs innumerable other questions of social psychology, of politics, of cultural history, and of the influence of science itself on human affairs. The particular archive from which I have sampled is certainly highly relevant and worth detailed study for evidence on just such questions, but needs to be supplemented from many other sources. But the working hypothesis of thematic continuity was not seriously disconfirmed, even on issues such as science and war, for which I had confidently expected to find a decisive change of tone. One could even pick up the traces of many particular controversies, such as the place of science in the primary school, or the value of numerical indicators of the fighting power of weapons, which seem to us to be so distinctly modern.

This selection of articles for attention, and of passages within them, has been a purely personal responsibility. Somebody else might read those volumes quite differently. For me, at least, it has been a valuable experience of the long time-scale of change in the inner ideology of a culture that seems to have been utterly transformed outwardly. I can clearly hear the characteristic thoughts and emotions, the enthusiasms and distastes, the concerns and indifferences, of my very own academic colleagues in the words of those nineteenth-century scientific worthies, drawn away from their laboratories for a little while to discuss education and employment, health and safety, peace and war, and other such issues of the 'outside' world. That could almost be heartening, for it suggests that *our* pessimism may be no more justified than *their* abounding optimism: if so few of *their* confident hopes for progress came to full fruition, need we be quite so dismal about *our* fears for the future, which could be equally without foundation?

DISCUSSION

A. G. KELLER

Department of History of Science, University of Leicester, Leicester, UK

Both Professor Schroeder-Gudehus and Professor Ziman reminded us how self-conscious scientists had become by the closing years of the nineteenth century; how aware of the social impact of their ideas; and how eager that their work be properly appreciated by the general public. To us, the age of Big Science often appears to have been inaugurated only by the terrible demands of the Second World War. But the criterion of Bigness must needs be based on a comparison, and, by the standards of 1900, scientists, who knew how private and modest science had been but could hardly imagine what science was soon to become, might well declare that their science was already a *Grossbetrieb* – big business, indeed *Grosswissenschaft* – big science (Schroeder-Gudehus cites the very phrase used by Mommsen in 1890). In several retrospects drawn up toward the end of the century, the same conclusion was reached: mighty are the deeds of science, which has transformed life to a greater extent during this wonderful nineteenth century than in many ages before. In the grand peroration to his Presidential Address to the British Association for the Advancement of Science in 1900, the anatomist Sir William Turner told over the triumphs of technology, inevitably ascribed to science. 'Great is Science,' he proclaimed 'and it will Prevail.'[1] Even if some scientists were less happy at this mental coupling, and regretted that ordinary people were more fascinated by marvellous engines and miraculous cures than by theoretical ideas, they had to acknowledge that technology was the justification of science. Future scientific progress would, moreover, cost increasingly more and demand of scientists more collaborative effort, to exploit expensive apparatus in expensive laboratories, which would make their work more like that of the industrial factory.

Most seem to have viewed this prospect quite cheerfully. In the 1899 Presidential Address to the British Association, Sir Michael Foster declared that 'the story of natural knowledge, of science in the nineteenth century . . . is, I repeat, a story of continued progress. There is in it not so much as a hint of falling back, not even of standing still'; and, furthermore, 'the material good which mankind has gained and is gaining through the advance of this science is so imposing as to be obvious to everyone, and the praises of this aspect of science are to be found in the mouths of all.'[2] Not quite all, perhaps. An uneasiness that had been voiced throughout the nineteenth century was not stilled: romanticists and idealists, those who rejoiced in the name of 'decadent' and those who complained at the decadence of their age, often accused science of being the source of the rationalism and materialism they abhorred. But such doubters were in the minority; the pride and self-confidence expressed by Turner and Foster would be echoed enthusiastically by far the greater number of their audience, and by the world at large. Only, this view could not be taken for granted; it had to be defended.

If science was at that time taking over, whereas nowadays it is rather defending ground already won, it remains understandable that the particular problems of the interaction of the scientific community with the rest of society have remained essentially similar over the past hundred years. Ziman claims a striking continuity even in detail: when discussing how science should be organized, how properly applied and properly understood, whether on issues of education or of industry, for peaceful expansion or military development, he finds almost the same arguments as might be used in comparable debates today. Presumably that is because as soon as science began to cost money, the public had to be shown that science would be good for them. Hence, the need for more widespread and effective scientific education, and for the employment of scientifically trained people to improve productivity, and to innovate in industry. The columns of *Nature* would be bound to produce evidence of these concerns, inasmuch as the journal had been founded by Norman Lockyer with this objective very much in mind, as Ziman notes. Lockyer remained the active publicist. In 1903 he devoted his own Presidential Address to the British Association to the 'Influence of Brain-Power on History', by which he really meant, the influence of science and scientific attitudes of mind. Then, in 1905, he was active in setting up the British Science Guild, because he felt these attitudes needed a more institutional projection and neither the Royal Society nor the British Association were doing their duty in this respect.[3]

When did these views first come to the fore, at least sufficiently so as to demand so much space in journals and affect the activities of scientific associations? In most parts of Europe, similar opinions could be heard from at least the 1830s, if not before: the founding of associations for the advancement of science, first in Germany, then in Britain, France and elsewhere, is evidence enough of that. But a general awareness of the way science could and would affect the organization of society – as cause and as effect of the role of science in industry – that, I suspect, dates from the 1850s and 1860s. The foundation of *Nature* in 1869 is in itself a clear sign; the previous year, Matthew Arnold, in a famous study of higher education in German-speaking lands, had pointed to the importance of science as a key to the success of their system. It was the visible prowess of the chemical industry, primarily in Germany, in the synthesis of new dyestuffs, in pharmaceuticals and in explosives, which made these developments public issues elsewhere. The progress of chemistry encouraged hopes of a revolution in public health through pharmacology and through the development of bacteriology (founded by a chemist turned microbiologist, Pasteur).

There were, of course, at the same time those who saw that this progress in chemistry could have a harmful side, for these new industries might introduce new forms of pollution. The first attempt to control pollution in Britain, the Alkali Act, dates from the same period – 1863. Chemistry might also introduce more terrible weapons of war. Even before the first number of *Nature* had appeared, a friend of Lockyer had written to him to beg that he would not entertain therein 'anything relating to war as breaking in upon the feeling of quiet and rapt contemplation which the name of Nature suggests'.[4]

The Franco-Prussian War in 1870, however, underlined the message of the past decade, convincing not only the defeated French but also neutral onlookers that science is the mainspring of industrial power and nourishes the sinews of war. Sir Henry Roscoe recalled how, when the British Royal Commission on Technical Education visited a trade school in Rouen in 1883, they noticed a Prussian soldier's helmet in the school museum. The principal explained that it had been picked up in the streets of Rouen during the German invasion. If any boy was slacking at school, he was summoned, found the helmet on the head's desk, and was told, 'Now, if you do not make progress and learn properly, this will happen to you again. . . .'[5] Roscoe had himself studied in Heidelberg before he became professor of chemistry at Owens College, Manchester (and subsequently for ten years an influential Liberal MP), so he took the point; he remarks that the members of the commission used the story many times as they travelled about Britain, 'preaching on the subject of educational progress'.

Still, by the 1880s and 1890s, most scientists and industrialists could believe that henceforth real shooting wars, with Uhlans riding through conquered cities, would be a thing of the past, at least in western Europe. Wars would be fought, if at all, outside the pale of civilization, in remote colonies or in the 'Near East'. Great Powers would compete more peacefully; industrial rivalry would be the bloodless battlefield of the future. The appalling prospect of what we would call a 'hot' war must deter any statesman from embarking on it. The language of military conflict was often used by the advocates of science in the service of nationalism; they could safely talk like that – it was only a metaphor.

The growth of the electrical industry during the last quarter of the century took physics along the road already travelled by chemistry. Electricity offered greatly improved communications, and a new heavy industry without the pollution caused by the discharge of smoke into the atmosphere. Electricity might even reverse the social disorder created when craftsmen were driven into factories by the Industrial Revolution. Physics profited from these exciting visions: one might point to the discoveries and inventions made in the year of Nobel's death, 1896, at the very frontier of physics: Röntgen's X-rays, Becquerel's uranic radioactivity, the first public demonstration of Marconi's wireless telegraphy. Understandably, not only fantasists like Robida, but serious commentators could predict that the beaming of pictures by wireless across the sea would soon follow.

The increasing ease of communication made international collaboration more practical than ever and led to, among other things, the growth of international meetings and associations, which really began only in the 1850s and 1860s. Schroeder-Gudehus suggests that the International Association of Academies was called into being in 1899 less to satisfy the call for an embodiment of the intellectual fraternity of scientists in a world union of their grand councils than to promote the academies themselves, so that narrower national objectives were not set aside. The academies feared that they would cease to be the central authorities for science in their states – the keystone of the arch of national science – since each scientific discipline was constructing its own arch, crowned by its own international body. Within each, there were practical needs for a fuller exchange of information, for cooperation in the containment of disease, and for the establishment of standards for chemical and electrical industries. Industry was developing multinational dimensions, and science would have to do likewise. But how?

The history of the International Association of Academies gives perhaps a somewhat different picture from that which might be displayed by one of the discipline-oriented bodies. The academies were the senates of science. Acceptance into their ranks had become a formal mark of honour. Usually, they were supported by the State and acted almost as part of the state apparatus. Always, they enjoyed a goodly measure of access to power and were called upon to advise ruling elites. Other associations with less restricted membership might be more democratic and more independent of the nation state. In any case, the practical need for collaboration across national boundaries was always justified by the rhetoric of scientific internationalism and free interchange of ideas. In orations on formal occasions, prominent scientists might feel entitled to use a grander and more idealistic language than in everyday business (which must certainly be borne in mind when reading those 'speeches from the throne of science', the presidential addresses to the British Association, or Mommsen's address, quoted by Schroeder-Gudehus). Nevertheless, they presumably said nothing they did not sincerely believe. Unless most of the audience shared the ideals, there would be little point in the orator's phrases. Just as in the early days of modern science, in the seventeenth century, science was seen as the conciliator, which would bring together those divided by their religious confession; so at the turn of the nineteenth-twentieth centuries, when jingoism was on the march in so many countries, scientists, whose work had often brought them into friendly contact with their foreign peers, could genuinely believe that their science would conciliate hostile nations.

The wider public might well share this hope. A full-page cartoon published in *Punch* in September 1899 celebrates a joint session of the British Association meeting at Dover with its French equivalent, the Association Française pour l'Avancement des Sciences, whose members had come over from Boulogne. The two associations are depicted as two girls collecting shells on a Dover beach, shaking hands with one another, with the comment, 'We, at least, can meet with neighbourly cordiality', while a smaller Belgium looks curiously at a crab (who could that be?).[6] The Fashoda incident, the Dreyfus case, the South African War had awoken political coolness; but science made friends. It is true that these associations found practical cooperation harder than the ideal, as Pancaldi has pointed out.[7] However, the joint meeting did serve to advertise the new form of international communications technology, when Mr Marconi beamed his wireless messages, for the first time, across the Channel. The following summer, no less than 126 international congresses took place at the Paris Exposition Universelle. Nearly half were devoted to scientific and technical cooperation, including congresses devoted to all the major scientific disciplines. It was at one of these, the physics congress, that the world first began to appreciate that research into the innermost recesses of nature had led not merely to the discovery of another new element, radium, but was foreshadowing powers in medicine and the control of energy that would leave behind anything achieved in the nineteenth century. Indeed, a special congress had already been devoted to 'Medical Electrology and Radiology'. In such times, the coming together of the central national councils of the Sages of Science could well seem more than a merely practical business. Turner, in his 1899 address to the British Association, welcomed the prospect of a World's Fair that would encourage 'international communing in the search for truth', and, above all, 'witness the first select witenagemot of the science of the world'.[8]

Notes

1. *Report of the Seventieth Meeting of the British Association for the Advancement of Science held at Bradford in September 1900*, London, 1900, p. 30.
2. *Report of the Sixty-ninth Meeting of the British Association for the Advancement of Science held at Dover in September 1899*, London, 1900, p. 15.
3. Meadows, A. J. (1972) *Science and Controversy. A Biography of Sir Norman Lockyer*, London. (Founding of *Nature*, pp. 25–29; his address to the British Association, pp. 265–267; the founding of the British Science Guild, pp. 272–277.)
4. J. J. Sylvester to N. Lockyer, cited in Meadows (note 3), p. 27.
5. Roscoe, H. E. (1906) *The Life and Experiences of Sir Henry Enfield Roscoe . . . Written by Himself*, London, p. 198. For the background to this Commission, see Cardwell, D. S. L. (1972) *The Organisation of Science in England*, revised ed., London, pp. 132–136.
6. *Punch*, 26 September 1899, Vol. CXVII, p. 134.
7. Pancaldi, G. (1981) *Scientific Internationalism and the British Association*. In: Macleod, R. & Collins, P., eds, *The Parliament of Science*, London, pp. 145–169. Compare the British Association's promotion of imperial links, discussed in an accompanying article: Worboys, K. *The British Association and Empire: Science and Social Imperialism, 1880–1940*, pp. 170–187.
8. Turner (note 2), p. 22.

II

New paths of scientific inquiry at the turn of the century: Physics, chemistry, biochemistry and physical chemistry

FIN-DE-SIECLE PHYSICS

J. L. HEILBRON

Office for History of Science and Technology, University of California, Berkeley, California, USA

During the last third of Alfred Nobel's life, Western countries began both to enjoy and to regret the fruits of high technology. While the telegraph and telephone bound the world tighter, the electric tram and subway helped enlarge the metropolis; the electric light caused day to encroach further on to night; new dynamos, hymning 'la déité du jour, l'électricité',[1] held the promise and threat of further transformations. No less novel than the scope and pace of change were its principal agents: discoveries made by academic scientists and exploited by engineers trained at a university or technical school.

The increasing tempo of civilized life and the squalor of the cities in which it was played out caused legitimate concern about uncontrolled technological progress and reservations about the sciences that were supposed to be responsible for it. To respond to these worries, and to fit their new prominence, physicists adjusted the image and even the substance of their discipline. Adjustment also appeared necessary to accommodate both recent advances and persistent difficulties in physical theory. The physicists of the *fin de siècle* sought to redefine their professional objectives so as simultaneously to achieve internal consensus and to secure their place in the wider society.

Their concern produced a large literature, in which speeches before the grand assemblies of scientists and fellow travellers, such as the American and the British Associations for the Advancement of Science (AAAS, BA) and the Gesellschaft Deutscher Naturforscher und Ärzte (GDNA), figured prominently. A common theme pervades this literature, despite the diversity in nationality, culture and general philosophy of its authors. I call the theme 'descriptionism', after a nineteenth-century word for a mere describer or reporter.[2] It signifies the least common denominator among conventionalism, instrumentalism, nominalism, etc.; and tolerance, even encouragement, of diverse, partial or complementary approaches to physical theory. Descriptionism recommends withdrawal from big questions and relaxation of claims to knowledge of truth.

In this paper I exhibit the state of physics around 1900 by describing the descriptionist literature, by identifying its objectives and estimating its success, and by confronting it with the material circumstances of the physicists. Descriptionism had some consequences the reverse of those intended and symptomatic of the compromises practised to obtain public subventions for private investigations.

I must emphasize that my primary historical problem is to determine the origin and purpose of a certain body of literature. Whether its authors believed fully in their rhetoric, or whether they overrated the strength of the opposition they hoped to placate, are questions beyond the reach of my method. They are also unimportant for my inquiry.

Descriptionism

Around 1870, spokesmen for physics taught that their science sought truth in the form of immutable principles, like Newton's gravitational attraction and the 'laws' of thermodynamics, or in the precise definition and measurement of a physical quantity. It went without saying that principles and theories, and entities to which they referred, like atoms and aethers, should constitute a unified and coherent whole. One looked forward to a time when all of physics might be subsumed under a 'single law, which will be the true one, accounting for everything'.[3] This optimistic and perhaps naïve opinion did not suit the up-to-date physicist of the 1890s. He endorsed instead the views of the mathematician and mathematical physicist, Henri Poincaré.

As keynote speaker at the grand international congress of physics, convened in Paris during the Universal Exposition of 1900, Poincaré recommended humility and did not hope for unity. Physicists, he said, are not decipherers of Nature's laws, but librarians and cataloguers of experience. The theorist draws up a catalogue; the experimentalist fills the gaps exposed with new facts bought from Nature with the modest means of his laboratory. The librarians of science may arrange and rearrange their collections as they please; what unity science possesses is secured by convention; theories and principles are not true or false, but more or less useful. The arbitrariness of general principle applies *a fortiori* to physical models, not excluding the atom and the aether. These are 'indifferent hypotheses', mere aids to the consultation and extension of the only items of permanent value in the library of science, phenomenological relations. 'If we know these relations, what does it matter if we judge it convenient to replace one model by another', or use conflicting or imperfect ones?[4]

Poincaré's talk was approved as 'one of the most perfect expressions of the state of mind of the masters of modern science'.[5] When his mind is in that state, the physicist holds that his aim is 'to describe in the shortest possible way how his senses have been, will be, or would be affected' by this or that experiment. So J. H. Poynting told his fellow physicists in his presidential address to the BA in 1899.[6] Repudiations of truth-seeking were already a commonplace when Poincaré mounted the podium in Paris. 'It is not the phenomenon itself one knows, but only a representation of it.' 'The philosophical scientist never sees absolute truth in a theory, but only a tool to ease his mental grip of phenomena.'[7] 'Our theories are not reality . . ., they are rather to be compared with a picture that is more or less similar to the original.' 'We recognize the futility of attempting an ultimate explanation of natural phenomena.'[8] Truth resides in a deep well, nay in a bottomless pit; we shall never reach it, and, if we did, would not recognize it.[9] 'The truly scientific mind [has] long been familiar with the truth that a so-called law of Nature is simply a convenient formula for the coordination of a certain range of phenomena.'[10]

Elements of descriptionism may be found in Kant and Comte, and in the instrumentalist physics of the late eighteenth century, to go back no further. Its proximate sources within physics in the 1890s were the practice of Kelvin and Maxwell, the scruples of Gustav Kirchhoff, and the jeremiads of Ernst Mach. To these perhaps should be added the philosophy of Emile Boutroux. It is characteristic that the British came to their position by doing physics, the continentals by talking about it.

Eschewing inquiry into the fundamental nature of force, Kirchhoff had declared in 1875 that the principles of mechanics, and even Newton's equations, were not laws laid down by Nature but descriptions invented by physicists.[11] The obligation of analytical mechanics, the most basic and secure of the sciences, is 'to describe in the simplest and most complete manner the motions occurring in nature . . ., not to find causes'; apparently it ranked no higher than natural history, which physicists since Newton had depreciated as unmanly gossip.[12] At the same time, Ernst Mach was warning his colleagues that by taking their theories too literally they had missed the basic truth of science, that it has nothing to do with truth, but only with precise, economical and simple description.[13] Poincaré's brother-in-law Boutroux, a religious man, found it convenient to stress the investigator's freedom of choice and the possibility of evolutionary change in phenomenological relations. He thus convicted science of being a large admixture of 'contingent' elements arising from peculiarities of time, place and taste.[14]

Mach's views at first met with indifference, Kirchhoff's with astonishment, Boutroux' with antagonism;[15] none prospered among continental physicists until Heinrich Hertz' demonstration of electromagnetic radiation, predicted by Maxwell, brought them to examine the British method in physics. Many professed to be shocked at what they found. Maxwell appeared to have no reverence for rigour. He employed contradictory mechanical analogies simultaneously, or jumped from one model to another. Poincaré noticed the 'feeling of discomfort, even of distress' with which Maxwell's acrobatics afflicted French readers, who, from 1889, could read his *Treatise* in their own language, with annotations by two distinguished *physiciens*. 'In passing from Maxwell's text to the notes of MM. Cornu and Potier one experiences the same feeling as in leaving a thicket for a well-arranged park.' As for Hertz, he acknowledged that he had never managed to follow the derivation of the equations from the model; while his countryman, J. W. Hittorf, unwilling to find physics unintelligible, fell into a depression over British electrodynamics.[16] But success brings its own recommendation. Continental physicists desired to know how Maxwell could erect an unshakeable structure from a rickety scaffolding of borrowings from the Victorian machine shop. Hertz, and Maxwell's stalking horse Poincaré, observed that since the models were intended only as descriptions or aids to description, they need not be exact or consistent, and could be chosen and discarded at will, without menacing the advances achieved with their aid.[17]

The British had invented *laissez-faire* physics. According to Ludwig Boltzmann, one of its continental admirers, the British method fitted much better with the *Geist* of science than the ordinary ideal, which required that all parts of a theory fit together. Theoretical entities conceived as analogies 'need not agree completely with nature'.[18] Maxwell's models could be defective. The kinetic theory of gases too had its faults. 'Analogies always break down somewhere or other.'[19] So what, if they bring 'a *soupçon* of truth', a few new relationships?[20] Since the physicist cannot attain to the essence of matter, he would be foolish to deprive himself of any aid his weak mind might suggest. He should be as free as his fellow citizen, 'à la liberté et la matière' should be his watchword.[21] He should be encouraged to change his mind, to be open and flexible. 'Only in adopting different and sometimes opposing viewpoints do the sciences progress', wrote the indulgent author of the official report on science for the Universal Exposition of 1900. 'Let us not mutilate the human spirit in the great task it has to accomplish.' Clarity and coherence are overrated, and even damaging. Heed the warning of Arthur Schuster: 'By purposely excluding everything that is vague from a physical treatise, we destroy all possibility of making the work useful in stimulating further research.'[22]

Meanwhile Mach began to prosper. Kirchhoff's teaching, endorsed by Hermann von Helmholtz, became a household word. Karl Pearson encountered it in Berlin in 1879, and made it the cornerstone of an aggressive descriptionism approved by Mach; and Mach himself, while emphasizing that he was the prior and greater radical, allowed that the sympathetic hearing of his own views owed much to Kirchhoff's authority.[23] In 1895, Mach was called to a new chair in history and philosophy of science from which to disseminate his epistemology. By the turn of the century he had a large following. Although both Pearson's *Grammar of Science* and Mach's empirico-criticism were too authoritarian to permit free use of analogy, around 1900 they seemed compatible with descriptionism. Their views were conveniently conflated with those of Boltzmann, von Helmholtz, Hertz, Kirchhoff and Poincaré,[24] and sometimes – in what appears to be a masterpiece of miscegenation – with those of that most resourceful analogist, J. J. Thomson.[25] The epistemologies of Kirchhoff and Maxwell mingled into the 'Fleisch und Blut der heutigen Naturwissenschaften'.[26]

What is convenient and simple to one mind can easily befuddle another. The wide descriptionist consensus of 1900 accordingly did not include agreement about the best mode of description. The great majority held the reduction of all physical phenomena to the principles of mechanics as the prime *desideratum,* and even as the definition of their subject. 'All physicists agree that the problem of physics consists in tracing the phenomena of nature back to the simple laws of mechanics.'[27] Nothing could be simpler, says Kirchhoff; according to Maxwell, anyone who has understood the concepts of mechanics — configuration, motion, mass, force — sees immediately that 'the ideas they represent are so elementary that they cannot be explained by means of anything else'.[28] And there was reason to believe that the reduction might be possible. The principle of energy conservation seemed to guarantee it; the unification of optics and electrodynamics had reduced the reductions required; the gas theory showed how to proceed. Reviewing these accomplishments, the President of the French Physical Society, Alfred Cornu, opened the Paris Congress with these ringing words: 'the spirit of Descartes hovers over modern physics, no, he is its beacon; the greater our knowledge of natural phenomena grows, the more developed and precise the bold Cartesian conception of the mechanism of the world becomes: "there is nothing in the physical world but matter and motion".'[29]

There is, however, a great variety of mechanical descriptions. Should we be content to state that the equations of electrodynamics and thermodynamics are formal consequences of an abstract formulation of general mechanics, such as the principle of least action? That is the recommendation of Poincaré and von Helmholtz. They observed, moreover, that if least action is satisfied, any number of descriptions can be found, with diligence. They consequently deemed it a waste of their time to look for any.[30] Or should our thoughts be illustrated by a general mechanical system, say a swarm of atoms or whirlpools in an aether? We would agree with Cornu, Marcel Brillouin, Woldemar Voigt and many more, and, in some of their moods, with Kelvin and Thomson. Or should we descend to detailed indifferent hypotheses, to the 'robust form and vivid colouring of a physical illustration', to 'full-bodied and pictorial imagery and mechanical models of what [is] going on'?[31] Here we regain the peculiarly British practice that established, though it did not define, descriptionism.[32]

The British recommended their style as good pedagogy. By imagining ourselves to be part of the machinery of Nature, we can feel its strains and anticipate its motions. Physics becomes easier. 'There is no doubt but that a concrete mechanism that we can distinctly picture the working of is enormously easier to reason about than one of whose structure we know nothing, but only know general laws of its action.'[33] 'In a word', writes the first historian of descriptionism, 'mechanism sums up the instinct of the race. We cannot think clearly without it.'[34] It also fits perfectly with our laboratory apparatus: facts obtained with the help of factory tools lend themselves to interpretation by mechanical analogy.[35]

A few decried the priority that was generally accorded to mechanical reduction. Some, for example Gabriel Lippmann, claimed that irreversible phenomena would forever elude reduction to the reversible laws of mechanics.[36] Mach and Duhem condemned the priority of mechanism as an historical accident. Opposition went furthest with Wilhelm Ostwald, a physical chemist allied with Mach.

In a famous speech he urged the GDNA to give up explicit pictures: 'thou shalt not make unto thee any graven image, or any likeness of anything'.[37] He recommended replacing matter with energy, of which the transformations constituted the sole legitimate study of the physicist. He suffered for his advice. His chemist friends told him they could not solve their problems without the concept of atoms.[38] German physicists had only contempt for his effort to replace matter with energy. The French referred him to Poincaré for lessons on method and worried for the souls in his professional care. G. F. Fitzgerald answered for the British: 'That may be allright for a German, who plods by instinct, but a Briton wants emotion in his science, something to raise enthusiasm, something with human interest', viz. a mechanical model.[39]

The apologetic air of descriptionism ill accords with the vigorous advocacy of mechanical reduction and the optimism of the Paris Congress. Nor did the doctrine correspond to physicists' attitudes toward their daily work; in their experiments and research reports, aethers and atoms and electrons enter on the same footing, as regards reality, with workbenches and colleagues. In this they resembled the philosophers derided by Balfour, who, 'knowing perfectly as an abstract truth that freedom is an absurdity, yet in moments of balance and deliberation invariably conceive themselves to possess it, just as if they were savages or idealists'. Several physicists remarked that descriptionism, though perhaps the only defensible philosophy, conflicts with the call of science: 'its introduction into the objective study of nature is misleading, and tends to confusion'.[40] Only when preparing public lectures can most physicists afford to be epistemologists.

One might seek to explain the insistence of *fin-de-siècle* physicists on a philosophy so little relevant to their practice as a consequence of the equivocal state of physical theory around 1900. Mechanism was then pressed not only at the outworks of its fortress, by Ostwald's ineffectual guerrillas, but also in its armoury, where Lord Kelvin, glancing up from the springs and rubber bands from which he was wont to fashion aethers, spied two clouds over the dawning twentieth century. He reminded physicists that they had no acceptable account of the interaction of aether and matter, and that an inescapable consequence of the gas model, the equipartition of energy, failed before the facts. These problems bothered many physicists, and, according to the Exposition historian, caused 'a kind of backing away in the boldness of scientific thought'.[41]

Kelvin's clouds presaged fundamental change. But around 1900 they did not appear to condemn the mechanical theory. Kelvin thought that he could dissipate the second cloud, and Planck, whose work did so, initially believed that his quantum brought nothing radically new.[42] As for the first cloud, it was knocked down by relativity; but Poincaré and H. A. Lorentz, who together played Moses to Einstein's Joshua, did not believe that they were leading the tribe of physicists to a new land. Malaise over the reach of mechanics or over the relations between mechanics and electrodynamics does not account plausibly for the constant sermonizing in favour of descriptionism. Physical theory had been in difficulty before, and would be again, without bringing its professors to professions of intellectual impotence. Why then?

The 'Fin de Siècle'

'The prevalent feeling is that of imminent perdition and extinction.' Thus Max Nordau opened his compendious catalogue of degeneracy in art and literature in the 1890s. He regarded degeneracy as a heritable nervous disease, brought on by use of alcohol, pollution and overwork and characterized by ennui, despair, pretentiousness, mysticism, pornography, diabolism, Ibsen and impressionist painting.[43] Writers and artists of the period did dwell on things they esteemed degenerate and decadent, including themselves. Nothing is more modern, said Nietzsche, than this general sickness, this oversensitivity of the nervous system, this lust for the crazy and incongruous. 'Modern is Paul Bourget [a naturalist writer turned Catholic] and Buddha, and the splitting of atoms.'[44]

Statistics appeared to prove that the lower classes were wasting away at least as quickly as their poetical betters. In 1895, the Austro-Hungarian Empire had trouble finding enough healthy recruits to fill its armies. Inquiry disclosed that in Germany and France, too, the rate of rejection of men unfit to be cannon fodder had increased steadily since the institution of compulsory military training. The British had to throw back many of the men fished up for service in the Boer War as too short, too light or too stupid. Less than a third of American inductees in the First World War scored above an intellectual age of fifteen on the standard intelligence test, and forty-five percent scored below age twelve. These facts, together with statistics about sharp increases in crime and suicide in cities, gave cause even to those indifferent to the perversion of literature to take seriously the proposition that the sun was setting on the West.[45]

Science was implicated in both spiritual and physical degeneracy. On the one hand, its indirect products – particularly improvements in manufacture, transportation and communication – encouraged the growth of cities. The resultant poverty, bad air, poor hygiene and nutrition, adulterated foods, and, above all, the frantic pace, pushed by the factory and aggravated by the daily press, produced that 'state of constant nervous excitement' remarked by all students of degeneracy.[46] Physical science and its technology had grown too rapidly for the human race to adapt. They are cancers, in the opinion of Paul Bourget, growing amok, menacing the life of the society that supports them. It occurred to some that society would do well to clamp down on its physicists, to enact a moratorium on research, to alter its suicidal course. 'Progress . . . may be too fast for endurance.'[47]

Secondly, science (by which its detractors meant scientism, naturalism, or scientific materialism) contributed to depression by claiming to remove mystery, spirit and choice from the world. The mechanistic picture derived from the principles of energy, evolution, chemical physiology and psycho-physics appeared to leave no room for faith or poetry.[48] The novelist, said Zola, should copy the scientist, showing how the actions of his characters are determined by their physical make-up and circumstances. The methods of physics and chemistry apply not only to physiology and medicine, but also to 'the investigation of human sufferings and feelings'.[49] The vague and unintelligible, the intuitive and spontaneous, shrivel before science, 'the eternal and immortal'. 'To her alone', says Zola, 'belongs mystery, for she marches continually to its conquest.'[50] The brain is no more spiritual than a kidney; the one excretes thought as the other does urine. The guardian of morality is not religion but science, which inculcates probity, accuracy, tolerance and modesty (!), and eliminates the error and superstition on which the Church bases its pathetic ethics. Science emancipates the human spirit; 'humanity knows no other guide'.[51]

These pretensions seemed to many not only inane but dangerous as the technical applications of science subverted the old order. Here, *avant-garde* degenerates, eager to free themselves from impending determinism, from the 'hard chain of cause and effect that science would succeed in untangling perfectly the day after tomorrow',[52] joined forces with their fierce opponents, the champions of yesterday, the spokesmen for established religion and education, the antirepublicans and ultramontanes. Science, they declared, is morally bankrupt. 'Bring up a woman in the positivist school, and you make of her a monster, the very type of ruthless cynicism, of all engrossing selfishness, of unbridled passion.' A man bred to science 'will view his mother's tears not as expressions of her sorrow, but as solutions of muriates and carbonates of soda, and of phosphates of lime; and he will reflect that they were caused not by his selfishness, but [by] cerebral pressure on her lachrymal glands'.[53] The teaching of science undermines not only the family, but also the values of allegiance and sacrifice on which the modern state depends; socialism and even anarchy are the likely results.[54] The threat of science does not stop at human life and community, at ethics, aesthetics, literature, music and good taste. By attacking the precepts of religion and the evidence of the soul, by destroying belief in a purposeful creation, it menaces the posthumous happiness of anyone taken in by its dogmatism.[55] 'What does it matter whether the conceptions men form of the universe are those of the nineteenth century or those of the ninth, in comparison with their acceptance of [religious] truths?'[56]

Critics complained that scientists in their turn had become dogmatic and arrogant, extravagant, superstitious and bombastic, and that their increasingly frequent meetings were so many occasions for parade and puffery.[57] In 1894, at one such gathering, the arrogance of 'denominational science' was thrown in the teeth of the BA by its president, the non-scientific Lord Salisbury; and in 1895 Ostwald courageously stood up before the GDNA to insist upon the extirpation of scientific materialism.[58] 'The new bigotry', as Karl Pearson called anti-scientism, derived satisfaction from the neighings of these Trojan horses. 'The theologically-minded have been hurrahing and throwing up their caps.'[59]

During the hurrahing, physicists discovered a new application of the epistemology of Kirchhoff and Maxwell. In science, descriptionism might legitimize shortcuts and fruitful incoherence; in the wider society, it could serve as a defence against the theologically minded. Its defensive role outside physics appears plainly in Pearson's reply to Salisbury: anyone who understands the methods and goals of science knows that, since it cannot attain to the essence of things, it cannot make the claims on which the accusations of arrogance are based. For several years after Salisbury's address the chief spokesmen for the BA, and especially the presidents of the division for physics and mathematics, insisted on the modest limits of scientific knowledge. Similar speeches were given before the GDNA; and, as we know, the organizers of the Paris Congress arranged to have the business aired in the keynote address. The tactic had some success. One noticed a decline in arrogance, if not an increase in humility, and a certain maturity, a growing out of the 'desire to announce propositions which should carry dismay and panic into the world of church and chapel'.[60]

Perceptive contemporaries understood the manoeuvre. The philosopher Edward von Hartmann: 'The more physics keeps its completely hypothetical character in mind, the better will be its scientific reputation in public opinion.' The politician Georges Sorel: 'It seems to me that [physicists] have adopted this sceptical attitude . . . because they believed that they would increase the confidence that men have in the results of science while freeing it from a compromising alliance.' The *littérateur* Paul Claudel: 'What a deliverance for the scientist himself, who henceforth will be able to devote himself in all freedom to the contemplation of things without having the nightmare of an "explanation" to maintain!'[61]

Senescence

If the human race were degenerating or stultifying, so very probably were its physicists, its senior tribe of scientists. This inference occurs in the reasoned pessimism of the historian Charles Pearson, who predicted in 1893 that the western nations, having reached the limits of colonial expansion, would stabilize internally through socialistic regimes; and that socialism would subvert family, character, ambition, competition, achievement, religion and science. Literature would die out; already the epic poem was dead, and the novel moribund: the lukewarm age to follow would do little worth celebrating. Science too was exhausting its topics. No doubt, much remained to be done in matters of detail, in diffusion, application and measurement, in the dry labour of specialists. Pearson expected their number to multiply, since their picayune occupations agreed perfectly with the enervation expected of the twentieth century. The world and the science that describes it were exhausted.[62]

Different roads lead to the same place. Oliver Lodge, a vigorous believer in the vigour of science, agreed that socialist states would destroy the possibility of excellence and with it 'the power of pursuing really advanced physical science'. Adolf Harnack, author of the official history of the Berlin Academy of Sciences in 1900, concluded that already at that time physicists had withdrawn from the big questions and, in their attention to insipid detail, had forfeited their intellectual ascendency. A. J. Balfour, nephew and successor to Salisbury, deduced the senescence of science from the teachings of evolution. Our senses, our only sources of information about the physical world, are few and imperfect. 'They were not, unfortunately, evolved for purposes of research.' We may be amazed at what has been accomplished. It will not continue.[63]

Evidence of devolution also came from the fabric of physics. Thermodynamics required the universe to end in homogeneous frigidity. To be sure, men would have ages of opportunity to exterminate themselves in other ways before the sun went out. Yet thermodynamic eschatology was hardly a happy subject. 'Science, with its record of glacial epochs and its forecast of vanishing heat; religion, with its warning that "the earth and the creatures that are therein shall be burned up", do not speak of life, but of death.'[64] Come together at last, to arrange the funeral of mankind, science and religion differed only over the temperature of the final dissolution.

Physicists did not find it easy to meet the charge – to which they frequently referred[65] – that their science was senescent. On the one hand, wishing to affirm their achievements, they could maintain that they were at the beginnings of their science; on the other hand, they could not allow that they knew enough for an adequate description of Nature. Firstly, it was not true; Kelvin's clouds obscured the electrodynamics of moving bodies and the kinetic elucidation of thermodynamics, and no more was known about the nature of gravity than when Newton stopped feigning hypotheses about it. Secondly, some physicists, worried about details of the mechanical theory of heat, concluded that, perhaps, it had given all it could.[66] To them, of course, the judgement that the theory had grown stale was an indication not of the senescence, but of the adolescence of science. Thirdly, those who paid for physics might have decided to diminish support of professors unlikely to produce more than uselessly refined measurements.

Descriptionism authorized no strong belief in the future progress of science. In the early 1890s spokesmen for physics were cautiously optimistic, but only on the soft ground of analogy. 'In view of the ever increasing pace of discovery in our time, it would be silly to want to answer this question [whether anything remained for the twentieth century to do] in the negative.' 'The progress made in the last century is not likely to cease or abate in the next.'[67] 'Who knows whether the limit has been reached?'[68] While awaiting the answer, physicists had much to do; there were 'many gaps yet to be filled, data to be accumulated, measurements to be made'.[69] In an address mocked by later generations of physicists, A. A. Michelson held that 'the future truths of Physical Science are to be looked for in the sixth place of decimals'. To Michelson, the most exact of measurers, the challenge and excitement of science had always been to determine the values of physical constants with exquisite accuracy. His statement should not be construed as pessimistic or complacent, but as an idiosyncratic expression of quiet optimism for the future of his profession.[70]

Suddenly, in 1896, the stock of physics inflated so sharply that the charge of stasis or bankruptcy lost its colour. The bearish defenders became bullish. 'Now we have Professor Röntgen's investigations.'[71] The effect on morale is neatly captured in a footnote to an address given early in 1896 on the occasion of the fiftieth anniversary of the founding of the Deutsche Physikalische Gesellschaft. The speaker had ended with the usual hedged optimism, with the hope that 'energy would not be lacking' for further advance. While his speech was being printed, he saw the first X-ray photographs. Had he known of them earlier, he said, he would have ended his talk on an entirely different note, 'in pride and joy that the second half of the life of the Society had begun as brilliantly as the first'.[72]

More joy was coming. In 1897, J. J. Thomson showed that physicists had for years been knocking chips off atoms, and he claimed that these fragments, or electrons, were the long-sought building blocks of matter. A year later, the Curies found a diamond in a dunghill, a grain of radium in a ton of pitchblend. The substance flaunted, if it did not violate, thermodynamics; its light, supposedly everlasting, rekindled embers of scientism and postponed indefinitely the heat death of the universe.[73] About the time of Röntgen's discovery, Lord Rayleigh and William Ramsay, attending to a discrepancy in the fifth place of decimals, found the noble gases, 'the most fundamental discovery in experimental chemistry since Davy'.[74] Who could hold that precise measurement did not pay off, or that the youthful creative period of physics had ended?[75]

In the image presented by the proceedings of the Paris Congress, the physicists of 1900 were proud of the progress of their mechanistic or materialistic science; refreshed by the discovery of novel phenomena; aware that their models had no value beyond pedagogy or heuristics; prepared, but not expecting, to alter their models and even their principles to accommodate the new material;[76] and optimistic that the descriptionist epistemology enunciated by Poincaré, and the discoveries of Röntgen and others, would disarm criticism of their discipline. As the president of the AAAS said in 1901: 'with respect to the future of scientific achievement, the consensus of expert opinion is cheerfully hopeful'.[77]

Material Considerations

The circumstances of physics professors at the end of the nineteenth century might not appear to be plausible grounds for their worry about reputation and support in the wider society. Why revert to descriptionist defences when the number of academic physicists, about one thousand worldwide, was increasing, at about three percent a year? And these academic physicists dominated the discipline; they produced between seventy and eighty percent of research papers in physics published in 1900, and nine papers in ten presented at the Paris Congress. The industrial or government physicist, as opposed to the technical expert, hardly existed. Around 1900 almost all the manpower, plant and money that were producing new physics, and giving intellectual direction to the discipline, were to be found in the universities and higher schools.

This academic physics was becoming expensive. The average cost of institutes built in Germany between 1905 and 1914 was almost double that of institutes built in the 1890s. Part of the increase came from sheer size: the principal lecture hall for elementary classes, which set the scale, accommodated 100 to 150 auditors in the earlier period, 200 to 400 in the later. Also, the number of advanced students requiring space for individual work began to go up. But perhaps as important as size in raising price was qualitative improvement.

Two considerations made for higher unit costs. First, the institute had to have all facilities necessary to demonstrate the major new phenomena of physics. Hence it required good vacuum lines and refrigerating mechanisms, gas and water conduits, and, above all, a plentiful electrical supply. As electricity became essential to all sorts of experiments, it was also exploited for other purposes: almost all institutes built after 1900 had electric light and lifts.

The second inflationary consideration was exact measurement. For students to learn the essence of physics, the quest for precision, by practice, they required work benches free from vibrations and rooms without electric or magnetic disturbance. Consequently institutes had special piers independent of the flooring to insulate instruments from the settling of the building, and rooms entirely free from iron, fastened with copper nails and supplied through copper pipes, to allow accurate magnetic measurements. The basic furnishings of the institutes – electrical installations, power supplies, meters, induction machines, spectroscopic gratings, motors, electro-magnets, X-ray tubes and radioactive substances – were procured from large commercial manufacturers. By 1900 physics had passed the age when the experimenter could rely on his wits and kitchen utensils.

The total being invested in academic physics in Western countries around 1900 was $2.5 million. The United States contributed the most – about 1.5 times the investment of the next biggest spender, the United Kingdom, twice that of Germany and thrice that of France. And the disparity was widening: American expenditures were growing at ten percent a year, British and German at five percent, French at two percent. Reckoned as a fraction of national income, however, the investment in each of the four countries was the same (0.005 percent); so was the fraction of the population engaged in academic physics (about three per million). At the end of the century of science, each of the main competitors had attained the same pace despite great differences in sources of funding and means of recruitment. This was, however, but a fleeting coincidence. France could not keep up; the United States, enjoying its progressive era and the philanthropy of nineteenth-century tycoons, went beyond, while the United Kingdom and Germany retained their relative positions.

It was not love of science but international industrial competition that had brought investment in physics to similar levels in the major belligerent countries. Physicists were supported to run nurseries for technicians, physicians, secondary-school science teachers, and, above all, electro-technologists, whose needs increasingly gave the tone to, and forced the growth of, the institutes. 'Next to Helmholtz stands W. Siemens.' Less than one percent of the 18 000 students in physics courses in the universities and higher schools in Europe in 1900 went on to graduate work in the subject; that year, the University of Berlin, with a total enrolment of 400 in lectures, granted ten doctorates in physics. There were correspondingly few advanced undergraduates or majors: in a representative case, Ernest Rutherford and his colleagues at McGill in 1902 taught over 300 students, of whom fewer than ten worked for honours in physics.[78]

The main purpose of the physics institutes of Europe and the United States was to help to train the 'Frontoffizieren' in the army of industrial warfare. The military metaphor became commonplace. The Humboldts, founders of the University of Berlin, were the 'Moltkes of scientific method'; 'the sinews of war come not only from the tax-payers' pockets but equally, if not to a greater extent, from our high schools of science'; 'the weapons which science places in the hands of those who engage in great rivalries of commerce leave those who are without them, however brave, as badly off as were the dervishes of Omdurman against the Maxims of Lord Kitchener'.[79] The equivalence of the technological and the military in national defence was institutionalized in Germany and France by remission of two years of military service for students of certain sciences and technologies.[80]

The French observed that their neighbours 'had prepared for economic warfare with the same ingenuity with which they had prepared previously for shooting wars'.[81] The Republic accordingly transformed its moribund faculties of science into vigorous universities, built schools of technology, and produced engineers good enough to alarm the Germans. The Reich for its part was 'as fully determined that their high schools of science shall be ahead of those of other countries as we [Britons] may be resolved that our fleet shall be equal to that of any two other nations'.[82] The determined Germans worried that Britain was building intermediate trade schools faster than the rest of Europe combined, and that nothing in Germany equalled the furnishings of the best American technical universities.[83] The British in turn envied the fittings and finances of German physics institutes. The Americans pointed to Europe and accelerated investment in staff and plant.

Everywhere, physicists hitched their wagon to 'Amerikanismus', to the star of practice. They delighted to hear the Kaiser praise physics for its part in the 'enormous industrial and financial upturn in Germany'.[84] They appealed to the connection to help them gainsay the nay-sayers of science: 'no one ever dared to speak of the bankruptcy of industrial progress'. They agitated for increased support from the state, industry and private donors as a patriotic necessity in the endemic war for markets. 'It is in the interests of the state economy that large sums of money should be judiciously disbursed by the Exchequer in furthering scientific pursuits.'[85] In some cases, particularly that of electricity, 'science and industry constitute but a single body of doctrine'. 'It is in the interests both of teaching and of research that physics as practised at the university be brought into the closest possible connection with electrotechnology.'[86]

As a confessed accomplice of technology, physics declared itself an enemy of the defenders of the old culture. And these defenders were not powerless. Particularly in Germany and Britain, their humanist and liberal education had won them positions in universities and ministries from which they could influence the flow of funds to academic physics. Their conception of its value could have a decisive effect upon a part of the institute's budget that had not increased in proportion to staff salaries and capital investment: the research provision. In Britain, for example, scientists complained that science did not receive a fair share from the endowments of the ancient universities, particularly in stipends for research workers. One tried to crack the monopoly of the humanists with such pedagogical arguments as 'physical science should be taught from the first as a branch of mental education, and not merely as useful knowledge'.[87] But the Oxbridge dons who controlled the scholarships and fellowships did not want to see the products of so novel an upbringing in their midst.[88] In Germany, administrators, legislators and professors who shared the sentiments expressed in the attack on scientism opposed raising science to a par with classical languages and did not favour applications for research support from academic physicists.[89]

A narrow research provision was nothing new to academic physicists. Their duties and their budgets had always been largely pedagogical.[90] But research, or the opportunity of making a great discovery, had become the fulcrum of professional ambition; and institute directors and department chairmen, whose morality had been perfected by the cultivation of science, became adept at diverting to research monies intended for teaching. The sums thus redirected might be a good fraction of the institute budget.[91] The necessity for subterfuge and occasional lack of apparatus became the more irksome as the size of the institutes and the opportunities for research increased. Naturally, physicists expected considerable improvements in research facilities along with enlargement of their clientele and workplace. Like other interest groups, they were (and are!) particularly sensitive to checks against their rising expectations. Funding that does not continue to increase appears to decline. The concern of academic physicists that expressed itself in the defensive literature of descriptionism may be understood as a reaction to the threat to their research provision posed by well-placed administrators educated as humanists and indifferent or antipathetic to the claims of science.

As it happened, funds for research in physics increased moderately during the first years of our century.[92] Physicists had their connections with industry, the great discoveries beginning with Röntgen, and a new democratic image, compounded of descriptionism and practicality, to thank for their good fortune. But, like the mechanical analogies on which theoretical physics depended, the adjusted image had negative aspects that grew more serious in time.

Two Clouds on the Image

With a slight change of emphasis, descriptionism could be turned against science by the very interests it was intended to placate. For example, in 1895 a French writer, Ferdinand Brunetière, inspired by a visit to the Vatican, charged science with moral and philosophical bankruptcy. His indictment was based on the broad premise that 'we cannot draw from the laws of physics or the results of physiology any way of knowing anything'. The scientific establishment found two answers to Brunetière. First, it agreed with the descriptionist premise.[93] Second, it rejected the conclusion by affirmation of the converse: the realization that there are no scientific dogmas brings not bankruptcy but liberation. Eight hundred affirmers, including forty senators and seventy deputies, ate a banquet in honour of Marcellin Berthelot, chemist and minister of state, and in homage of science, 'source de l'affranchissement de la pensée'. Berthelot explained that everything good – morality, art, literature, the Revolution, the Republic – derived from the scientific Spirit. The president of the chamber of deputies replied, simultaneously recalling the tie between science and the Republic and condemning as political Brunetière's appropriation of the physicists' epistemology. 'The formula "the bankruptcy of science" is, above all, a phrase of the political order, a means of reactivating clerical reaction.'[94]

Karl Pearson perceived a similar threat in Salisbury's remarks to the BA. Salisbury too proceeded from a descriptionist premise – 'we live in a small bright oasis of knowledge surrounded by a vast unexplored region of inscrutable mystery' – and appropriately exemplified the penumbra of ignorance by reference to the atom, the aether and natural selection. What bothered Pearson was Salisbury's move from confession of present ignorance to affirmation of continued ignorance, from 'ignoramus' to 'ignorabimus', in the vocabulary of the arch-mechanist Emil Du Bois-Reymond. Insistence on future impotence could only encourage the inference that, since science did not have the answers, religion must be allowed its say. (The same argument is used in California by fundamentalists trying to insert the biblical account of creation into public school texts.) According to Pearson, Salisbury intended to court the 'new bigotry which is likely to prove such a powerful engine of political warfare in the days to come'.[95] If so, the danger was real, for Salisbury was the Prime Minister of England.

The poet too discovered the utility of descriptionism. Apparently his vision was no worse than the scientist's: together, in the twilight of the nineteenth century, they looked forward to 'la sainte nuit, la bienheureuse ignorance'.[96] In ignorance we are all equal: 'Das Licht wird leuchten, weil es leuchten muss/ drum knurrt nur immer: Ignorabimus!/Transzendental ist nichts in der Natur/ transzendental ist unsere Dummheit nur!'[97] From descriptionism to distrust or depreciation of reason is a small step. Few besides Pearson and Planck saw the danger, which took dramatic form in Gottfried Benn's literary murder of a pedantic descriptionist professor by students wanting something warm and magical in their studies. As the poor man died he cried out – *Ignorabimus*![98] The depreciation of the physicists' product and the reach of reason played a part, perhaps a decisive part, in the creation and interpretation of quantum mechanics, where we find 'Ignorabimus!' at the core of physical theory.[99]

The negative side of identification with technology is scarcely news today. The military image of the *fin de siècle* has become a reality; modern warfare has required the services of physicists and enhanced the prestige of their discipline. Domestic claims of relative inferiority in international commercial competition, used to promote science and technology around 1900, have modern counterparts in the Sputnik scare and the missile gap, from which emerged the space programme, with literally infinite opportunities for expenditures in physics. In the matter of pollution, another concern of the old antiscientists, physicists have gone beyond chemical engineers in their contributions to the radioactive contamination of the planet, and have suffered in reputation if not in conscience. As for the pace of life that bothered the doctors of degeneracy, television, the jet plane, communication satellites, and the computer have quickened the tempo beyond the imaginings of even so wild a writer as Nordau, who thought humans had only to adapt to 'read[ing] a dozen square yards of newspaper daily, to be[ing] constantly called to the telephone, to . . . thinking simultaneously of the five continents of the earth, to liv[ing] half their time in a railroad carriage or in a flying machine, and to satisfy[ing] the demands of a circle of ten thousand acquaintances, associates, and friends'.[100]

Physicists of 1900 who resisted unabated technology argued not that it might ruin the world, but that it menaced their discipline. Like descriptionism, it lowered status. The menace appeared most serious at the extremes of academic culture, in Germany and in the United States, in the one because the prestige of the professor had been so high, in the other because it had been so low. The German physicist who considered himself an incarnation and conveyor of the highest aspirations of the German people, and who ranked on the civil list with ministers of state, had something to lose by identification with 'Materialismus und Amerikanismus'. Physicists who emphasized the humanistic education that they shared with high government officials and that separated them from most commercial and industrial men were among the most bitter opponents of efforts to upgrade technical schools and to introduce technical facilities into the universities.[101]

In the United States, prominent physicists tried to distance themselves from a technology that compromised their struggle to obtain public recognition of the value of disinterested or 'pure' science. The *Popular Science Monthly* complained of those who would degrade science to a 'low, money-making level'; the AAAS heard about the wastage of American intellect 'in the pursuit of so-called practical science'; memorialists praised defunct physicists who had resisted the blandishments of industry.[102] The threshold of resistance measured the scientist's culture and morality, and even his chances of professional success: 'Nature turns a forbidding face to those who pay her court with the hope of gain, and is responsive only to those suitors whose love for her is pure and undefiled.'[103] The electric telegraph, apparently the greatest recommendation for physical science – a 'phenomenon to which nothing in the history of our planet presents anything equal or similar' – was in fact its greatest danger. For, as Henry Rowland observed in a plea for pure physics, the public had inconveniently mixed up the telegraph and the electric light with science.[104]

Technology also appeared to present a material threat to established academic physics. Electrotechnics may have begun in discoveries made in research laboratories. But once discovered, the relevant principles could be exploited by engineers; and the research required to nourish the next round of exploitation would continue to trickle, as it had done, without lavish support. Those who made this argument recommended that public and industrial monies be directed not to academic research, but to higher technical training. Electrotechnologists, not professors, are the 'born confederates of the minister of education'.[105] In Britain, the technical schools were growing and multiplying far faster than university facilities for science. In France, the Sorbonne professoriate, recognizing a threat to their subventions as well as to their culture, rejected proposals to add technical courses to the curriculum.[106] But such opposition only helped to drive students to technical schools. Everywhere by 1900 enrolments in university physics courses were stagnating or declining under competition from the technical schools, whose graduates industry preferred; while industry itself, by offering better opportunities to instructors, helped to keep down the number of postgraduate students.[107]

The great interests at stake are mirrored clearly in the long fight to obtain for the German *Technische Hochschulen* the right to grant the doctor's degree. The universities, including many science professors, opposed the concession on the grounds that it would cheapen the degree intellectually and socially, admit the upstart technical schools to equality with the ancient seats of culture, and diminish teaching and research funds previously reserved to the universities for the monopolistic production of PhDs.[108] The Kaiser intervened. He announced the award of the coveted right in 1899, at the artificially computed centennial of the Berlin Technische Hochschule. The ceremony began with the unveiling of monuments to those heroes of high technology, Siemens and Krupp. The Imperial Family, generals and admirals, captains of industry and the Minister of War attended. The Kaiser, wearing the uniform of an officer of engineers, praised the school for its part in securing Germany's prosperity, and for its 'social role', now abetted by the prize of the doctorate, of bringing sons of good families to useful ways of life. The Rector went further than the Emperor. He understood the social programme of his school to be the stamping out of humanism from those 'trade schools for teachers', the universities, to whose incompetence and puffery the Reich was indebted for whatever difficulties it still faced in international commercial warfare.[109]

The uncomfortable adjustment to the new circumstances in which physicists found themselves around 1900 continued in force until the First World War. Some alleviation occurred through increases in research support, progress in physical theory and redefinition of the roles of the university, technical school, government bureau and industrial laboratory. The Nobel prize probably also helped to protect against the 'drowning out of the gentle music of natural laws by the trumpet blasts of technical success'.[110] That at least was the assessment of the first laureate in chemistry, J. H. van't Hoff, who applauded the decision to extend rewards to discoveries of use in science as well as in practice. 'In our time of industrial competition among the various nations the adverse side had already acquired a great predominance.'[111]

Acknowledgements

I am most indebted to Paul Forman for the example of his work on the ideology of Weimar physicists and for his generosity in putting his notes on *fin-de-siècle* physics at my disposal. I am also obliged to Ms Holly Murray for information about German literary responses to scientism, and to the United States National Science Foundation for support for work a few results of which are presented here.

Notes

1. Perrier, E. (1895) Talk at Berthelot banquet. *Rev. sci.*, 3, 470.
2. Cf. 'beschreibende Naturwissenschaft'. In: Kirchhoff, G. (1876) *Vorlesungen über mathematische Physik*, Vol. 1, Vorrede.
3. *Grand dictionnaire universel* (1874), Vol. 12, p. 291.
4. Poincaré, H. (1900) *Relations entre la physique expérimentale et la physique mathématique.* In: *Congrès International de Physique, Rapports,* Vol. 1, Paris, pp. 1–29.
5. Guillaume, C. E. (1900) The international physics congress. *Nature*, 62, 425–428.
6. Poynting, J. H. (1899) Address to Section A. *British Association for the Advancement of Science Report*, pp. 615–624.
7. Respectively, Brillouin, M. (1895) Pour la matière. *Rev. gén. sci.*, 6, 1022–1024; and Abegg, R. (1899) Review of W. Ostwald's *Grundriss. Phys. Z.*, 1, 136.
8. Respectively, Hagenbach, A. (1902) Die Entwicklung und der heutige Stand der

Kathoden- und Röntgenstrahlen. *Dtsch. Rev.*, **27** (3), 334–341, on 334; and Merritt, E. (1900) On kathode rays. *Science*, **12**, 41–48, 98–104, on 99.

9. Respectively, J. Larmor, introduction to Poincaré, H. (1905) *Science and Hypothesis*, London, p. xx; Heaviside, O. (1894) *Electromagnetic Theory*, Vol. 1, London, p. 1; and Du Bois-Reymond, P. (1890) *Über die Grundlagen der Erkenntnistheorie*, Tübingen, pp. 120–121.
10. Knott, C. G. (1892) Review of Pearson's *Grammar. Nature*, **46**, 97–99, 221–222.
11. Kirchhoff (note 2).
12. Weber, H. M. (1881) *Über Causalität*, Leipzig, p. 14; Willbois, J. (1899) La méthode des sciences physiques. *Rev. Métaphys. Moral*, **7**, 579–615, on 581–582.
13. For example, Mach, E. (1943) *On the economical nature of physical theory*. In: *Popular Scientific Lectures*, La Salle, Ill., pp. 186–213; Mach, E. (1919) *The Science of Mechanics*, 4th ed., Chicago, p. 555.
14. Nye, M. J. (1979) The Boutroux circle and Poincaré's conventionalism. *J. Hist. Ideas*, **40**, 107–120.
15. Mach, E. (1910) Die Leitgedanken meiner naturwissenschaftlichen Erkenntnistheorie. *Phys. Z.*, **11**, 599–606; Kleiner, A. (1902) Über die Wandlungen in den physikalischen Grundanschauungen. *Schweiz. Naturforsch. Ges. Verh.*, **84**, 113–141, on 126–127; Classen, J. W. (1908) *Vorlesungen über moderne Naturphilosophen*, Hamburg, pp. 109–110.
16. Poincaré, H. (1890) *Electricité et Optique*, Vol. 1, Paris, p. viii; Duhem, P. (1954) *The Aim and Structure of Physical Theory* [1906], Princeton, p. 85; Gilbert, P., review of French translation of Maxwell, quoted by Paul, H. (1980) *The role and reception of the monograph in 19th century French science*. In: Meadows, A. J., ed., *Development of Science Publishing in Europe*, Amsterdam, pp. 123–148, on 128; Hertz, H. (1893) *Electric Waves*, London, pp. 20, 27; von Meyenn, K. (1981) Paulis Weg zum Ausschliessungsprinzip. *Phys. Bl.*, **37**, 13–19, on 15.
17. Duhem (note 16), pp. 87–91, 101–102, 294–319; cf. Langevin, P. (1914) In: *Henri Poincaré*, Paris, pp. 122, 131–134, 142, and Boutroux, E., *ibid.*, pp. 220–221.
18. Boltzmann, L. (1892) *Über die Methoden der theoretischen Physik*. In: Dyck, W., ed., *Katalog mathematisch-physikalischer Modelle*, Munich, pp. 89–98, on 97–98; Boltzmann, L. (1899) Über die Entwicklung der Methoden der theoretischen Physik in neurer Zeit. *Phys. Z.*, **1**, 60–62, 77–79, 84–87, 92–98, on 78, 86–87.
19. O. Lodge to W. H. Bragg, 5 March 1891. In: Caroe, G. M. (1978) *William Henry Bragg*, Cambridge, p. 37.
20. Bouasse, H. (1898) La rôle des principes dans les sciences. *Rev. gén. sci.*, **9**, 561–569, on 561.
21. Brillouin (note 7), p. 1024.
22. Picard, E. (1904) *'Sciences,' Paris, Exposition Universelle, Jury International, Rapports, Introduction générale*, Vol. 2, pp. vii-ix, 1–114, on 27–29, 33–34, 113–114.
23. Classen (note 15), pp. 109–110; Pearson, K. (1892) The grammar of science. *Nature*, **46**, 199–200; Mach to Pearson, 16 January 1900, and to H. Hertz, 25 September 1890. In: Thiele, J. (1978) *Wissenschaftliche Kommunikation. Die Korrespondenz Ernst Machs*, Kestellaun, pp. 190, 55; Mach (note 13), p. 556.
24. Classen (note 15), pp. 108, 115–116; Kleiner (note 15), pp. 129–130; Kleinpeter, H. (1913) *Der Phänomenalismus*, Leipzig, p. 247.
25. As in Cameron, F. K. (1900) Some objections to the atomic theory. *Science*, **11**, 608–612; and Kimball, A. L. (1906) *The relations of the science of physics of matter to other branches of learning*. In: Rogers, H. J., ed., *Congress of Arts and Science . . ., St Louis, 1904*, Vol. 4, New York, pp. 69–86, on 77.
26. Kleinpeter (note 24), pp. 137, 220, 244; Klein, F. (1904) *Philosophische Fakultät*. In: Lexis, W., ed., *Die Universitäten im deutschen Reich*, Berlin, pp. 243–266, on 247–248.
27. Hertz, H. (1899) Preface. *The Principles of Mechanics* [1894], London. Cf. *La Grande encyclopédie* (circa 1896), p. 822.
28. Kirchhoff, G. (1894) In: Planck, M., ed., *Theorie der Wärme*, Leipzig, pp. 1–2; Maxwell, J. C. (1890) *Scientific Papers*, Vol. 2, Cambridge, p. 418; Picard (note 22), p. 20.

29. Cornu, A. (1900) Discours d'ouverture. In: *Congrès International de Physique, Rapports*, Vol. 1, pp. 5–8 (*Rev. gén. sci.*, **11**, 919–920).
30. Poincaré (note 4), pp. 25–26. Cf. Garbasso, A. (1895) Review of Hertz' *Mechanik. Nuovo cimento*, **1**, 40–59, on 42. The principle that no mechanical representation can be unique was also stressed by J. J. Thomson and by Maxwell.
31. Respectively, Maxwell, quoted by Thomson, J. J. (1931) in *James Clerk Maxwell, a Commemoration Volume*, Cambridge, p. 31; Lodge, O. (1902) George Francis Fitzgerald. *Royal Society of London, Yearbook, 1902*, pp. 251–259.
32. That the style was peculiarly English was emphasized by Duhem (note 16); Thompson, S. P. (1901) Address. *Phys. Soc. London, Proc.*, pp. 12–25, on 18; Poincaré, L. (1906) L'evolution de la physique. *Rev. sci.*, **5**, 481–486, on 484; Houllevigue, L. (1910) *The Evolution of the Sciences*, New York, p. xviii.
33. Fitzgerald, G. F. (1902) *Scientific Writings*, Dublin and London, p. 311; cf. Poynting, J. H. (1920) *Collected Scientific Papers* [1893], Cambridge, p. 264.
34. Rey, A. (1907) *La Théorie de physique chez les physiciens contemporains*, Paris, p. 178.
35. Sorel, G. (1905) Les préoccupations métaphysiques des physiciens modernes. *Rev. Métaphys. Moral*, **13**, 859–889, on 875, 879; Hagenbach (note 8), p. 6.
36. Lippmann, G. (1900) *La Théorie cinétique des gaz*. In: *Congrès International de Physique, Rapports*, Vol. 1, Paris, pp. 546–550; cf. Picard (note 22), pp. 27–28, 32.
37. Ostwald, W. (1926) *Lebenslinien*, Vol. 2, Berlin, pp. 149–157, 310; Thiele (note 23), p. 174; Ostwald, W. (1895) Die Überwindung des wissenschaftlichen Materialismus. *Ges. dtsch. Naturforsch. Ärzte Verh.*, **67**, 155–168 (*Rev. gén. sci.*, **6**, 953–958).
38. Ostwald, *Lebenslinien* (note 37), p. 179; 'Überwindung' (note 37), pp. 32–33; Körber, H. G. (1961) *Aus dem wissenschaftlichen Briefwechsel Wilhelm Ostwalds*, Vol. 1, Berlin, pp. 118–120.
39. Sommerfeld, A. (1944) Das Werk Boltzmanns, *Wiener Chem. Ztg.*, **47**, 25–28; Cornu, A. (1895) Quelques mots en réponse à [Ostwald]. *Rev. gén. sci.*, **6**, 1030–1031; editorial, *ibid.*, p. 1070; Fitzgerald (note 33), p. 388.
40. Balfour, A. J. (1918) *The Mind of Alfred James Balfour*, New York, p. 234; Morgan, C. L. (1892) Review of Pearson's *Grammar. Nat. Sci.*, **1**, 300–308, on 308; cf. Epstein, S. S. (1896) Hermann von Helmholtz. *Dtsch. Rev.*, **21**, 31–41, 192–202, 328–339, on 201–202; Picard (note 22), p. 25.
41. Kelvin, Lord (1904) *Baltimore Lectures* [1900], London, pp. 486–527; Picard (note 22), pp. 36, 41.
42. Kelvin (note 41), pp. 492, 527; Kuhn, T. S. (1978) *Blackbody Theory*, Oxford and New York, pp. 115–120.
43. Nordau, M. (1895) *Degeneration*, New York, pp. 2, 9–21, 31, 34–35.
44. Nietzsche, F. (1966) *Werke*, Vol. 2, Munich, p. 913; von Hofmannsthal, H. (1966) *Ausgewählte Werke*, Vol. 2, Frankfurt/Main, pp. 293–294.
45. J. Donath in Kende, M. (1901) *Die Entartung des Menschengeschlechts*, Halle; Kende, *ibid.*, pp. 28, 49–51, 60–63; Smyth, A. W. (1904) *Physical Deterioration*, London, pp. 13–26; Coblentz, S. A. (1925) *The Decline of Man*, New York, p. 68; Rasch, W. (1977) *Fin de siècle als Ende und Neubeginn*. In: Bauer, R. *et al.*, eds, *Fin de siècle. Zu Literatur und Kunst der Jahrhundertwende*, Frankfurt/Main, pp. 30–49.
46. Nordau (note 43), pp. 34–39; Kende (note 45), pp. 79–81, 104–119.
47. Bourget [1883–85], quoted by W. Wiora, '*Die Kultur kann sterben*'. In: Bauer *et al.* (note 45), pp. 50–72, on 63. Coblentz (note 45), pp. 3–8, 144; Zweifel, P. (1900) Pläne und Hoffnungen für das neue Jahrhundert. *Dtsch. Rev.*, **25** (1), 108–120, on 118; Sokal, E. (1896) Mechanistische und energetische Naturforschung. *Dtsch. Rev.*, **21**, (3), 374–377; quote from Crookes, W. (1891) Electricity in relation to science. *Nature*, **45**, 63–64.
48. Cf. Balfour (note 40), p. 248; Brooks Adams (1943) *The Law of Civilization and Decay* [1896], New York, Beard, C. A., ed., p. 308; Aliotta, A. (1914) *The Idealistic Reaction against Science*, London, pp. xv–xvi; and Gregory, F. (1977) *Scientific Materialism in 19th-century Germany*, Dordrecht.
49. Respectively, Zola, E. (1971) *Le Roman expérimental* [1880], Paris, p. 63; Hart, H. &

Hart, J. (1962) Dichtung und Naturwissenschaft [1880], quoted in Ruprecht, E., ed., *Literarische Manifeste des Naturalismus 1880–1892*, Stuttgart, pp. 29–30; and Bölsche, W. (1962) Charles Darwin und die moderne Aesthetik [1888], quoted *ibid.*, pp. 103–104. Cf. Bölsche, W. (1887) *Die naturwissenschaftlichen Grundlagen der Poesie*, Leipzig, pp. 6–7.

50. Zola, E. (1895) Speech at Berthelot banquet. *Rev. sci.*, **3**, 473.
51. Vogt, C., quoted by Pictet, R. (1896) *Etude critique du Matérialisme*, Geneva, p. 24; Richet, C. (1895) La science a-t-elle fait banqueroute? *Rev. sci.*, **3**, 39; Berthelot, M. (1897) La rôle de la science dans les progrès des sociétés modernes. *Rev. gén. sci.*, **7**, 641–643.
52. Claudel, P. (1947) *Contacts et Circonstances*, Paris, p. 12; cf. Pictet (note 51), p. 451 and Gide, A. (1953) *Paludes* [1895], tr. G. D. Painter, London, p. 40: 'It's enough that it might be different, and isn't. All our acts are so well-known.'
53. Respectively, Lilly, W. S. (1886) Materialism and morality. *Fortn. Rev.*, **46**, 575–594, on 589; Cobbe, F. P. (1888) The scientific spirit of the age. *Contemp. Rev.*, **54**, 126–139, on 130. Cf. Berthelot (note 51), p. 642; and Balfour (note 40), pp. 233–234.
54. For example, Mivart, St G. (1895) Denominational science. *Fortn. Rev.*, **64**, 423–468, on 438.
55. Lilly (note 53), p. 538; Burnell, A. (1900) Science and religion. *Westminister Rev.*, **154**, 440–445, on 445; Balfour (note 40), pp. 238, 241; Kaibel, G. (1900) Die neue Bildung. *Dtsch. Rev.*, **25** (1), 57–67, on 67.
56. Anon. (1895) A century of science. *Q. Rev.*, **180**, 381–405, on 392.
57. Mallock, W. H. (1889) Science and the revolution. *Fortn. Rev.*, **52**, 600–619, on 619; Anon. (1891) *Spectator*, **67**, 524, 723, and Anon. (1896) *Spectator*, 77, 138. Cf. Anon. (1894) *Saturday Rev.*, **78**, 150–151.
58. Mivart (note 54); Salisbury, Lord (1894) *Evolution*, London, p. 53. Cf. Balfour (note 40), pp. 321, 326, 328.
59. Pearson, K. (1894) Politics and science. *Fortn. Rev.*, **62**, 334–351; Anon. (1894) *Saturday Rev.* **78**, 150–151; Anon. (1895) *Gaea*, **31**, 720; Mivart (note 54), p. 434; Anon. (note 56), pp. 402–403; H. Spencer to T. H. Huxley, 10 August 1894. In: Duncan, D. (1908) *Life and Letters of Herbert Spencer*, Vol. 2, New York, p. 74.
60. Langley, S. P. (1902) The laws of nature. *Science*, **15**, 921–927, on 924; Anon. (1900) *Saturday Rev.*, **90**, 321, re BA.
61. von Hartmann, E. (1902) *Die Weltanschauung der modernen Physik*, Leipzig, p. 219; Sorel (note 35), p. 864; P. Claudel to A. Gide, 7 August 1903. In: Claudel, P. & Gide, A. (1949) *Correspondance*, Paris, p. 48. Cf. Brush, S. G. (1967) Thermodynamics and history. *Grad. J.*, 7 (2), 477–565, on 523–527; Duhem (note 16), pp. 275–282, 288.
62. Pearson, C. H. (1893) *National Life and Character*, London, pp. 261, 267, 273, 286–290, 293, 298–302, 312–320, 344. Cf. Badash, L. (1972) The completeness of 19th-century science. *Isis*, **63**, 43–58.
63. Lodge, O. (1898/1899) The scientific work of Lord Rayleigh. *Natl Rev.*, **32**, 89–102, on 90; Harnack, A. (1900) *Geschichte der Königlich-Preussischen Akademie der Wissenschaft*, Vol. 1, Berlin, pp. 792, 979; Balfour (note 40), pp. 263–264, 292.
64. Pearson (note 62), p. 342; cf. Brush (note 61), pp. 505 ff., and Hiebert, E. (1966) The uses and abuses of thermodynamics in religion. *Daedalus*, **95** (4), 1046–1080.
65. For example, Berthelot (note 51), p. 643; Lévy, M. (1904) *'Industrie,' Paris, Exposition Universelle, Jury International, Rapports*, Vol. 2, Paris, pp. 115–392, on 117; Planck, M. (1910) *Acht Vorlesungen über theoretische Physik*, Leipzig, p. 2; Roberts, W. (1897) Science and modern civilization. *Nature*, **56**, 621–624, on 623.
66. For example, Planck, at GDNA, 1891. In: *Physikalische Abhandlungen und Vorträge*, Vol. 1, Braunschweig, p. 372; Picard (note 22), p. 29.
67. Gerland, E. (1899) Die Physik im 19. Jahrhundert und ihre Aufgaben für das 20. Jahrhundert. *Dtsch. Rev.*, **24** (4), 240–249, on 248; Thomson, E. (1899) The field of experimental research. *Science*, **10**, 236–245, on 236.
68. E. Gray, welcoming address. *International Electrical Congress, 1893, Proceedings*, pp. 1–4.

69. Respectively, Thomson (note 67), p. 236; and Carhart, H. S. (1900) The Imperial Physico-Technical Institution in Charlottenburg. *Science,* **12**, 697–708, on 707; cf. Elsdale, H. (1894) Scientific problems of the future. *Contemp. Rev.*, **65**, 362–375.
70. Quoted in *Physics Today,* April 1968, p. 56; cf. Badash (note 62), p. 52. On the profound importance of measurement, see Planck (note 65), p. 3; and K. B. Hasselberg to G. E. Hale, 29 December 1907. In: Livingston, D. M. (1973) *The Master of Light,* New York, pp. 238–239.
71. Münsterberg, H. (1896) The X-rays. *Science,* **3**, 161–163.
72. von Bezold, W. (1896) Festrede. *Dtsch. phys. Ges. Verh.*, **15**, 19–25, on 24.
73. Guillaume (note 5), p. 427; cf. Burchfield, J. D. (1975) *Lord Kelvin and the Age of the Earth,* New York, pp. 163–170.
74. Lodge (note 63), p. 91.
75. Riecke, E. (1899) Strahlende Materie. *Dtsch. Rev.*, **24**, (1), 44–60, on 60.
76. Cf. Boltzmann (note 18), p. 86; Picard (note 22), p. 50.
77. Woodward, R. S. (1901) The progress of science. *Science,* **14**, 305–315, on 306.
78. Harnack (note 63), p. 979; *Jahres-Verzeichniss der an den deutschen Universitäten erschienen Schriften,* **15** (1899/1900); Heilbron, J. L. (1979) *Physics at McGill in Rutherford's time.* In: Bunge, M. & Shea, W. B., eds, *Rutherford and Physics at the Turn of the Century*, New York, pp. 42–73, on 45. The data on the physics enterprise come from Forman, P., Heilbron, J. L. & Weart, S. (1975) *Physics circa 1900: Personnel, Funding, and Productivity of the Academic Establishments (Historical Studies in the Physical Sciences, No. 5),* Princeton.
79. 'Frontoffizieren' is F. Klein's slogan, as quoted in Manegold, K. H. (1970) *Universität, Technische Hochschule, und Industrie*, Berlin, pp. 123, 132; 'Humboldts': Armstrong, H. E. (1897) The need of organizing scientific opinion. *Nature*, **55**, 409–411, 433–435, on 434; 'sinews': Magnus, P. (1898) Technical high schools, a comparison. *Nature,* **58**, 52–54; 'Maxims': Haldane, Viscount (1903) quoted in *Nature*, **68**, 338.
80. Cf. von Diefenbach, J. (1897) *Technische Erziehung in Württemberg.* In: *International Congress on Technical Education, 1897, Report,* London, pp. 42–46; *ibid.* (1900) *Rapports,* Paris, pp. 703–704.
81. Lauth, C. (1900) L'École municipale de physique et chimie industrielles. *Rev. sci.*, **13**, 780–786, on 781.
82. Liard, L. (1890) *Universités et Facultés,* Paris, pp. 25–132; Klein's opinion of French excellence, quoted by Manegold (note 79), p. 104; on the fleet, Lot, F. (1903) quoted in *Nature*, **67**, 433, and Magnus (note 79), p. 52.
83. Brooks, C. P. (1897) *Technical Education in Europe,* Lowell, Mass., pp. 3–5; Manegold (note 79), pp. 116–117, 147.
84. 'Amerikanismus', Manegold (note 79), p. 114; cf. Planck (note 65), p. 2; and Witt, O. N. (1897) In: *International Congress, 1897* (note 80), pp. 4–8.
85. Frankland, P. F. (1889) Scientific education in relation to industrial prosperity. *Natl Rev.*, **13**, 343–353, on 344; various commentators in *International Congress, 1897* (note 80), pp. 15, 22, 47; Lauth (note 81), pp. 780–781.
86. Respectively, Lévy (note 65) and Riecke, E. (1906) *Die Physikalische Institut der Universität Göttingen,* Göttingen, pp. 25–26; cf. Manegold (note 79), pp. 101, 152–153.
87. BA (1889), quoted by Armstrong in *International Congress, 1897* (note 80), p. 9.
88. Cf. the exchange between G. C. Brodrick and E. R. Lanksester (1900) in *Saturday Rev.*, **89**, 331–332, 394–395, 459–460, 527–528; and Heilbron, J. L. (1974) *H. G. J. Moseley, The Life and Letters of an English Physicist,* Berkeley, California, pp. 20, 28.
89. Hart, H. 'Neue Welt' [1878], quoted in Ruprecht (note 49), pp. 13–14; Hillebrand, J., 'Naturalismus schlechtweg!' *ibid.*, p. 65; Ringer, F. (1969) *The Decline of the German Mandarins,* Cambridge, Mass., pp. 25–58.
90. Kevles, D. (1978) *The Physicists,* New York, p. 36; Manegold (note 79), pp. 150–151.
91. Forman, Heilbron & Weart (note 78), pp. 73–81.
92. Forman, Heilbron & Weart (note 78), pp. 75–83; Kevles (note 90), pp. 69–71; Manegold (note 79), p. 214.

93. Richet (note 51), p. 33; cf. Paul, H. (1968) The debate over the bankruptcy of science in 1895. *Fr. hist. Stud.*, **5**, 299–327.
94. Anon. (1895) *Rev. sci.*, **3**, 474.
95. Salisbury (note 58), pp. 15–16, 21–22, 27, 31–33; Pearson (note 59), pp. 339–340.
96. Respectively, Hofmannsthal as represented in Janik, A. & Toulmin, S. (1973) *Wittgenstein's Vienna*, New York, pp. 113–115; and Claudel to Gide (note 61).
97. Holz, A. (1962) *Buch der Zeit* [1885], *Werke*, Vol. 5, Neuried am R./ Berlin, p. 56.
98. Planck (note 65), pp. 4–7; Benn: In Ritchie, J. M. (1972) *Gottfried Benn*, London, pp. 73–80. Cf. McCormmach, R. (1974) On academic scientists in Wilhelmian Germany. *Daedalus*, **103** (3), 157–171, on 164; and Aliotta (note 48), pp. xv, xx, 4, 395–398.
99. Cf. Forman, P. (1971) Weimar culture, causality, and quantum theory. *Hist. Stud. phys. Sci.*, **3**, 1–115.
100. Nordau (note 43), p. 541.
101. McCormmach (note 98), pp. 161–163; Manegold (note 79), pp. 87, 114, 131–133, 155–156.
102. Respectively, Kevles (note 90), p. 45; and Rowland, H. (1899) Highest aim of the physicist. *Science*, **10**, 825–833; a sample obituary is Nichols, E. L. (1903) Ogden Nichols Rood. *Phys. Rev.*, **16**, 311–313.
103. Newcomb, S. (1904) *Evolution of the scientific investigator.* In: *Smithsonian Institution, Annual Report, 1904*, pp. 221–233, on 223.
104. Rowland [1883], quoted in Kevles (note 90), p. 43; Salisbury [1889], quoted in Appleyard, R. (1930) *The History of the Institution of Electrical Engineers*, London, p. 113.
105. See *Z. Elektrochem.*, **6**, 1–3.
106. Paul, H. (1972) *The Sorcerer's Apprentice*, Gainsville, Fla., pp. 23–27; Shinn, T. (1979) The French science faculty system, 1808–1914: Institutional change and research potential in mathematics and the physical sciences. *Hist. Stud. phys. Sci.*, **10**, 271–332, especially 313–325.
107. Forman, Heilbron & Weart (note 78), p. 29.
108. Manegold (note 79), pp. 145–147, 289–298.
109. Manegold (note 79), pp. 301, 304–305; Kaibel (note 55), p. 58.
110. Simon, T., Göttingen, quoted by Manegold (note 79), p. 142.
111. van't Hoff, J. H. (1902) Die Nobel-Stiftung. *Dtsch. Rev.*, **27**, 80–86, on 86.

THE INTERPLAY OF CHEMISTRY AND BIOLOGY AT THE TURN OF THE CENTURY

JOSEPH S. FRUTON

Yale University, New Haven, Connecticut, USA

Introduction

Around 1900, the area of science now called biochemistry, although flourishing, was marked by prominent differences of style in empirical investigation and theoretical speculation. In part, the differences were related to the uncertain institutional status of biochemistry, for only a few universities, notably Strasbourg and Yale, had independent chairs in 'physiological chemistry', the more common name for this field at that time.[1,2] Instead, research on chemical problems posed by living organisms was largely conducted in chemistry departments, schools of medicine and pharmacy, some agricultural experiment stations and several new institutes, in particular the Carlsberg, the Pasteur, the Lister and the Rockefeller. As a consequence, there was wide diversity in the training brought to biochemical work and in the purpose of the university departments and the institutes where such research was being done.

This dispersal of biochemical effort among laboratories with different objectives, and staffed by people whose formal education ranged from physical chemistry to clinical medicine, also reflected the coexistence of different directions of scientific thought. At the risk of oversimplification, it may be suggested that three points of view characterized biochemical work at the turn of the century. The salient feature of each of these attitudes was to consider a living organism as (1) an assembly of chemical substances whose individual properties and mutual interactions could explain biological phenomena; or (2) a physico-chemical system whose structure and dynamics could be described in terms of molecular physics, thermodynamics and kinetics; or (3) an assembly of cellular units whose chemical behaviour could be adequately understood only in terms of the properties of whole organisms, intact cells or protoplasmic entities which lost their biological character upon chemical dissection. These three ways of looking at the chemistry of life represented traditions inherited from the nineteenth century, and indeed had their historical roots in earlier scientific thought. As has been shown by the historical development of biochemistry during the twentieth century, all of them proved to be fruitful, and they were not mutually exclusive.

The aim of this essay is to consider the ways in which leading biochemical investigators responded to these three approaches. In keeping with the theme of this symposium, I shall limit the discussion largely to the two decades that preceded the First World War, and I propose to consider research on some of the biochemical problems that were deemed to be important at that time. Before doing so, however, it may be useful to sketch the trends in chemistry and biology that most influenced such research, and to identify some of the leading personalities.

By 1900, chemistry had undergone a bifurcation, with the emergence of physical chemistry as a branch separate from the strongly established organic chemistry. The latter had burst into flower during the second half of the nineteenth century, after the development of concepts of chemical structure based on valence and stereochemistry. Through achievements that led to the synthesis of new dyes, new drugs and new explosives, organic chemistry provided proof of the value of pure science to modern industry, and in return received generous support, especially in Germany. Organic chemistry was no longer what Berzelius had called it in 1806, namely 'the part of physiology that describes the composition of living bodies, together with the chemical processes that occur in them'.[3] Nevertheless, its ties to biology, agriculture and medicine, although increasingly tenuous, remained unbroken, and were implicit in the research of Emil Fischer, the unquestioned leader of organic chemistry during the first decade of this century. His work on the structure and synthesis of sugars and purines indicated his bias toward chemical problems of biological importance, and its value was recognized by the award of the 1902 Nobel Prize in Chemistry, just after he had embarked on his studies on amino acids, peptides and proteins. Throughout the nineteenth century, the development of the field that came to be called 'physiological chemistry' was dependent on advances in organic chemistry. This dependence has continued to the present, as is evident from the influence on modern biochemical research of the many organic chemists who received Nobel prizes for elucidating the structure of important chemical constituents of living organisms.[4]

At the turn of the century, the role of organic chemistry in the development of biochemistry was matched by the newly emergent branch of physical chemistry, whose leading figures were Arrhenius, van't Hoff and Ostwald, all of whom were to receive Nobel prizes before 1910. The work of the physical chemists on electrolytic dissociation, reaction kinetics, chemical thermodynamics and surface phenomena profoundly affected biochemical thought and experiment, and continues to do so today.

If one looks at the biological side of the interplay of chemistry and biology around 1900, a noteworthy feature was the institutional dependence of biochemical research on branches of medicine, above all physiology, but also pathology, pharmacology and bacteriology. In part, this came from the historical development of organic chemistry that separated it, after 1850, from the animal chemistry of Berzelius and von Liebig. More importantly, at mid-century, leading physiologists, notably Bernard and Ludwig, and the great pathologist Virchow, emphasized the importance of chemical methods for the study of biological and medical problems. The chief protagonist of an independent physiological chemistry, Hoppe-Seyler (a disciple of Virchow), directed the research of his laboratory first at Tübingen, then at Strasbourg, to problems of medical interest. The same may be said of Hofmeister, who succeeded him at the latter university in 1896, and of Chittenden, who at Yale headed the other major independent department of physiological chemistry at that time. Many of the significant biochemical achievements during the latter part of the nineteenth century, for example Bernard's discovery of glycogen, came from studies on animal metabolism, which he and other leading physiologists considered to belong to their domain. In addition to departments of medical physiology, those in pharmacology, notably that of Schmiedeberg at Strasbourg, were important centres of biochemical research. Toward the end of the century, after the work of Pasteur, Koch and von Behring had led to the establishment of bacteriology as an independent subject, problems such as the nature of bacterial toxins and of the immune response were added to those being studied by chemists associated with leading physiologists, pathologists and pharmacologists. Also, in some hospitals acute observations by chemically minded pathologists and clinicians provided important clues to metabolic pathways.

Another development at the turn of the century that influenced biochemical work and thought was the efflorescence of cytology, due to the achievements of zoologists like van Beneden and Boveri, anatomists like Oskar Hertwig and Flemming, and botanists like Strasburger. By 1900, their discoveries had not only illuminated intracellular structure and the dynamics of fertilization and cell division, but also raised new questions about the chemical nature and function of protoplasm. Because these investigators depended heavily on microscopy and the use of fixatives and stains, some biochemists emphasized the possibility of artefacts, and the interaction of the new cytology with biochemistry was less pronounced than it became some thirty years later. Also, the rediscovery of Mendelian genetics appears to have had little immediate impact on the course of biochemical work, with the notable exception of Garrod's studies on inborn errors of human metabolism. Nor did the chemists associated with medical physiologists and bacteriologists appreciate fully the importance of the microbiological studies of men like Winogradsky and Beijerinck, whose work entered the mainstream of biochemical research only during the 1920s, largely through the efforts of Kluyver and Stephenson. Nevertheless, the groundwork for the later emergence of a 'general biochemistry' was laid during the first two decades of this century by Loeb and Bayliss, among others, who advocated the study of a 'general physiology' that gave attention to chemical studies on plants and invertebrate organisms, as well as on higher animals.[5,6]

Role of Enzymes in Metabolism

Metabolism

Among the biochemical problems under active investigation and discussion at the turn of the century, perhaps the most challenging ones were those concerned with the chemical events in what had been variously called '*Stoffwechsel*', 'metamorphosis', 'nutrition' or 'metabolism'. During the latter half of the nineteenth century, considerable empirical knowledge had been obtained about respiratory gas exchange and heat production in higher animals, a line of research that had begun with Lavoisier in the 1780s. By 1900, a peak had been reached, in the achievements of Rubner, and later of Atwater and Rosa, who showed that, as a combustion apparatus, the animal organism obeys the law of conservation of energy (more accurately, Hess' law of constant heat summation). A significant stimulus to this development had been provided during the early 1840s by von Liebig, whose ideas about animal metabolism were based on the assumption that carbohydrates and fats represent the fuel of the animal body, and that proteins are oxidized during muscular activity or starvation. Although his ideas about the oxidation of proteins required correction during the 1860s, the assumption that measurements of nutritional balance and respiratory exchange should bear a direct relation to the metabolic heat output also characterized the work of a succession of investigators — von Liebig's Munich colleague von Pettenkofer, whose student Voit in turn taught Rubner.

Such animal calorimetry, and the nutritional studies with which it was associated, were not universally admired by physiologists. In 1865, Bernard wrote:

> 'One can undoubtedly establish the balance between the food consumed by a living organism and what it excretes, but these would be nothing but purely statistical results unable to throw light on the intimate phenomena of the nutrition of living things. It would be, according to the phrase of a Dutch chemist [Mulder], like trying to tell what happens in a house by watching what goes in the door and what leaves by the chimney. One can determine exactly the two extreme limits of nutrition, but if one then wishes to construe the intermediates between them, one finds oneself in an unknown region largely created by the imagination, all the more easily because numbers often lend themselves admirably to the proof of the most diverse hypotheses.'[7]

The concept of intermediary metabolism had been formulated by the 1850s;[8] but the experimental difficulties in the study of such step-by-step metabolic conversion were formidable, and until about 1880 few systematic efforts were made to overcome them. In comparison with the achievements of Voit and Rubner in the quantitative measurement of heat output, respiratory exchange and the balance of nutrients and excretory products, those interested in intermediary metabolism had relatively little to offer, except with regard to such processes as the transformation of dietary proteins to peptones or of starch to sugar in the animal digestive tract, or the conversion by microorganisms of nutrients (e.g. glucose) to other organic compounds. By about 1875, it had been widely accepted that the oxidative metabolism of animals occurs in the tissues, rather than in the blood; but no intermediates leading to carbon dioxide and water had been identified as such, although there was considerable discussion about the mechanism of intracellular oxidations. Prominent among the questions raised in this discussion was the one related to the nature of the agents involved, not only in such biological oxidations, but also in other metabolic changes effected by living organisms, notably fermentations and the synthesis of complex constituents (e.g. proteins, glycogen) from simpler compounds. At the turn of the century, the question acquired new urgency when, in 1897, Eduard Buchner reported that he had succeeded in preparing from brewer's yeast a cell-free extract that could ferment glucose.[9] This finding gave strong support to the idea advocated during the nineteenth century by some chemists that living cells contain discrete catalytic substances (which we now call enzymes) whose action is important in intracellular metabolism.

The ferments

An important factor in drawing attention to Buchner's achievement was the controversy generated by the forthright stand Pasteur took in 1860 in favour of an organismic theory of alcoholic fermentation. After the demonstration, during the 1830s, by Caignard-Latour, Schwann and Kützing that the fermentation of glucose to alcohol and carbon dioxide is caused by living microorganisms, and not by the decomposition of dead albuminoid matter, many chemists of the 1840s and 1850s accepted the view later adopted by Pasteur. The question then arose as to the relation between the ferments that cause the breakdown of glucose and the known agents (also termed 'ferments') that could be extracted from living organisms and which effect the conversion of starch to sugar (named 'diastase' by Payen and Persoz in 1833) or the dissolution of coagulated egg white (named 'pepsin' by Schwann in 1836). Subsequently, other such 'soluble ferments' were identified, including one that converts sucrose to glucose and fructose. For Berthelot, who obtained this agent from yeast, and named it 'invertin', there was no fundamental distinction between such soluble (or 'unorganized') ferments and the 'insoluble ferments' present in the microorganisms (or 'organized ferments') that produce alcohol or other products, such as lactic acid, from glucose. In his view, they are all chemical reagents, and his professed aim was 'to banish life from all explanations relative to organic chemistry'.[10]

The question of the relation between the soluble ferments and the microbial agents of fermentation reappeared often in succeeding years, and especially in 1878 during the famous Pasteur-Berthelot debate over some experimental notes of Bernard on alcoholic fermentation. Shortly before his death, Bernard had apparently convinced himself that he had demonstrated the conversion of sugar to alcohol by agents separable from living yeast, in support of his long-held view that only synthetic processes (such as the formation of glycogen) are correlative with life. For the physiologist Bernard, 'vital destruction is on the contrary of a physico-chemical order, most often the result of a combustion, of a fermentation, of a putrefaction, in a word, of an action comparable to a large number of chemical decompositions or cleavages'.[11] During the course of the acerbic but inconclusive debate with Berthelot, Pasteur stated:

> 'I must add finally that it is always an enigma to me that one could believe that I would be disturbed by the discovery of soluble ferments in the fermentations properly designated as such, or by the formation of alcohol from sugar, independently of living cells. Certainly, I must confess it without hesitation, and if one wishes, I am ready to explain myself on this point at greater length, I do not now see either the necessity for the existence of these ferments or the utility of their function in this kind of fermentation.'[12]

I have quoted the last passage because it underlines two themes that were reiterated during the succeeding two decades. The first is the experimental failure to demonstrate cell-free alcoholic fermentation, for Pasteur cast sufficient doubt on the validity of Bernard's experiments to leave the question open; and indeed, before 1897, cell-free alcoholic fermentation had not been demonstrated in a manner that could be reproduced by other investigators. Pasteur's second point, that the assumption of intracellular agents of fermentation was unnecessary, is perhaps of greater historical interest, because it recurred frequently in the writings of physiologists of that period. In 1878, the noted medical physiologist Pflüger stated that this assumption 'is not only unnecessary, but indeed highly implausible';[13] and in the following year, the influential plant physiologist Nägeli wrote:

> 'The agent of fermentation is inseparable from the substance of the living cell, i.e., it is linked to plasma [Nägeli's word for protoplasm]. Fermentation occurs only in immediate contact with plasma in so far as its molecular action extends. If the organism wishes to exert an effect on chemical processes in places or at distances where the molecular forces of living matter are without power, it excretes ferments. The latter are especially active in the cavities of the animal body, in the water where moulds live, and in the plasma-poor cells of plants. It is even doubtful whether the organism ever makes ferments that are intended to function within the plasma; since here it does not need them, because it has available to it in

> the molecular forces of living matter much more energetic means for chemical action.'

For Nägeli, fermentation 'is the transfer of the states of motion of molecules, atomic groups and atoms of the compounds that compose living plasma (which remain chemically unaltered) to the fermentable material, whereby the equilibrium in the molecules of the latter is destroyed and they are brought to decomposition'.[14] In particular, Nägeli linked such states of motion with the micellar structure of protoplasm. This physicalist conception of the chemical action of protoplasm was echoed near the turn of the century by many biologists.

'Living' proteins

In addition to Nägeli's 'molecular-physiological' speculations, by 1900 much attention was being given to ideas that stemmed from Pflüger's more specific descriptions of what he called 'living proteins'. For Pflüger,

> 'an albumin molecule, which in the brain cooperates in the production of thought, which in the spinal cord mediates sensation, which in the muscles performs mechanical work, . . . is derived from the same albumin, but in the cells it becomes something else, . . . It commences to respire, to live. . . . The albumin of the blood, I may say, is dead . . . so long as it has not become cellular material.'[13]

According to Pflüger, the distinctive chemical character of living proteins is their instability, arising from the presence of their nitrogen, not in the amide groups of 'dead proteins', but in the form of reactive cyanogen groups, whose reactions lead to 'a continuous series of small explosions whose force increases molecular vibrations'.[15]

The respectful attention given Pflüger's views encouraged others to produce variants. In 1881, Loew and Bokorny claimed that the chemical feature which characterizes living plant protein is the presence of aldehyde groups. Although this idea was derided by some leading physiological chemists, notably Baumann and Hoppe-Seyler, others, for example Nencki, gave it approval:

> 'If we wish to approach the phenomena associated with the word "life", research on the chemistry of the albuminoid bodies must take a new direction. As Pflüger stated rightly over 10 years ago, the protein of the living cell must have an entirely different molecular structure from that of dead tissues, and the evidence that the protein of living cells had a labile aldehyde structure increases daily.'[16]

This 'increase' in evidence was in large part a consequence of Loew's persistence in reiterating his claim.[17]

To this heritage of ideas prevalent at the turn of the century about the distinctive chemical character of protoplasmic proteins must be added the

1885 article on biological oxidations by Paul Ehrlich, in which he introduced the idea that the living protein of protoplasm attracts to itself 'side chains' that represent the agents of metabolic processes. Although he stated that 'speculations would as yet be premature concerning the nature and origin of these binding groups, which presumably differ with their function', he added that 'I shall content myself with observing that possibly the aldehyde groups, the existence of which is assumed by Loew and Bokorny, may play an important role in this connection'.[18] By 1900, however, the centre of the stage in the field of 'energy-rich' proteins had been taken by Verworn, who added the catchword 'biogen' to the multiplicity of presumed fundamental units of life that had proliferated during the last decades of the nineteenth century. He ascribed to biogen the explosive properties that Pflüger gave his 'living proteins' and the same kind of side chains that Ehrlich gave to protoplasm.[19]

In the historical background of Buchner's discovery, there were also the ideas and experiments of the nineteenth-century chemist Moritz Traube, who did not believe that the assumption of intracellular ferments was either unnecessary or implausible. I mention them last because they appear to have been less highly regarded by his contemporaries than those of Pflüger and Nägeli. In 1878, Traube reiterated views he had expressed twenty years earlier:

> 'The ferments . . . are chemical substances related to the albuminoid bodies which, although not accessible in pure form, have like all other substances a definite chemical composition and evoke changes in other substances through definite chemical affinities. . . . Schwann's hypothesis (later adopted by Pasteur), according to which fermentations are to be regarded as the expressions of the vital forces of lower organisms is unsatisfactory. . . . The reverse of Schwann's hypothesis is correct: Ferments are the causes of the most important vital-chemical processes, and not only in lower organisms, but in higher organisms as well.'[20]

Traube was most concerned with the chemical mechanisms of biological oxidation, and offered a unified theory that explained both intracellular oxidation and fermentation in terms of discrete catalytic entities. Traube's view that biological oxidations and fermentations involve the action of chemically defined catalysts was echoed in the writings of Hoppe-Seyler, the leading physiological chemist of the 1880s, although the two differed on the relative importance of the transfer of oxygen atoms and of the utilization of the hydrogen atoms of water. Soon afterward, reports began to appear on the occurrence of intracellular oxidative ferments in plant and animal tissues, and by 1894 it was clear that 'oxidases' of the type envisioned by Traube actually existed.

Enzymes

It was during the course of the discussion about the biological role of intracellular ferments that the term 'enzyme' appeared in the scientific literature. This word was used in 1876 by Kühne, who took a dim view of Hoppe-Seyler's description of various metabolic processes, including oxidations and reductions, as chemical reactions catalysed by ferments. In order to 'avoid misunderstandings and laborious circumlocutions', Kühne emphasized the difference between the action of agents like pepsin and the metabolic capacities of protoplasm by defining enzymes as 'the unformed or unorganized ferments whose action can occur without the presence of organisms and outside the latter'.[21] It should be noted that Kühne did not coin the word, for it had been used long before in medieval theological disputes about whether the Eucharist should be celebrated with leavened (*enzyme*) or unleavened (*azyme*) bread. Also, it is ironic that Kühne has recently been hailed[22] for the introduction of a word that assumed great importance in twentieth-century biochemistry to denote intracellular agents whose action he explicitly excluded from his definition, and ascribed to living protoplasm.

To turn to the circumstances surrounding Buchner's discovery, it is of interest that he had been trained as an organic chemist: during the early 1890s he was an assistant in von Baeyer's institute in Munich. To say that von Baeyer was primarily responsible for the pre-eminence of German organic chemistry during the first decade of this century may be an overstatement; his important role is indicated, however, by the fact that among his pupils and assistants were the future Nobel prize winners Emil Fischer, Richard Willstätter and Heinrich Wieland, as well as many others of lesser renown, but whose chemical achievements and influence were considerable. Apparently, Buchner did not show the promise expected of the best of this group, for much later Willstätter reported that after Buchner's discovery of cell-free fermentation, von Baeyer said: 'This will bring him fame, even though he has no chemical talent.'[23] Nor is this judgement inconsistent with the fact that, no matter how significant that discovery was in relation to the debates that came before and the experimental advances that ensued, it was a matter of improved technique rather than new theoretical insight. Also, the happy circumstance that Eduard Buchner was given the opportunity to collaborate with his brother Hans, a noted bacteriologist, and his associate Hahn on the preparation of cell-free microbial extracts that might contain antitoxins,[24] and the fact that Eduard's subsequent contributions to the development of the research field he opened were modest, cannot be omitted from the historical record. Nevertheless, the Nobel Prize in Chemistry, which he received in 1907, was richly merited; and the fact that he was chosen underlines the historical importance of a single great discovery, as compared with a lifetime of sustained achievement.

In announcing his discovery in the principal German journal of organic chemistry Buchner wrote:

'. . . the initiation of the fermentation process does not require so complicated an apparatus as is represented by the yeast cell. The agent responsible for the fermenting action of the press juice is rather to be regarded as a dissolved substance, doubtless a protein; this will be denoted *zymase*.'[9]

This statement was in the tradition derived from the writings of Moritz Traube and Hoppe-Seyler, and a blow to the upholders of the view that fermentation is not only correlative with life but is also the expression of the activity of an organized protoplasm not accessible to chemical dissection without loss of that activity. Although the chemist Ahrens triumphantly announced in 1902: 'The ancient conflict over the question "What is fermentation" had ended: fermentation is a chemical process',[25] the debate about the role of intracellular enzymes in metabolism was still in progress. The recent suggestion that by 1903, 'the protoplasm theory in general was being swept out of biology'[26] glosses over the continued adherence of many leading biologists and biochemists to this theory. Indeed, the uncertain status of the so-called 'enzyme theory of life' continued into the 1920s, for Otto Warburg not only dismissed the metabolic role of the dehydrogenases studied by Thunberg and Wieland, but also that of the oxidases found in cell extracts:

> 'If the extract-oxidases had been preformed in the cell, a single type of cell would contain innumerable oxidases. But the multiplicity of oxidases in the living cell would be in opposition to a sovereign principle in the living substance. . . . Therefore, if many oxidases have been found in extracts of a cell type, these were not ferments that were already present in the living cell, but rather products of the transformation and decomposition of a single homogeneous substance present in life.'[27]

Further evidence of this attitude may be found in Kluyver's statement in 1925 that

> 'It is no longer necessary to have recourse to the assumption of a large number of separate enzymes to explain the partial reactions of the dissimilation process . . . the changes which are ascribed to separate enzymes such as catalase, reductase, zymase [he lists seven more] are actually only manifestations of a definite degree of the affinity of protoplasm for hydrogen.'[28]

Intracellular oxidases

The background of such views of the 1920s may be found in the discussion earlier in this century about the nature of intracellular oxidation. Thus, in 1897, Spitzer proposed that 'the known oxidation processes by animal cells . . . are due to the presence in these cells of specifically active nucleoproteins or, more precisely, are attributable to oxygen transfer mediated by organically-bound iron'.[29] At that time, the cell nucleus was widely considered to be the site of intracellular oxidation, and the 'nuclein' found there years before by Miescher was a popular subject of biochemical investigation. In the succeeding decade, important work by Battelli and Stern led them initially to conclude that the 'main respiration' (corresponding to seventy-five percent of the oxygen uptake) of minced liver occurs only in the presence of intact cells, and they noted that

> 'there is no basis for ascribing the main respiration . . . to the action of enzymes. In fact, no one has yet succeeded in achieving main respiration in the absence of cells or cell fragments, and it is possible that the physical structure of the cell is unavoidably essential for the occurrence of the main respiration.'[30]

Subsequent work by Battelli and Stern and by Warburg showed that the main respiration was associated with subcellular granules. The question of the biological importance of the low respiration of cell extracts reflected a parallel discussion about the significance of Buchner's discovery. For example, Rubner argued that the true process of alcoholic fermentation depends on the integrity of the yeast cell, because Buchner's press juice was so much less active than an equivalent amount of yeast;[31] and in subsequent years, it was not uncommon to find in the biochemical literature the statement that a metabolic process is 'linked to life' because it could not be demonstrated in cell-free extracts.

Enzymes as proteins

To return to the period around 1900, if the organic chemist Eduard Buchner gave a decisive impetus to enzyme research through his discovery of cell-free alcoholic fermentation, a no less significant contribution was made by the greater organic chemist, Emil Fischer, through his work on the specificity of enzyme action. In 1894, Fischer reported that α-methyl glucoside is readily hydrolysed in the presence of a yeast extract, which he called 'invertin', but not by a preparation of emulsin obtained from almonds. On the other hand, the isomeric β-methyl glucoside was cleaved by emulsin but not by invertin. In his words,

> 'As is well known, invertin and emulsin have many similarities to the proteins, and undoubtedly also possess an asymmetrically constructed molecule. Their restricted action on the glucosides may therefore be explained on the basis of the assumption that only with a similar geometrical structure can the molecules approach each other closely, and thus initiate the chemical reaction. To use a picture, I would say that the enzyme and glucoside must fit each other like a lock and key, in order to effect a chemical action on each other. . . . The finding that the activity of enzymes is limited by molecular geometry to so marked a degree should be of some use in physiological research. Even more

> important for such research seems to me the demonstration that the difference frequently assumed in the past to exist between the chemical activity of living cells and of chemical reagents, in regard to molecular asymmetry, is non-existent.'[32]

In these sentences, there is not only the denial of the validity of Pasteur's view that molecular asymmetry is uniquely the expression of biological processes, but also the demonstration of the power of the new synthetic organic chemistry to illuminate the mode of enzyme action by the creation of artificial substrates whose chemical structure can be modified systematically. Fischer first used this approach to study the enzymic hydrolysis of glycosides because by the 1890s he had largely established the glycosidic nature of natural sugars such as sucrose, maltose and lactose, already known to be cleaved by soluble ferments. In the succeeding decade, after he had developed a method for the laboratory synthesis of peptides, he initiated studies on the hydrolysis of such compounds by enzymes known to act on proteins and on products of their physiological breakdown.

But it was not enough to have substrates of known structure. Shortly after the appearance of Fischer's 1894 paper, Bourquelot rightly called attention to the inhomogeneity of the invertin preparation Fischer had used, and noted that 'the ingenious hypothesis he advances may perhaps correspond to the facts . . . but it will not be possible to study it until a means has been found to prepare a chemically-pure soluble ferment, something that it has not been possible to do thus far'.[33] The problem of the purification of enzymes was to bedevil biochemical research for several decades, and at the heart of the problem was the question whether enzymes are proteins. Fischer and Buchner believed they were, but the issue was in doubt. It was not settled until the 1930s, when the claims of Sumner and Northrop to have obtained urease and pepsin in the form of crystalline proteins were generally accepted. Before then, many chemists who worked on the problem believed that enzyme purification should yield protein-free material. The most famous among them was Willstätter, who wrote in 1933 that while his laboratory 'aimed at gradual and, if possible, complete liberation of enzymes from protein, these American colleagues proceeded exactly in the opposite direction'.[34] It should be noted that well-defined proteins such as egg albumin, serum albumin, haemoglobin and several seed globulins had all been crystallized before 1900, by means of procedures not very different from those used later by Sumner and Northrop to purify urease and pepsin. The delay in the application of such methods to enzymes may be ascribed in part to the belief that, like the known hormones adrenaline and thyroxine, they should be small molecules, and that the presence of proteins in enzyme preparations is a consequence of the ready adsorption of such molecules by colloids.

In the face of this uncertainty about the organic-chemical nature of enzymes, many biochemists concerned with intracellular metabolism during the first decade of this century were receptive to ideas offered by the new physical chemistry, whose leading spokesman was Wilhelm Ostwald. In a lecture on catalysis given in 1901, he stated:

> 'We will see in the enzymes catalysts which are formed in the organism during cellular life and through whose action the living thing accomplishes most of its tasks. Not only digestion and assimilation are governed from beginning to end by enzymes, but also the supply of the needed chemical energy through combustion with atmospheric oxygen comes from the decisive cooperation of enzymes and would be impossible without them.'[35]

The relation of the new physical chemistry to biology had been established around 1880 by Pfeffer's measurements of osmotic pressure in plants and the stimulus they provided to the formulation of van't Hoff's osmotic pressure equation. The subsequent development of a theory of solutions founded on thermodynamics and buttressed by Arrhenius' theory of electrolytic dissociation in turn affected biological thought about such problems as the selective permeability of cellular membranes. So also van't Hoff's theory of chemical equilibria and reaction rate assumed biological importance when Ostwald proposed that the function of catalysts, including enzymes, is to accelerate the attainment of equilibrium in a chemical reaction that proceeds immeasurably slowly in their absence at a given temperature. These developments had a profound influence on biochemical thought; and in their advocacy of a 'general physiology' during the first decade of this century, Loeb and Bayliss enthusiastically adopted the physical-chemical approach to the explanation of biological phenomena. It is perhaps understandable, therefore, that significant progress was made in the study of the kinetics of enzymic catalysis around 1902, when Brown and Henri independently concluded that an enzyme-substrate compound is a kinetically important intermediate. Ten years later, after Fernbach and Sørensen had shown the importance of controlling the hydrogen ion activity (pH) in the measurement of enzyme kinetics, the Henri equation was further developed by Michaelis, and the idea of a rate-limiting enzyme-substrate complex (of the kind postulated by Fischer) became a basic concept in enzymology.

The ideas offered by van't Hoff and Ostwald about the reversibility of chemical reactions catalysed by enzymes were readily accepted by some biochemists at the turn of the century, for physical chemistry appeared to offer a solution to the problem of the mechanism of the biosynthesis of cellular constituents and of other products of animal metabolism. By the end of the nineteenth century, several metabolic processes were recognized to involve chemical synthesis; among these were the conversion of glucose to glycogen (in animals) or to starch (in plants), and it was also believed that blood proteins are formed from the breakdown products (peptones) arising from the digestive degradation of food proteins. Around 1900, such biosyntheses were considered to involve condensation reactions and, together with biochemical reductions, were associated with the phase of metabolism long denoted 'assimilation'; in keeping with the then current physiological thought they were defined as endothermic processes. The degradative metabolic reactions involving hydrolysis, oxidation or fermentation (collected under the term 'dissimilation') were defined as exothermic processes. Wide acceptance was given to Bernard's view, mentioned earlier, that only the degradations, such as the metabolic breakdown of proteins, fats or starch and glycogen, can be effected by soluble ferments, whereas assimilation was correlative with life and required the integrity of cellular structure. Indeed, in the first edition of his well-known book on enzymes, published in 1900, Oppenheimer stated that they could only effect chemical breakdown and not chemical synthesis.[36] At about the same time, however, experiments were reported that indicated the reversibility of the enzymic hydrolysis of disaccharides, fats and proteins. Thus, around 1900, Arthur Croft Hill claimed that he had demonstrated the synthesis of maltose from glucose by the enzyme 'maltase', Kastle and Loevenhart reported the enzymic synthesis of fatty acid esters, and several investigators described the formation by proteolytic enzymes of protein-like 'plasteins' from peptones. The attention given such reports is suggested by Loeb's comment that 'it is no exaggeration to say that Hill's paper entirely changed the conceptions of the physiology of metabolism'.[5] 'Synthetases' that catalysed the reversal of well-known hydrolytic reactions were consequently considered by many biochemists to be the intracellular catalytic agents in the biosynthesis of polysaccharides, fats and proteins. Similarly, the new physical chemistry, in providing a theoretical basis for electrochemical oxidation and reduction, gradually led to revision of the assumption, made on the basis of cytochemical studies, that separate 'reductases' existed in cells. By around 1910, it became clear to biochemists that an 'oxidation place' in the cell can also be the site of reduction, for the concept of oxidation-reduction potentials had been well developed. It was not applied rigorously to biochemical processes, however, until the 1920s, notably by Clark and Michaelis.

Colloid chemistry

Another branch of physical chemistry that attracted the attention of biochemists during the first decade of this century was colloid chemistry, founded during the 1860s by Thomas Graham. Among the non-crystalline materials of large molecular size which he called colloids were the albuminoid substances. Not only were colloids retained by membranes that allowed the passage of water and salts, but like protoplasm they imbibed water and adsorbed substances on their surface. Such adsorption phenomena had been studied qualitatively at mid-century, but the new physical chemistry provided a consistent theory to explain them. As a consequence, many investigators tended to view enzymic catalysis as the result of the adsorption of substrates on colloidal surfaces rather than of chemical combination with specific groups in a defined enzyme molecule. This attitude was clearly stated by Bayliss in the five editions of his book on the nature of enzyme action published between 1908 and 1925. In particular, he criticized the conclusion many had drawn from Fischer's experiments with invertin and emulsin:

> 'The subject needs attacking rather from a dynamic than from a static point of view; rates of reaction need more investigation than fitting of locks and keys. . . . Much of the recent work on the nature of enzyme action tends to show that the point of view of pure structural chemistry gives little help in the difficult problem . . . of the nature of the intermediate chemical compound between enzyme and substrate, which is supposed to be formed and afterward decomposed, if indeed, such a compound actually exists.'[37]

In Bayliss' view, which was shared by many of his contemporaries, enzymic catalysis takes place exclusively at the surfaces of colloidal particles suspended in a solution of a substrate, and not between substances that are all in true solution.

The period early in this century when biologists were attracted to colloid chemistry has recently been termed 'the dark age of biocolloidology'.[38] This disparagement is regrettable, for it obscures important contributions of colloid chemistry to biochemistry, as in Svedberg's demonstration during the 1920s that proteins are large molecules of defined particle weight. It also overlooks the fact that the chemical structure of proteins, long considered to be the most important protoplasmic constituents, was largely unknown. At that time, several noted organic chemists (e.g. Bergmann, Karrer) did not accept Staudinger's view that macromolecules linked by covalent bonds can exist, and various theories of protein structure were offered in which the physical association of low-molecular units such as diketopiperazines was invoked to explain the high particle weight of proteins. Indeed, the peptide theory of protein structure, advanced by Fischer and Hofmeister at the turn of the century, was questioned well into the 1930s.[39] Although, during the four decades before the Second World War, organic chemists were notably successful in elucidating the structure of many important chemical constituents of biological systems, the validity of the view that proteins are long-chain polypeptides was not generally accepted until after Sanger established the amino acid sequence of insulin during the 1950s. Moreover, colloid chemistry offered satisfying explanations of several physiological phenomena, notably the selective permeability of cell membranes. It is perhaps not surprising, therefore, that the physical-chemical approach to the study of the colloidal nature of protoplasm should have been attractive to many biologists and biochemists.

Intermediary Metabolism

From the foregoing it is evident that during the first decade of this century enzymes were widely considered to be colloidal catalysts, possibly associated with proteins, which accelerate the attainment of equilibrium in chemical processes and exhibit specificity in their action. Although some biochemists rejected Ostwald's view that enzymes are the principal agents of metabolism, influential voices were raised in support of an enzyme theory of life. In 1901, Hofmeister wrote:

> 'With the help of energy transfer, as well as release and inhibition mechanisms, one can easily build a complicated machine and it may be imagined that one could achieve, by a skillful combination of reciprocally-controlling chemical processes, and without many mechanical aids, an automatic mechanism which in regular sequence forms certain chemical products and converts them to others. The cell resembles . . . such an automatically-functioning chemical machine, and it is a noteworthy coincidence that the most important parts of this machine, the ferments, are particularly subject to release and inhibition effects and offer great possibilities for the combination of such effects.'

In the same article, he also stated;

> 'In the protoplasm the synthesis and breakdown of various substances proceeds by way of a series of intermediate steps, not always involving the same kind of chemical reaction, but rather a series of reactions of different kinds. . . . A regular sequence of chemical reactions in the cell requires, however, separate function of the individual agents and a definite direction of movement of the products, in short a chemical organization which is incompatible with a ubiquitous uniformity of protoplasm.'[40]

Although, as mentioned before, ideas about intermediary metabolism had been offered as early as the 1850s, it is probably fair to say that the first major contribution to the systematic study of metabolic pathways in animal organisms came in 1904, with Knoop's experiments on the oxidation of fatty acids,[41] and his proposal that this process involves the successive removal of two carbon atoms at a time from the carboxyl end. This work marked the introduction of a labelling technique for the study of intermediary metabolism. After stable and radioactive isotopes became available to biochemists during the 1930s, Schoenheimer and de Hevesy showed the power of such techniques in the elucidation of metabolic pathways.

Knoop's results, obtained in Hofmeister's Strasbourg institute, provided striking evidence for the latter's view that metabolic synthesis and breakdown 'proceeds by way of a series of intermediate steps'. Soon afterward, Embden, also a member of Hofmeister's laboratory, performed notable experiments which, together with the parallel work of Dakin, strongly buttressed Knoop's theory.

These achievements at the turn of the century are recounted in the books by Leathes[42] and Dakin,[43] and form the immediate background for Hopkins' oft-quoted statement in 1913 that it was a matter of studying the metabolism of 'simple substances undergoing comprehensible reactions',[44] rather than speculating about large protoplasmic molecules, derived from Pflüger's 'living proteins' or Verworn's 'biogen'. It also appears to have been a matter of reliance on organic chemistry to provide models of 'comprehensible reactions' that might occur in metabolic processes, and avoidance of the question of the nature of enzymes, except to note their colloidal nature.

These advances in the study of intermediary metabolism followed closely upon the demonstration, during the preceding decade, that animal organisms conform to the law of the conservation of energy. It should be added, however, that many chemists and biologists equated the heat liberated in a metabolic process with the work the process might be expected to effect in a living organism. For example, in discussions of muscular contraction, insufficient attention was paid to de Fick's insistence during the 1870s that this is a 'chemical-dynamic' process in which chemical energy is converted to work in ways other than through heat, and cannot be compared with the operation of a heat engine, which depends on differences in temperature. Indeed, as late as 1912, A. V. Hill was impelled to write:

> 'Unfortunately there have been, even among physiologists, many and grievous misconceptions as to the application of the laws of thermodynamics; these have been due partly to the desire to make over-hastily a complete picture of the muscle machine, partly to the completely erroneous belief that the laws of thermodynamics apply only to heat engines and not to chemical engines, and that no information can be obtained from the Second Law as to the working of a chemical machine at uniform constant temperature. The muscle fibre has been treated as a heat-engine when it is inconceivable that there are finite differences of temperature in it. . . . The muscle is undoubtedly a chemical machine working at constant temperature.'[45]

Thus, together with the partial acceptance of the enzyme theory of life, and progress in the study of intermediary metabolism, the turn of the century also witnessed the introduction of chemical thermodynamics into biology. For example, Wilhelm Ostwald proposed that in the transfer of energy between chemical processes, labile intermediates are formed in coupled chemical reactions.[46] But in 1900, the nature of the reactants, labile intermediates and products of the individual steps in biological oxidations and fermentations were largely unknown. Only when it became possible during the succeeding decades to define these components as chemical entities, and to dissect the metabolic pathways in which they appear, did it become evident how much more limited is the information about metabolism given by respiratory calorimetry. At the turn of the century, however, when the discussion of the role of enzymes in intracellular metabolism had been spurred by Buchner's discovery but was still inconclusive, those who continued to do research along the line set by von Liebig, Voit and Rubner affirmed their adherence to an organismic and physicalist view of animal metabolism. I have already noted Rubner's criticism of the conclusions drawn from Buchner's work. And in 1906, Lusk, one of the younger leaders in the field of animal calorimetry, wrote:

> 'The constancy of the energy requirement in metabolism makes difficult the explanation of the various ferments found in the body. These are of three varieties: hydrolytic, synthetic and oxidizing, but these from the very principles of our knowledge must be subservient to the requirement of the living cells, and not themselves masters of the situation, as, for example, they become in the autolysis of dead tissues. It seems to be the requirement of the mechanism of cell activity which determines metabolism, and not primarily the action of enzymes, whose influence appears to be only intermediary. . . . [M]etabolism does not depend on the satisfaction of chemical affinities, but rather upon a definite law of utilization of energy equivalents. However clearly formulated the laws of metabolism may be, and many of them are as fixed and definite as are any laws of physics and chemistry, still the primary cause of metabolism remains a hidden secret of the living bioplasm.'[47]

I have dwelt on the problem of the role of enzymes in intracellular metabolism because it was deemed to be of fundamental importance at the turn of the century, and because its history offers an example of the competition among the various biochemical attitudes to which I referred at the beginning of this paper. This problem was not the only one, however, that elicited lively discussion among chemists and biologists around 1900, and many others could be mentioned that illustrate to a greater or lesser degree the competition between styles of biochemical speculation. Among these other facets of the interplay of chemistry and biology at that time, the discussion between Ehrlich and Arrhenius about the nature of the immune response is of special historical interest; an excellent account has been published recently.[48]

Immunochemistry

Ehrlich had entered the new field of immunology during the 1890s, shortly after Emile Roux and von Behring had discovered antitoxins in the sera of animals infected with pathogenic bacteria. By 1900 it was evident that such 'antisera' could counteract the lethal effect of the toxins present in the bacterial cultures, and the question arose as to the nature of the toxin-antitoxin (or antigen-antibody) interaction. In that year, Ehrlich presented his views in a Croonian lecture,[49] where he proposed a 'receptor' theory for the action of bacterial toxins, as an outgrowth of his early suggestion (in 1885) that protoplasm is equipped with certain atomic groups ('side chains') whose function is to attach themselves to food materials. In the later elaboration of this theory, as applied to immunity, the toxin was thought to unite specifically and covalently with a particular protoplasmic 'haptophore' group and to exert its toxic action by virtue of its interaction with a 'toxophore' group. The immune response to a bacterial infection was, according to Ehrlich, the secretion into the blood of a 'receptor' derived from protoplasmic side chains and containing one or two haptophore groups. The latter receptors ('amboceptors') were thought to combine both with the toxin and with complement (a heat-labile serum factor just discovered by Bordet) and thus to act as a bridge between the two. In explaining his theory, Ehrlich drew analogies from organic chemistry; this organic-chemical orientation is evident also in his earlier work on the use of synthetic dyes to study cellular structure and metabolism and in his later excursion into chemotherapy. In particular, Ehrlich's side-chain theory of immunity appears to have been influenced by Fischer's lock-and-key hypothesis of enzyme action, presented a few years earlier. In turn, Ehrlich's ideas may have influenced Langley to propose in 1906 that nerve cells have 'receptive substances' that interact specifically with toxins such as nicotine or curare. It may also be noted that Ehrlich's idea later reappeared in the 'two-affinity' theory of enzyme action, proposed in 1923 by Hans von Euler: he suggested that an enzyme has one group that binds the substrate and another that is responsible for the catalytic action.

Ehrlich's theory was opposed by Arrhenius, who entered the field of immunology in 1901, largely as a consequence of his association with the bacteriologist Madsen. At that time, the importance of Arrhenius' theory of electrolytic dissociation was almost universally accepted (he received the Nobel Prize in Chemistry in 1903). In his study of the stoichiometry and kinetics of the immune response, Arrhenius convinced himself that the antigen-antibody reaction involved equilibria of the kind encountered in the interaction of weak acids and weak bases. He consequently applied the laws of reversible chemical equilibria to immune processes, denied the formation of the covalent bonds postulated by Ehrlich, and criticized Ehrlich's penchant for postulating new substances to explain every new phenomenon in antigen-antibody reactions. His attitude was later expressed pungently as follows:

> 'I am convinced that biological chemistry cannot develop into a real science without the aid of the exact methods offered by physical chemistry. The aversion shown by biochemists, who have in most cases a medical education, to exact methods is very easily understood. . . . The physical chemists have found that the biochemical theories, which are still accepted in medical circles, are founded on an absolutely unreliable basis and must be replaced by other notions agreeing with the fundamental laws of general chemistry.'[50]

On the other hand, Ehrlich stated:

> 'The natural aim of physical chemistry must always be to produce as few factors as possible for purposes of calculation whereas biological analysis always seeks to pay due regard to the wonderful multiplicity of organic matter. However, I believe that these two methods can readily be combined and that this will be very desirable. The biologist will have to content himself in so far as yielding to the economy of the mathematical view that he restricts his assumptions to the smallest number possible. The physical chemist, on the other hand, cannot escape the obligation of paying due heed to this minimal multiplicity, the result of experimental research.'[51]

The competition between those who shared Ehrlich's biological orientation and those who subscribed to Arrhenius' commitment to physical chemistry continued long afterward; an example may be found in an article by Northrop, in which he espouses Arrhenius' attitude and the principle of Occam's Razor.[52] As the subsequent development of the study of immune reactions showed, however, neither Ehrlich nor Arrhenius was as decisive a figure in the emergence of modern immunochemistry as their younger contemporary Landsteiner, who noted in 1905 that 'because of the lack of a sharp dividing line between so-called physical and chemical processes, the discussion as to whether these reactions are one or the other is meaningless'.[53] Other voices were also raised to counsel moderation; thus Bechhold responded to Pauli's claim that colloid chemistry had overcome Ehrlich's side-chain theory as follows:

> 'Physical chemistry and especially colloid chemistry offer a series of highly valuable methods for the study of biological and pathological processes, and I am certain that in the future they will make possible great progress; I remain convinced, however, that we should not disdain other methods, especially structural chemistry, and the structural-chemical attitude toward processes in living matter is at least of equal importance.'[54]

If the Arrhenius-Ehrlich debate may be seen as a confrontation between a physical chemist and a biologist who drew on the organic chemistry of his time, both were opposed by Metchnikoff who had developed, during the preceding two decades, a cellular theory of immunity based on phagocytosis. He staunchly defended this theory in a classic book *L'Immunité dans les Maladies Infectieuses*, published in 1901. When, four years later, Wright and Douglas proposed that immune sera contain agents (opsonins) that help leucocytes digest bacteria, the contradictions between the 'chemical' and the 'physiological' explanations of immunity became less sharp. The importance of Metchnikoff's achievements was widely recognized; he and Ehrlich shared the 1908 Nobel Prize for Physiology or Medicine for their work on immunity.

As in the discussion of the nature of enzymes at the turn of the century, the question of the nature of the bacterial toxins and the serum antitoxins was unresolved, and the view was widely held that they are all breakdown products of protoplasmic proteins of unknown chemical structure. Although Rudolf Kraus had shown in 1897 that purified proteins can act as antigens and can induce the formation of specific antibodies, the increasing emphasis on the colloidal properties of proteins led to interpretations of such findings in terms of adsorption phenomena.[55] In retrospect, we can see the importance of the 1906 report by Obermayer and Pick, who showed that the injection of a chemically-modified protein leads to the production of two types of antibodies – one directed against the modified groups in the antigenic protein and the other against the rest of the antigen. This represents the immediate background of Landsteiner's use of chemically-labelled antigens, and his decisive role in the emergence of modern immunochemistry.[56]

Conclusion

I take the liberty of concluding this paper with a quotation from an earlier article:

> 'There is little evidence of a linear progression within a single scientific discipline toward the so-called mature biochemistry of today, and the continuity of the biochemical enterprise may be seen rather in the competition among attitudes and approaches derived from different parts of chemistry and biology. Inevitably, such competition is attended by tensions among the participants. I venture to suggest that this competition and these tensions are the principal source of the vitality of biochemistry and are likely to lead to unexpected and exciting novelties in the future, as they have in the past.'[57]

May I also add an expression of gratitude for the opportunity of preparing this paper, because I find in the current biochemical literature, albeit on a larger scale, much of the optimism and dogmatism, as well as competition and confusion, that characterized it at the turn of the century. If the interplay of biology and chemistry is allowed to continue for another eighty years, perhaps at that time there will be another conference such as this one, and someone will describe the various attitudes that characterized biochemical work and thought during the 1980s in language not too different from mine, although based on a vastly greater body of scientific knowledge and in relation to problems we now deem to be important.

Notes

1. Eulner, H. H. (1970) *Die Entwicklung der Medizinischen Spezialfächer an den Universitäten des Deutschen Sprachgebietes*, Stuttgart, Enke.
2. Chittenden, R. H. (1930) *The Development of Physiological Chemistry in the United States*, New York, Chemical Catalog Co.

3. Berzelius, J. J. (1806) *Föreläsningar i Djurkemien*, Stockholm, Delén, p. 6.
4. Ruzicka, L. (1971) Nobelpreise und Chemie des Lebens. *Naturwiss. Rundsch.*, **24**, 50–56.
5. Loeb, J. (1906) *The Dynamics of Living Matter*, New York, Columbia University Press.
6. Bayliss, W. M. (1914) *The Principles of General Physiology*, London, Longmans Green.
7. Bernard, C. (1865) *Introduction à l'Étude de la Médecine Expérimentale*, Paris, Baillière, p. 228.
8. Lehmann, C. G. (1855) *Physiological Chemistry* (translated by G. E. Day and edited by R. E. Rogers), Vol. 2, Philadelphia, Blanchard & Lea, p. 354.
9. Buchner, E. (1897) Alkoholische Gärung ohne Hefezellen (Vorläufige Mittheilung). *Ber. chem. Ges.*, **30**, 117–124.
10. Berthelot, M. (1860) *Chimie Organique Fondée sur la Synthèse*, Paris, Mallet-Bachelier, p. 656.
11. Bernard, C. (1879) *Leçons sur les Phénomènes de la Vie Communs aux Animaux et aux Végétaux*, Vol. 1, Paris, Baillière, p. 40.
12. Pasteur, L. (1878) Première réponse à M. Berthelot. *C. R. Acad. Sci. (Paris)*, **87**, 1053–1058.
13. Pflüger, E. (1878) Ueber Wärme und Oxydation der lebendigen Materie. *Pflügers Arch.*, **18**, 247–380.
14. Nägeli, C. (1879) Theorie von Gärung. *Abh. k. Akad. Wiss. München*, **13**(2), 77–205.
15. Pflüger, E. (1875) Beiträge zur Lehre von der Respiration. I. Ueber die Physiologische Verbrennung in den lebendigen Organismen. *Pflügers Arch.*, **10**, 251–369, 641–644.
16. Nencki, M. (1885) Ueber das Parahämoglobin. *Arch. exp. Pathol. Pharmakol.*, **20**, 332–343.
17. Loew, O. (1896) *The Energy of Living Protoplasm*, London, Kegan Paul.
18. Ehrlich, P. (1885) *Das Sauerstoff-Bedürfnis des Organismus. Eine Farben-analytische Studie*, Berlin, Hirschwald.
19. Verworn, M. (1903) *Die Biogenhypothese*, Jena, Fischer.
20. Traube, M. (1878) Die chemische Theorie der Fermentwirkungen und der Chemismus der Respiration. *Ber. chem. Ges.*, **11**, 1984–1992.
21. Kühne, W. (1876) Ueber das Verhalten verschiedener organisierten und sog. ungeformter Fermente. *Verhandlungen Heidelberger Naturhist.-Med. Vereinsbl.*, **NS 1**, 2–5.
22. Gutfreund, H. (1976) Wilhelm Friedrich Kühne: an appreciation. *FEBS Lett.*, **62**, Suppl., E1-E2.
23. Willstätter, R. (1949) *Aus Meinem Leben* (edited by A. Stoll), Weinheim, Verlag Chemie, p. 63.
24. Kohler, R. E. (1971) The background to Eduard Buchner's discovery of cell-free fermentation. *J. Hist. Biol.*, **4**, 35–61.
25. Ahrens, F. B. (1902) Das Gärungsproblem. *Samml. chem. chem.-tech. Vortr.*, **7**, 445–495.
26. Kohler, R. E. (1973) The enzyme theory and the origin of biochemistry. *Isis*, **64**, 181–196.
27. Warburg, O. (1929) Atmungsferment und Oxydasen. *Biochem. Z.*, **214**, 1–3.
28. Kluyver, A. J. & Donker, H. J. L. (1925) The catalytic transference of hydrogen as the basis of the chemistry of dissimilation processes. *Proc. R. Acad. Sci. Amsterdam*, **28**, 605–618.
29. Spitzer, W. (1897) Die Bedeutung gewisser Nucleoproteide für die oxydative Leistung der Zelle. *Pflügers Arch.*, **67**, 615–656.
30. Battelli, F. & Stern, L. (1911) Die Oxydation der Citronen-, Äpfel-, und Fumarsäure durch Tiergewebe. *Biochem. Z.*, **31**, 478–505.
31. Rubner, M. (1913) Die Ernährungsphysiologie der Hefezelle bei der alkoholischer Gärung. *Arch. Physiol.*, **Suppl.**, 1–392.
32. Fischer, E. (1894) Einfluss der Konfiguration auf die Wirkung der Enzyme. *Ber. chem. Ges.*, **27**, 2985–2993.
33. Bourquelot, E. (1896) *Les Ferments solubles (Diastases-enzymes)*, Paris, Société Editions Scientifiques, pp. 133–134.
34. Willstätter, R. (1933) Problems of modern enzyme chemistry. *Chem. Rev.*, **13**, 501–512.

35. Ostwald, W. (1901) Über Katalyse. *Z. Elektrochem.*, **1**, 995–1004.
36. Oppenheimer, C. (1900) *Die Fermente und ihre Wirkungen*, Leipzig, Vogel, p. 22.
37. Bayliss, W. M. (1925) *The Nature of Enzyme Action*, 5th ed., London, Longmans Green, pp. 143–144.
38. Florkin, M. (1972) *A History of Biochemistry*, Amsterdam, Elsevier, p. 279.
39. Fruton, J. S. (1979) Early theories of protein structure. *Ann. N.Y. Acad. Sci.*, **325**, 1–18.
40. Hofmeister, F. (1901) *Die Chemische Organisation der Zelle*, Braunschweig, Vieweg, pp. 19, 26.
41. Knoop, F. (1904) *Der Abbau aromatischer Fettsäuren im Tierkörper*, Freiburg, Kuttruff.
42. Leathes, J. B. (1906) *Problems in Animal Metabolism*, Philadelphia, Blakiston.
43. Dakin, H. D. (1912) *Oxidations and Reductions in the Animal Body*, London, Longmans Green.
44. Hopkins, F. G. (1913) The dynamic side of biochemistry. *Nature*, **92**, 213–223.
45. Hill, A. V. (1912) The heat production of surviving amphibian muscles during rest, activity, and rigor. *J. Physiol.*, **44**, 466–513.
46. Ostwald, W. (1900) Über Oxydationen mittels freien Sauerstoffs. *Z. phys. Chem.*, **34**, 248–252.
47. Lusk, G. (1906) *Elements of the Science of Nutrition*, Philadelphia, Saunders, pp. 296–297.
48. Rubin, L. P. (1980) Styles in scientific speculation: Paul Ehrlich and Svante Arrhenius on immunochemistry. *J. Hist. Med.*, **35**, 397–425.
49. Ehrlich, P. (1900) On immunity, with special reference to cell life. *Proc. R. Soc.*, **66**, 424–448.
50. Arrhenius, S. (1915) *Quantitative Laws in Biological Chemistry*, London, Bell, p. vi.
51. Ehrlich, P. (1910) *Studies in Immunity*, 2nd ed., New York, Wiley, p. 500.
52. Northrop, J. H. (1961) Biochemists, biologists, and William of Occam. *Ann. Rev. Biochem.*, **30**, 1–10.
53. Landsteiner, K. & Reich, M. (1905) Ueber die Verbindungen der Immunkörper. *Centralbl. Bakteriol.*, **39**, 83–93.
54. Bechhold, H. (1905) Report on address by Wolfgang Pauli at meeting of Gesellschaft der Aerzte in Wien. *Wiener klin. Wochenschr.*, **18**, 550–551.
55. Pauli, W. (1907) *Physical Chemistry in the Service of Medicine*, New York, Wiley.
56. Landsteiner, K. (1933) *Die Spezifizät der serologischen Reaktionen*, Berlin, Springer.
57. Fruton, J. S. (1976) The emergence of biochemistry. *Science*, **192**, 327–334.

DEVELOPMENTS IN PHYSICAL CHEMISTRY AT THE TURN OF THE CENTURY*

ERWIN N. HIEBERT

Department of History of Science, Harvard University, Cambridge, Massachusetts, USA

During the late 1880s a number of chemists took it upon themselves to launch physical chemistry as a constituent branch of the discipline within the larger profession of chemistry. The objective of these chemists was to establish this new domain as an autonomous, academic and industrially significant profession, primarily alongside organic and inorganic chemistry, but also within the context of the growing professionalization of other related sub-domains such as pharmaceutical, medical and analytical chemistry. The newly self-proclaimed physical chemists also were intent upon forging a strong and academically acceptable link to connect the various domains of physics with those of chemistry. While no 'Newton of chemistry' had been identified by the end of the nineteenth century, the link between physics and chemistry had been cultivated with considerable success for many years by investigators, both physicists and chemists, such as Davy, Faraday, Bunsen, von Helmholtz, Clausius, Kirchhoff, Landolt, Gibbs, Boltzmann, Planck and Nernst.

Wilhelm Ostwald was the most enthusiastic and enterprising proponent of the new thrust.[1] Over a period of two decades, as Professor of Physical Chemistry at the University of Leipzig, Ostwald managed, by virtue of being able to take full advantage of his own strong native abilities and constitution, in addition to a variety of favourable environmental circumstances, to emerge as a powerful and cunning spokesman for the intellectual and professional legitimacy of people who called themselves physical chemists. An excellent experimentalist, Ostwald skilfully organized and administered an efficient and exciting experimental research programme. He was acclaimed at home and abroad as a persuasive and resourceful lecturer and teacher. He was a lucid, speculative and provocative expositor of chemistry, its theory, its history, its philosophy and its organizational structure. His literary-scientific output was immense: he wrote

* This paper is dedicated to my teacher, friend and colleague Aaron J. Ihde. An earlier version was delivered at the 23rd Annual Meeting of the Midwest Junto of the History of Science Society in April of 1980 on the occasion of Aaron's retirement from thirty-seven years of teaching chemistry and the history of chemistry at the University of Wisconsin in Madison.

some forty-five books, 500 scientific papers and 4000 reviews, besides, at one time or another, being involved in the editing of six journals.

Ostwald's imposing disposition was further enhanced by his being a romantic artist, who put considerable effort into painting and into the study of the science of colours. While his reputation as an entrepreneur for the new theories of chemistry was unique among chemical colleagues of his own generation, the breadth of his literary and public interests put him into contact with an international constellation of scientists, philosophers and humanists. It goes without saying that his ideas often engendered controversy and stiff opposition. The German expression that most cogently captures what I think Ostwald would have wanted to be for his time was 'der grosse Kulturträger'.

The American physical chemist Wilder Bancroft, who received his doctorate under Ostwald in 1892, and who was one of the most critical of his students, wrote in 1933:

> 'We can distinguish three groups of scientific men. In the first and very small group we have the men who discover fundamental relations. Among these are van't Hoff, Arrhenius and Nernst. In the second group we have the men who do not make the great discovery but who see the importance and bearing of it, and who preach the gospel to the heathen. Ostwald stands absolutely at the head of this group. The last group contains the rest of us, the men who have to have things explained to us. . . . Ostwald was a great protagonist and an inspiring teacher. He had the gift of saying the right thing in the right way. When we consider the development of chemistry as a whole, Ostwald's name like Abou ben Adhem's leads all the rest. . . . Ostwald was absolutely the right man in the right place. He was loved and followed by more people than any chemist of our time.'[2]

Perhaps a single example may help to demonstrate the awe-inspiring magnetism with which Ostwald and the Leipzig environment drew young scientists into its orbit. Einstein in 1901 published his first scientific paper on capillary phenomena.[3] He sent a copy to Ostwald in Leipzig with the following comment:

> 'Since it was your work on general chemistry [Ostwald's *Lehrbuch*] that furnished the stimulus for the enclosed paper, I have taken the liberty of sending you a copy. On this occasion I shall also allow myself to ask whether you could perhaps make use of a mathematical physicist who is familiar with absolute measurements. I allow myself the freedom of making such an inquiry only because I am without means, and only such a position would offer me the possibility for additional training.'

Two weeks later Einstein wrote again, this time from his parents' home in Milan, repeating the address at which he could be reached, and telling Ostwald that a response to his article on capillary phenomena would be of great importance to him.[4]

Apparently, Ostwald did not answer, because ten days later Einstein's father wrote apologetically to Ostwald saying that his son, although only twenty-two years old, was very talented, diligent, in love with his studies and had passed brilliantly the Diplom-examination at the Zurich Polytechnikum, but was deeply unhappy not to have been able to secure an assistantship that would enable him to extend his education in theoretical and experimental physics. The father continued:

> 'Since among all of the scholars who are active in physics today my son respects and esteems you most highly, Hochgeehrter Herr Professor, I venture to turn to you to ask you politely please to read the article that he published in the *Annalen der Physik*, and eventually to send him a few lines of encouragement so that he again may achieve the joy of living and working [*Lebens- & Schaffensfreudigkeit*]. . . . Once more I beg you please to excuse me for my forwardness; and allow me to add that my son has no suspicion about this unusual request on my part.'

There is no record to show whether Ostwald responded to these letters of 1901. Einstein was not invited to Leipzig as Ostwald's assistant. Instead, as we know, the following year he managed to secure employment in the Berne patent office. While there, between 1905 and 1906, he published four papers that contributed conspicuously to establishing the direction of twentieth century theoretical physics. What is of note here is that Einstein's fascination with a person of Ostwald's type was no youthful mercurial streak. He wrote to Ostwald in 1916 that he had 'twice with enchantment' read his treatise on colour theory.[5] In 1929, when Ostwald delivered a lecture on colour theory at the Prussian Academy of Sciences in Berlin, Einstein was the liveliest discussant. Ostwald was pleased, as he wrote to Einstein, that the excitement raised the temperature of his colleagues by twenty-five degrees.[6]

We may note that in Ostwald's day, a good number of physicists had made fundamental contributions to both experimental and theoretical chemistry. It had been recognized, besides, that chemists in increasing numbers were engaged in activities that were contiguous with various domains of physics, especially in the study of the thermal, optical and electrical properties of matter. This in itself is hardly surprising since, for example, heat and electricity were often treated more as a part of chemistry than of physics up to the middle of the nineteenth century.

Pre-ionist Developments in Physical Chemistry

This is not the place to discuss in any detail the pre-Ostwaldian rise and development of physical chemistry. Nevertheless, in order to acquire some perspective on the movement of the 1880s, it may be appropriate to isolate the growth of those dominant experimental and theoretical accomplishments and traditions that laid the foundations for the legitimation of physical chemistry as a special domain within the larger outlook of chemistry. Certain it was that physical chemistry was not conceived in 1887 within a vacuum.

Soon after the atomic-molecular conceptions of Dalton and Avogadro had taken root, Dulong and Petit announced, in 1819, the principle of constancy of atomic heats. The idea was later extended, by F. E. Neumann and Hermann Kopp, to include the molecular heat of solid compounds as an additive property of the atomic components of molecules. The importance of the study of specific heats as a function of composition, constitution and temperature runs like a guiding thread through the theoretical deliberations of chemical thermodynamics that culminated in the discovery of the puzzling and theoretically significant characteristics of matter that show up as one approaches the absolute zero of temperature.

The oldest and perhaps most diligently pursued branch of physical chemistry was electrochemistry[7] and the chemistry of electrolytes. Already in 1834, Faraday was able to show how electrochemical equivalents can be calculated from the laws of electrolytes that he had laid down. The history of nineteenth-century electrochemistry – which we shall not explore here – provides a veritable treasure of insights into the chemistry of solutions. Most important was the recognition, around 1883, that the electromotive force values for reversible galvanic cells are directly proportional to the free energy of the process, and therefore serve as a measure of chemical affinity. At about the same time, in 1882, Raoult's experiments led to the correlation of molecular weights with the depression of freezing-points and vapour pressure of solutions. The anomalous behaviour of salt solutions (volume relationships, refractive indices, electrical conductivity, vapour pressure, and colligative properties in general) were all explained convincingly – at least for dilute solutions – by the more general theory of solutions that van't Hoff published the following year.

In another area we recognize that by mid-century the acquisition of an impressive amount of information about organic compounds, chemical composition and molecular constitution suggested the importance of carrying out a comprehensive study of the physical properties of compounds. For half a century, chemists were kept occupied exploring the additive as well as the anomalous physical properties of molecular substances. These studies focused on specific heat, volume, refractivity, the rotation of plane polarized light, the selective absorption of light, the dielectric constant, the parachor, surface energy, heats of vaporization, viscosity of liquids, etc. Thus, early nineteenth-century physical chemistry was predominantly oriented toward the systematic study of physical properties as a function of chemical composition and constitution.

During the second half of the century, the beginnings of colloid chemistry, chemical catalysis, photochemistry, the spatial conception of molecular constitution (i.e. stereochemistry), and the elucidation of optical and geometrical isomerism were seen. Even more central to the future focus of physical chemistry was the study of chemical change (extent, position of equilibrium, and speed of chemical processes) and the factors that influence chemical change.

With the enunciation of the law of mass action and the first and second laws of thermodynamics, the old notion of chemical affinity took on a precise meaning. Thus, the basis was laid for a chemical thermodynamics that focused on methods for establishing the criteria for the equilibrium conditions, chemical reactivity, and the stability of chemical compounds. Nothing contributed more to changing the whole complexion and theoretical priorities of physical chemistry than the recognition that the course of spontaneous processes is governed by both the first and second laws of thermodynamics, and not just the first, as the Thomsen-Berthelot principle implies.

van't Hoff, in 1884, demonstrated that second-law considerations provided the clue to determining the work of chemical affinity for a reversible isothermal reaction. From within thermodynamics, too, came the correlation of osmotic pressure with gaseous pressure and with the other colligative properties that figure so prominently in the theory of dilute solutions. All said, however, physical chemistry at this stage had not yet taken on the stamp of its own professional awareness as a branch of chemistry with its own, unique cognitive character. The important point to emphasize in this context is that by 1887, when the inner driving force and physics-oriented emphasis on chemistry received special acclamation, there had been considerable activity, which, by any meaningful definition of boundary conditions for the discipline, would have to be labelled 'physical chemistry', although the term was not yet used.

In order to examine the nature of physical chemistry at that time, it would be informative to identify scientists of Ostwald's generation who were singled out by the leading contemporary physicists as chemists whose research activities merited being looked upon as a recognized component of physics. Fortunately, we have access to records – covering the period from 1870 to 1929 – which reveal the names of the chemists who were elected to the Berlin Academy of Sciences on the recommendation of physicists.[8] We see that over a period of fifty-nine years, ten out of the seventy-three members elected were chemists: Hans Landolt, Lothar Meyer, J. H. van't Hoff, William Ramsay, Dimitri Mendeleev, Wilhelm Ostwald, Walther Nernst, T. W. Richards, Fritz Haber and Otto Hahn.

In his *laudatio* for Ostwald, who was elected to the Berlin Academy in 1905, van't Hoff said: 'Among contemporary chemists Ostwald is absolutely one of the most prominent figures in all parts of the world; the magnitude of his work has seldom been attained by his predecessors.'[9] The following year, in 1906, at age fifty-three, Ostwald retired from teaching immediately after his return from a German exchange professorship at Harvard. He moved without delay – family, library, apparatus, artist's easel and cat – to Grossbothen, Saxony, there to spend the rest of his life in his country home which he had baptized 'Landhaus Energie'. Isolated, and free to do as he pleased, he continued to champion energetics, the international brotherhood of man, Esperanto, colour harmonics, and the importance of organizational activity as the greatest tasks of the twentieth century. In 1909 he received the Nobel Prize in Chemistry 'in recognition of his work on catalysis and for his investigations with the fundamental principles governing chemical equilibria and rates of reaction'.[10]

The Pioneers: Ostwald, van't Hoff and Arrhenius

The *annus mirabilis* of physical chemistry was 1887. That year saw the completion of Ostwald's monumental two-volume *Lehrbuch der allgemeinen Chemie* (*Stöchiometrie* and *Verwandtschaftslehre*). In the second edition, in 1910, he added a comparable volume entitled *Chemische Energie*. Ostwald's *Lehrbuch* was based on a fifty-year survey and exposition of the literature of physics and chemistry and their interconnection.

In February of that year, with van't Hoff of Amsterdam as co-editor, Ostwald published the first issue of a journal exclusively devoted to physical chemistry. The full title of the publication was: *Zeitschrift für physikalische Chemie, Stöchiometrie und Verwandtschaftslehre*. Volume One contained articles by, among others, van't Hoff, Ostwald, Arrhenius, Lothar Meyer, Raoult, Guldberg, Ramsay, Mendeleev, Julius Thomsen, Le Chatelier and Planck. Ostwald explicitly emphasized that the new journal would be devoted to 'general chemistry', meaning the foundations of all chemistry – by way of contrast to the preoccupation of chemists with the specialized chemistry of individual substances. He also wanted to appeal to the interests of physicists, to stress new experimental investigations, and to provide an open platform for speculations and theoretical discussions.

In his editorial, Ostwald reiterated the earlier message (1882) of Emil Du Bois-Reymond who, while proclaiming physical chemistry as the chemistry of the future, had reminded his readers that chemistry had come a long way toward silencing the slanderous remarks that Immanuel Kant had dropped a century earlier. Kant had explained that chemistry was a systematic art rather than a true science because it could not provide knowledge about nature in the way that dynamics does. Kant had maintained that chemistry could attain to being a science in this highest sense only if man could divine the causes of elasticity, velocity and the labile and stabile positions of equilibrium for particles, as we do for the stars. Accordingly, the 'astronomy of chemistry', in Kant's judgement, had not yet progressed to the stage of astronomy of the time of Copernicus and Kepler.[11]

In September of 1887, Ostwald took up his new post as Professor of Physical Chemistry at the University of Leipzig. Actually this post – the first in Germany – had been created for Gustav Wiedemann in 1871. Ostwald inherited the chair when, on the retirement of Wilhelm Hankel, in 1887, Wiedemann accepted the Leipzig professorship of physics. However, it should be mentioned that the position had been offered first to van't Hoff, who decided to stay with his post in Amsterdam.[12]

Beginning in 1887, and for a period of two decades, Ostwald became an apostle for the new ideas stemming from the electrolytic dissociation theory of Arrhenius, the osmotic theory of solutions of van't Hoff, and the general thermodynamic research programme connected with the study of chemical processes: mass action, chemical affinity and the conditions and criteria for chemical equilibrium and spontaneity. Ostwald's laboratory soon became the hub of experimental investigations and theoretical deliberations for venturesome graduate students from far and wide – including an abundance of Americans.

The principal master-architects who, early on, were responsible for promoting the establishment of physical chemistry were four: viz. Arrhenius, van't Hoff, Ostwald and Nernst. The geographical centres from which the strategies of these investigators were directed were, respectively, Stockholm, Amsterdam, Leipzig and Berlin. It will be informative to make a few remarks about the collaborative efforts of these physical chemists. Arrhenius (b. 1859), van't Hoff (b. 1852) and Ostwald (b. 1853) were all in communication with one another by the mid-1880s. Nernst (b. 1864), who was younger by ten years or so, joined the Leipzig axis of *Ioner*, as they came to be called, several years later. In a number of significant ways, as we shall see, Nernst endorsed and nourished an interpretation of physical chemistry more from the perspective of a physicist than that of a chemist.

Arrhenius in his doctoral dissertation in 1884 had spelled out the basic components of an electrolytic theory of dissociation. In June of that year he sent Ostwald a copy of his *Recherches sur la conductibilité galvanique des électrolytes*.[13] Ostwald, who was then at the Polytechnikum in Riga, enthusiastically accepted the theory, visited Arrhenius in Stockholm, and, while there, offered him the position of Dozent in Riga. Reluctant as his Swedish countrymen had been about adopting the theory of ionization, from that time on Arrhenius received some recognition for his contributions in various quarters. With support from Ostwald and the Swedish Academy of Sciences, Arrhenius undertook a travelling scholarship that began in Ostwald's laboratory in Riga (1886–1887) and ended in Ostwald's Physical Chemical Institute in Leipzig (1889–1890). In between those periods, Arrhenius worked with Friedrich Kohlrausch in Würzburg, Boltzmann in Graz and van't Hoff in Amsterdam.

The scientific collaboration between Ostwald and Arrhenius, as reflected in their correspondence over a period of forty-three years, gives us a vivid picture of how diligently they struggled to put their new branch of chemistry on a par with the high level that organic chemistry had attained in Germany by the end of the nineteenth century.[14] We may note here that Ostwald was born in Riga, in Latvian Russia, and had received his chemical training there at the University of Dorpat (now Tartu). He remarked that the intense preoccupation with organic chemistry began only at the German border and that he would probably have become an organic chemist if he had studied chemistry in Germany.[15]

Early in 1886, Arrhenius was making plans to visit Ostwald in Riga, more, as he said, to learn something about his methodology (*Arbeitsmethoden*) than to carry out joint experimental investigations. Unfortunately, his plans to get to Riga via Finland and St Petersburg (where he had wanted to search out Mendeleev and Menschutkin) were foiled because the ships were frozen in.[16] By the end of 1886, Arrhenius was in Würzburg and wrote that the person who was engaged in the most exciting work there was a young man by the name of Nernst.[17] Arrhenius was also captivated by the theoretical ideas of Boltzmann in Graz. With van't Hoff in Amsterdam, he became involved in investigations on the lowering of vapour pressure and the depression of the freezing-point of salt solutions.[18]

We learn that an increasing number of prominent scientists were lending their support to Arrhenius' ionic theory; besides Ostwald and van't Hoff, we could cite Nernst, Planck, Boltzmann, Guldberg, August Kundt, Friedrich Kohlrausch and Hans Landolt. As long as Arrhenius was on his travelling scholarship (1886–1890), he was confident that his dissociation theory, and the ionists in general, would be given a favourable hearing. However, a more hostile reception awaited him on his return to Stockholm in 1891. It was becoming apparent, besides, that the scientists associated with Wiedemann in Berlin were beginning to band together into an outspokenly hostile anti-ionist front.[19] In addition, scientists elsewhere were examining these issues critically and, at most, provided their own reconstructed versions of ionic chemistry: Rudorff in Berlin, Ladenburg in Breslau, Henry E. Armstrong and Pickering in London, Traube in Hanover, and so on.[20] J. J. Thomson in Cambridge had been arguing that all of these so-called ionizations could be explained on the basis of the attractions between solute and solvent.[21] Lothar Meyer from Tübingen warned Ostwald that a too caustic approach towards his critics would backfire, and that the *Zeitschrift* should avoid airing personal issues and stick to the substantive aspects of ionization theory.[22] That, however, did not prevent Ostwald from making plans – never carried out – to publish a farcical discourse involving an ionist, a physicist and an organic chemist. Writing to Arrhenius, Ostwald said: 'Just think how neat it would be if we could have Traube and Pickering come on the scene in person!'[23]

It is evident that Arrhenius was notably sensitive about the opinions of scientists and that he felt the need to establish and maintain his scientific image. Determined to gain a reputable status among his colleagues, so as to become eligible for an academic position, he undertook experiments to demonstrate conclusively that it is the radiation in air that renders it electrically conductive. Apparently he believed that the establishment of a kinetic interpretation of ions both in gases and in solutions would strengthen his much criticized theory of electrolytic dissociation. Arrhenius correctly sized up the situation by realizing that physical chemistry was not really recognized in Sweden, and that he probably could improve his image there as a scientist only by moving further away from chemistry in the direction of physics.[24]

Matters improved to some extent for Arrhenius in 1891 when he acquired the position of lecturer in physics at the Stockholm Högskola. He now spoke with pride about the three 'temples for physical chemistry' that had been established: Ostwald in Leipzig, Nernst in Göttingen and Arrhenius in Stockholm.[25] However, by 1895, Ostwald was becoming fed up with teaching, as his interest in chemistry began to wane, just when he was inheriting the new laboratory that he had waited for so long.[26] In that year, with the help of letters of strong recommendation to the rector of the University of Stockholm from Ostwald, Boltzmann and Planck,[27] Arrhenius attained his full professorship in physics at the Högskola. But we may note that by this time Arrhenius' interests had also shifted rather radically in the direction of cosmic physics, immunology and serum physiology. In 1903, he received the Nobel prize in chemistry 'in recognition of the extraordinary services he has rendered to the advancement of chemistry by his electrolytic theory of dissociation'.[28] From 1905 until he died in 1927, he was director of the Nobel Institute for Physical Chemistry of the Swedish Academy of Sciences in Stockholm.

While there can be no doubt about the unique roles that Ostwald and Arrhenius occupied as entrepreneurs in the establishment of physical chemistry as a professional discipline, it was undoubtedly van't Hoff who exerted the most profound influence on the theoretical status of this new branch of chemistry. During his lifetime he received wide acclaim for his epoch-making theoretical ideas on the tetrahedral carbon atom, chemical kinetics, thermodynamics and the theory of solutions. Thus, it is not surprising that he was the recipient, in 1901, of the first Nobel Prize for Chemistry, 'in recognition of the extraordinary services he has rendered by the discovery of the laws of chemical dynamics and osmotic pressure in solutions'.[29] Ostwald wrote at the time to Arrhenius:

> 'That van't Hoff has received the Nobel prize is a very good thing for physical chemistry in Germany; because now the organic chemists here are beginning to fear for their hegemony and try everywhere to repress us. So they have been offended that neither [Adolf von] Baeyer nor [Emil] Fischer were favoured. How happy I was about this, because of my personal friendship with van't Hoff, I need not tell you.'[30]

van't Hoff (unlike Arrhenius) secured high-level teaching posts in his native Holland early on, at the age of twenty-six. True, he began in 1876 as lecturer in physics at the State Veterinary School in Utrecht and because of his enunciation of the tetrahedral structure of the carbon atom became the butt of Kolbe's vitriolic attacks; but by 1878 he was Professor of Chemistry, Mineralogy and Geology at the University of Amsterdam. In 1896, after his election to the Prussian Academy in Berlin, he served as Professor of Chemistry at the University in Berlin.

Like Ostwald, van't Hoff possessed a poetic, visionary and romantic temperament. His favourite author seems to have been Lord Byron; his favourite scientist, Sir Humphry Davy. In a somewhat disjunct but remarkably perceptive philosophical inaugural lecture in 1878 on the role of imagination in science, we are exposed to the view that 'it is sometimes easier [in approaching a scientific problem] to *circumvent* prevailing difficulties rather than to *attack* them'.[31] Scientific discovery, said van't Hoff, does not resemble 'the shooting down of a fortress from different sides, the cautious scaling of the ruins, and the battle to raise the flag on top after the arrival of all the forces'. Rather, quoting H. T. Buckle (with Davy as discoverer in mind), he says: 'There is a spiritual, a poetic, and for aught we know a spontaneous and uncaused element in the human mind, which ever and anon, suddenly and without warning, gives us a glimpse and a forecast of the future, and urges us to seize the truth as it were by anticipation.'[32] Small wonder that van't Hoff and Ostwald got along so well and laboured together for so many years to establish the reputation of physical chemistry at a level *par excellence*. The relationship between Ostwald and Arrhenius was far less congenial.

Ostwald was first exposed to van't Hoff's work in 1886. It was in Riga that he discovered for himself the great wealth of ideas that were contained in van't Hoff's *Etudes de dynamique chimique* of 1884. This treatise, and the extension of its ideas to the colligative properties of ionic and non-ionic solutions, opened up for Ostwald the new world of chemical thermodynamics and chemical kinetics: the crucial role of free energy in the analysis of the equilibrium conditions for chemical processes, reaction rates as a function of temperature and the general expression that supplies the analogies that can be drawn in studying gases, non-electrolytes and dilute ionic solutions. Ostwald mastered these matters very well and was able to build upon the profound implications that the concept of electrolytic dissociation furnishes for studies of chemical affinity. It soon became evident that van't Hoff would be the third member of a triumvirate of physical chemists whose organizational centre in Leipzig drew much of its spiritual support from its intellectual outposts in Stockholm and Amsterdam.

The Ostwald-van't Hoff correspondence over a period of twenty-five years (1886–1911) deals mainly with organizational and publishing matters connected with their co-editorship of the *Zeitschrift*.[33] Together, they set up a truly international forum for chemical contributions that admirably laid bare the borderland where physics and chemistry overlap. In a reference to Duhem's support for the dissociation theory, van't Hoff wrote to Ostwald that the dynamics faction (*Dynamik-partei*) was on its way to turning chemistry upside down: 'God have mercy on our souls.'[34] With profound satisfaction, he wrote:

'You will be interested to learn that Boltzmann has derived the law of osmotic pressure, and likewise the law of diffusion from the kinetic theory of gases.'[35] Aware of the tremendous administrative burdens that Ostwald encountered in his multifarious publication projects, van't Hoff wrote: 'May the rapid awakening of physical chemistry grow up above your work like a huge laurel plant.'[36] Perhaps it is not too surprising that Ostwald was anxious to retire by age fifty in order to gain a new freedom outside of chemistry. van't Hoff wrote to Ostwald: 'You know that I hold your longing for this absolute freedom as something abnormal. . . . There is something tragic in this, namely that the person who has attained the best [that exists] still does not find peace therein. It is the fate of the everlasting wanderer.'[37] By 1910 Ostwald wrote: 'My interests wander ever further away from chemistry. Internationalism, pacifism and cultural energetics are now my problems.'[38]

Ostwald and the Physicists: Boltzmann, Planck and Nernst

Having examined to some extent the mode of interaction of the three pioneers of physical chemistry – Ostwald, Arrhenius and van't Hoff – I would like to make some comments about the way in which a number of contemporary physicists got involved in physical chemistry. For this purpose I shall limit myself primarily to the interchange of ideas between Ostwald and Ludwig Boltzmann, Max Planck and Walther Nernst.[39]

The Ostwald-Boltzmann interchange of ideas began in 1890 when Boltzmann published a paper in Ostwald's *Zeitschrift* that analysed van't Hoff's hypothesis of osmotic pressure from the point of view of the kinetic theory of gases.[40] While Boltzmann followed closely and in fact contributed substantially to the clarification of the theory of solutions and electrolytic dissociation, the major and recurrent theme that he explored with Ostwald was weighing the relative merits of the mechanistic versus the energetic interpretation of natural phenomena. Ostwald adamantly defended the energetic view. Boltzmann consistently championed the mechanical view, but was constantly asking Ostwald – in a most sympathetic and open way – to explain his position so that physicists and mathematicians might understand his point of view.

At times, Boltzmann seemed to bend over backwards to accommodate Ostwald's original conceptions. For example, in a letter to Ostwald in 1892, Boltzmann compared the electrical and magnetic components of radiation with the kinetic and potential components of vibrating mechanical systems. This seemed to him to suggest that radiant energy adheres to the matter of the aether as mechanical energy adheres to material bodies.[41] This point of view would have appealed to Ostwald, who was determined to show that mechanics can be reduced to energetics.

Boltzmann was not convinced by Ostwald's energetic point of view and wrote in 1892: 'Since I do not in any way take a negative stand toward your ideas, but rather am an advocate of them, you all the more should not take my doubts amiss when I tell you that today at least I will not trade them off for the old mechanics.'[42] In the same communication, Boltzmann discussed some mathematical errors in Ostwald's second edition of the *Lehrbuch*.[43]

The controversy surrounding the energetic philosophy of Ostwald and his henchmen, Georg Helm of Dresden and Franticek Wald of Prague, came to a head at the sixty-seventh annual meeting of the German Society of Scientists and Physicians in Lübeck in 1895.[44] Ostwald's lecture on 'The Conquest of Scientific Materialism' stirred much debate. Sommerfeld wrote:

> 'Helm . . . gave the paper on energetics. He was supported by Ostwald, and both were supported by the philosophy of Mach, who was not present. The fight between Boltzmann and Ostwald resembled, externally and internally, the fight of the bull with the lithe swordsman. But this time, in spite of all his swordsmanship, the toreador [Ostwald] was defeated by the bull. The arguments of Boltzmann broke through. At the time, we mathematicians all stood on the side of Boltzmann.'[45]

Shortly thereafter, Boltzmann published a detailed critical analysis of the papers of Ostwald and Helm.[46] Still, he mentioned that he considered their scientific contributions to be truly outstanding; and he hoped to continue to count them among his personal friends.[47] What we learn, however, from the extant Boltzmann-Ostwald correspondence was that from this time on there was both a falling off in the frequency of letters and a deterioration in the level of scientific interchange.

To complicate matters, Boltzmann was not at all happy with his situation in Vienna. He claimed that he was lonely, melancholy and without access to good students. Accordingly, Ostwald in 1900 arranged a professorship in theoretical physics for him in Leipzig. This did not work out as well as Boltzmann had hoped; and so he returned to Vienna two years later, where he worked until his death in 1906. Undoubtedly, Ostwald's anti-atomistic energeticism contributed substantially to the cooling off of a friendship which at an earlier date had been warm and genuine.

Like Boltzmann, Max Planck got involved with Ostwald by virtue of his enthusiasm for the theory of dilute solutions and electrolytic dissociation. Planck felt that these were just as much the property of physicists as chemists.[48] In 1890, he published an important article in the *Annalen der Physik* on the electric potential between two dilute solutions of a binary electrolyte.[49] Another article, in Ostwald's *Zeitschrift*, dealt with van't Hoff's thermodynamic investigations relating the osmotic pressure of solutions to their vapour pressure.[50] Shortly thereafter, Planck requested from Ostwald some clarifications about various aspects of energetics – and especially about Ostwald's claim that the dissipation of energy and the increase of entropy had nothing to do with the second law of thermodynamics.[51]

Apparently, Ostwald was not clear in his own mind about the concept of irreversibility and had tended to side with people like G. Zeuner and W. Meyerhoffer who looked upon the second law as an alternative form of the principle of conservation of energy. Planck obviously attempted to push Ostwald as far as possible in the direction of what he judged to be rather empty conclusions.[52] Planck, it turns out, was extremely knowledgeable about the chemical aspects of thermodynamics. There is ample evidence for this in the *Grundriss der allgemeinen Thermochemie* which he published in Breslau in 1893. It was a work that had been commissioned by Ladenburg for inclusion in his *Handwörterbuch der gesammten chemischen Wissenschaft.*

In retrospect, the correspondence shows that there is far more to the Planck-Ostwald interchange of ideas than might be suspected from Planck's severe public criticisms of Ostwald's thermodynamics. Planck seems to have valued Ostwald for the manner in which he was able to stimulate provocative and fruitful discussions. In 1893, Planck wrote to Ostwald:

> 'On one single point I must agree with you completely. That your ideas furnish an inspiration for experimental investigations, and that [they do so] more than many others whom I consider to be more correct is a demonstration of your worth under all conditions. But you will have to admit that even the greatest discoveries sometimes have not confirmed the ideas that led to them. And a great danger lies in mixing the discovered facts with those ideas, which then often in another way become damaging so that they can retard [the advancement] of science.'[53]

Ostwald replied that he basically was not interested in the question of truth or falsity but rather in what was appropriate and what inappropriate in science. He went on to remark that if Planck would grant him – as he seemed to do – that none of their differences could be proven or disproven, but that they merely could be reduced to the question of suitability, then he suggested that Planck also would have to grant him that his ideas not only served the function of a stimulus but that they were also correct – insofar as it was meaningful to speak at all about 'correctness'.[54] The only substantive point that Planck seems to have conceded to Ostwald in the realm of thermodynamics was that he had drawn a cogent distinction between the first and second laws, classifying them as *perpetuum mobile* arguments of the first and second kinds.

After the meeting of the Lübeck German Society in 1895, Planck decided that he could no longer keep his criticisms of Ostwald's energeticism to himself. In a paper of 1896, he confined his remarks to an analysis of the mathematical imperfections of the new energetics, instead of more generally defending the mechanistic view of nature.[55] Planck's two main criticisms were that the proponents of energetics overestimated the substantive relevance of energetics for mechanics, and that they were quite inept at coming to grips with the fundamental significance of the entropy function for irreversible phenomena. Planck and Ostwald had not resolved their differences; and that, essentially, is where they left the subject hanging.

Boltzmann and Planck made very substantial contributions to the direction that physical chemistry would take in the twentieth century. To the extent that chemistry is a molecular science, and since we now know that the properties of molecules are determined significantly by quantum considerations, it is no exaggeration to say that quantum mechanics has turned out to be at least as important for chemistry as for physics. In fact, there are undoubtedly far more domains within modern physics than in modern chemistry that can be treated without reference to quantum mechanics. Planck recognized early on that the correlation of dynamics with heat theory was the last of the great and enigmatic theoretical problems in need of attack, at least for his time.

Since thermal and chemical processes differ from both dynamic and electrodynamic phenomena, in that they proceed in one direction only, Planck and Boltzmann, like Clausius earlier, correctly perceived the fundamental theoretical difference between reversible and irreversible phenomena. In 1914, in his essay on the relation of physical theories, Planck wrote that the phenomenon of irreversibility

> 'finds its expression in the second law of thermodynamics which states that for every thermo-chemical process the total entropy of the bodies increases and remains constant only for the ideal limiting case of reversible processes. The enormous fertility of this theorem for heat theory and physical chemistry was seen to lie in the peculiar contrast to the apparently insuperable difficulties to understand this from a dynamic point of view.'

The solution was

> 'to abandon any purely dynamical exploration of the second law and substitute an exclusively statistical law to embrace all the results of thermal and chemical measurements. . . .'[56]

Ostwald, Arrhenius and van't Hoff never managed to build this probabilistic, statistical and quantum interpretation of irreversible processes into their world view. Nernst and his school of physical chemists in Göttingen and Berlin became more deeply involved in all of these physical interpretations of chemical phenomena than did the Leipzig-Stockholm-Amsterdam axis.

It remains, therefore, to say something about how Nernst, who was trained as a physicist in Zurich, Berlin, Graz and Würzburg, came to focus his attention on the application of physics to chemical problems.[57] Arrhenius and Nernst learned to know one another in Kohlrausch's laboratory in Würzburg in 1886. While visiting Boltzmann in Graz, Arrhenius introduced Nernst to Ostwald, with whom he had worked at the Polytechnikum in Riga. The outcome of this meeting was that Nernst joined Ostwald in Leipzig in 1887 as an assistant in his laboratory. By 1891, Nernst had accepted a professorship in physics at the

University of Göttingen, and there it was that he assembled an international group of scholars to cooperate in an intensive and comprehensive investigation of experimental and theoretical physicochemical problems, and especially those connected with electrochemistry and thermochemistry.

Upon the retirement of Hans Landolt in 1905, Nernst was called to the Chair of Physical Chemistry at the University of Berlin. Planck strongly supported this appointment because he felt that Nernst was the only chemist in Europe at that time who was in a position to lead Berlin out of its chemical doldrums. During his thirty years of active research in Berlin, Nernst managed to stay in remarkably close contact with most of the foremost physicists and chemists. These included Fritz Haber, Max Bodenstein, Max Volmer, Friedrich Bonhoeffer, Franz Simon, Walter Noddack, J. H. van't Hoff, Otto Warburg, Otto Hahn, Herbert Freundlich and Michael Polanyi. Nernst's contacts and collaborators among physicists were likewise impressive: Max Planck, Max von Laue, Erwin Schrödinger, Fritz London, Friedrich Paschen, Emil Warburg, Gustav Hertz, Albert Einstein and Rudolf Ladenburg. It is evident that Nernst preferred to be recognized as a physicist engaged in chemistry rather than as a physical chemist in the Ostwald sense.

In his first Silliman lecture at Yale in 1906, Nernst remarked:

> 'Natural changes have long been grouped into physical and chemical. In the former the composition of matter usually plays an unimportant part, whereas in the latter it is the chief object of consideration. . . . This classification has real value, as is shown by the customary separation of physics and chemistry, not only in teaching, but also in methods of research – a fact that is all the more striking as both sciences deal with the same fundamental problem, that of reducing to the simplest rules the complicated phenomena of the world. But this separation is not altogether advantageous, and is especially embarrassing in exploring the boundary region where physicists and chemists need to work in concert.'[58]

Before the turn of the century, Berlin was one of the last bulwarks of opposition to the ionic theory of dissociation. Chemists essentially ignored, as well, the new thermodynamics, and more or less confined their attention to the old thermochemical programme associated with emphasis on heats of reaction for chemical processes. Although van't Hoff, in his professorial post and by virtue of his prominence in the Royal Prussian Academy of Sciences, lent his support to the new physical chemistry, he had by then shifted his own research interests in the directions of solid solutions and double salts as related to phase-rule phenomena and the origins and conditions of oceanic deposits.

von Helmholtz and Planck had already in the early 1890s endorsed both the ionic theory and Nernst's electrochemical investigations. Nevertheless, the general response to such views among the Berlin chemists at that time was so manifestly reticent that von Helmholtz concluded that whereas thermodynamics was of central importance within chemistry, the chemists in Berlin were poorly informed about the latest activities. By the time that Nernst arrived in Berlin in 1905, the situation in physical chemistry could be characterized as one of mere tolerance for the work of the *Ioner*. Chemists were just beginning to take seriously what by then had become the old physical chemistry of Ostwald, van't Hoff and Arrhenius, with its somewhat exclusive focus on traditional thermochemistry, the ionic theory of dissociation and the colligative properties of solutions.

Until about 1904, Nernst's main efforts were directed predominantly toward electrochemistry and the refinement of methods to explore principles already current among Ostwald's group and the *Ioner*. After moving to Berlin, and notably during the decade prior to the First World War, Nernst was breaking new ground in the exploration of the implication of his heat theorem and in the study of low-temperature, specific-heat measurements as a solution to the century-old search for the criteria guiding chemical equilibrium and spontaneity.

The twentieth-century developments associated with the thermodynamics of chemical processes and statistical thermodynamics were connected largely with scientific investigations, both experimental and theoretical, initiated by scientists like Boltzmann, Planck, Einstein and Nernst. Simultaneously with his contributions to the third law of thermodynamics, Nernst became involved in exploring problems connected with chemical kinetics. This was a branch of chemistry that was extraordinarily far removed – especially in its early phases – from thermodynamic models and the thermodynamic mode of reasoning. In classical thermodynamics one ignores questions concerning the atomic-molecular structure of matter, the constitutional nature of the reactants and products, intermediate states, the mechanism of chemical reactions, and the rate at which reactions proceed. All of those matters, which are associated with the development of chemical kinetics, the genesis and formulation of the principles of statistical thermodynamics and the stringent testing of the third law, reach further into the twentieth century than it is our goal to pursue here. Their histories are complex and allied in a fascinating way with the development of quantum chemistry.

Conclusion

By way of conclusion, I shall briefly reiterate the main points that I have tried to bring out in this paper. First, I tried to show that physical chemistry was inaugurated as a new discipline in 1887 primarily through the work and cooperation of three investigators, viz. Ostwald, Arrhenius and van't Hoff. The glue that held the programme together, so to speak, was the new theory of solutions, the electrolytic dissociation theory and the correlation of those matters with a thermodynamics that had not yet embraced either the statistical thermodynamics of irreversible processes, the full implications of the second law, or third law considerations.

Second, I wanted to emphasize that the investigation of various chemical problems – which obviously belonged to physical chemistry even before the term was introduced – had been explored in considerable depth prior to the time that physical chemistry became a professional discipline within the context of chemistry proper. Certainly, the work of all of the people we have singled out for special consideration was built upon foundations that were laid mostly during the nineteenth century. By 1887, the stage was set for some enterprising person like Ostwald to provide the organizational base and the stimulus for a new, distinctive discipline perspective.

Third, my aim was to show that a number of physicists became very deeply involved in early physical chemistry. At the levels of both experiment and theory, their contributions to the establishment of the domain of modern physical chemistry were central. Physicists and chemists who approached their subject matter basically from the point of view of physics provided very important intellectual models on which to build the new physical chemistry. These were models without which the whole chemical programme could not really have been extended to include what lies at the heart of all of modern chemistry, namely, atomic and molecular states, quantum theory, chemical process and chemical kinetics.

Notes

1. The secondary literature on the life and times and contributions of Ostwald, both laudatory and critical, is extensive. For a concise biographical statement and interpretive evaluation of Ostwald's career, see Hiebert, E. & Körber, H.-G. (1978) *Wilhelm Ostwald (1853–1932).* In: *Dictionary of Scientific Biography*, **15**, 455–469. A recent biography by Rodnyj, N. I. & Solowjew, Ju. I. (1977) *Wilhelm Ostwald*, Leipzig (German translation from the Russian) is recommended for its inclusion of hitherto unpublished documents from the Wilhelm-Ostwald-Archiv, Grossbothen/Saxony in the German Democratic Republic.
2. Bancroft, W. D. (1933) Wilhelm Ostwald. The great protagonist. *J. chem. Educ.*, **10**, 539–542, 609–613; quotation on p. 612.
3. Einstein, A. (1901) Folgerungen aus den Capillaritätserscheinungen. *Ann. Phys. Ser. 4*, **4**, 513–523.
4. These two letters and the one by Einstein's father cited below, all written in 1901, are reproduced in Körber, H.-G. (1964) Zur Biographie des jungen Albert Einstein. *Forsch. Fortschr. Nachrichtenbl. dtsch. Wiss. Tech.*, **38**, 74–78. All translations from foreign languages used within the text are those of the author.
5. The letter is reproduced in Herneck, F. (1957) Wilhelm Ostwald. Zum 25. Todestag des grossen Chemikers. *Wiss. Fortschr. Pop. Monatsz.*, **7**, 69–72.
6. Letter from Ostwald, Grossbothen (Sa.), to Einstein, dated 10.2.29, Princeton, NJ, Einstein Archive of the Institute for Advanced Study. See also Ostwald, G. (1953) *Wilhelm Ostwald. Mein Vater*, Stuttgart, pp. 244–245.
7. For a discussion of *Selected Topics in the History of Electrochemistry*, see Dubpernell, G. & Westbrook, J. H., eds (1978) *Proceedings of The Electrochemical Society*, Vol. 78-6, Princeton, NJ.

8. Kirsten, C. & Körber, H.-G., eds (1975) *Physiker über Physiker. Wahlvorschläge zur Aufnahme von Physikern in die Berliner Akademie 1870 bis 1929 von Hermann v. Helmholtz bis Erwin Schrödinger*, Berlin.
9. *Ibid.*, p. 168.
10. *Nobel Lectures in Chemistry, 1901–1921*, Amsterdam, 1966, pp. 145–172.
11. Ostwald, W. (1887) An die Leser. *Z. phys. Chem.*, 1, 1–4.
12. For a discussion of these transactions in the correspondence of Ostwald and van't Hoff see Körber, H.-G., ed. (1969) *Aus dem wissenschaftlichen Briefwechsel Wilhelm Ostwalds*, Vol. II, Berlin, pp. 208–210.
13. There were two parts to the communication (1884): La conductibilité des solutions aqueuses extrêmement diluées déterminée au moyen du dépolarisateur. *Bih. Sven. Vetenskapsakad. Förh.*, 8, (13), 1–63; and Théories chimiques des électrolytes. *Ibid.*, 8 (14), 1–89.
14. Körber (note 12), pp. 3–197, 329–331. There are hints in this correspondence that the relationship between Ostwald and Arrhenius became ever more strained with advancing years. An examination of the correspondence of Ostwald and Arrhenius with other scientists reveals important differences in their approaches to physical chemistry.
15. See Hiebert & Körber (note 1), pp. 455–457.
16. Körber (note 12), pp. 19–23.
17. Körber (note 12), p. 26.
18. Körber (note 12), pp. 42–43.
19. Körber (note 12), pp. 61–66, 85–90, 109.
20. Körber (note 12), pp. 88–94. The discussions about the status of the theory of solutions, with special reference to Britain and America, are dealt with by Dolby, R. G. A. (1976) Debates over the theory of solutions: a study of dissent in physical chemistry in the English speaking world in the late nineteenth and early twentieth centuries. *Hist. Stud. phys. Sci.*, 7, 297–404.
21. Körber (note 12), pp. 87–88.
22. Körber (note 12), p. 94.
23. Körber (note 12), p. 16.
24. Körber (note 12), pp. 74–75, 111.
25. Körber (note 12), pp. 99–100.
26. Körber (note 12), pp. 133, 136.
27. Körber, H.-G., ed. (1961) *Aus dem wissenschaftlichen Briefwechsel Wilhelm Ostwalds*, Vol. I, Berlin, p. 21.
28. *Nobel Lectures in Chemistry, 1901–1921*, Amsterdam, 1966, pp. 41–61.
29. *Ibid.*, pp. 1–14.
30. Körber (note 12), p. 169.
31. Springer, G. F., ed. (1967) *Jacobus Henrichs van't Hoff. Imagination in Science (1878)*, New York; quotation on p. 8.
32. *Ibid.*, p. 18.
33. Körber (note 12), pp. 199–325.
34. Körber (note 12), p. 214.
35. Körber (note 12), p. 231.
36. Körber (note 12), p. 236.
37. Körber (note 12), p. 308.
38. Körber (note 12), p. 320.
39. The correspondence of Ostwald with Boltzmann (1890–1904), Planck (1890–1902) and Gibbs (1887–1896) is given by Körber (note 27), pp. 3–69, 89–104.
40. Boltzmann, L. (1909) Die Hypothese van't Hoff's über den osmotischen Druck vom Standpunkte der kinetischen Theorie. *Wiss. Abh.*, 3, 386–397.
41. Körber (note 27), pp. 7–8.
42. Körber (note 27), pp. 13–14.
43. Körber (note 27), pp. 14–15.
44. See Hiebert, E. (1971) The energetics controversy and the new thermodynamics. In:

Roller, D. H. D., ed., *Perspectives in the History of Science and Technology*, Norman, Oklahoma, pp. 67–86.

45. Quoted in Körber (note 27), p. 22 from *Chem. Ztg.*, 47, 25.
46. Boltzmann, L. (1905) Ein Wort der Mathematik an die Energetik (1896). *Populäre Schriften*, Leipzig, pp. 104–136.
47. *Ibid.*, p. 105.
48. Körber (note 27), pp. 52, 60.
49. Planck, M. (1958) Ueber die Potentialdifferenz zwischen zwei verdünnten Lösungen binärer Elektrolyte [1890]. *Phys. Abh. Vortr. (Braunschweig)*, **1**, 356–371.
50. Planck, M. (1958) Ueber den osmotischen Druck [1890]. *Phys. Abh. Vortr. (Braunschweig)*, **1**, 327–329.
51. Körber (note 27), pp. 34–35.
52. Körber (note 27), pp. 36–47.
53. Körber (note 27), pp. 49–50.
54. Körber (note 27), p. 52.
55. Planck, M. (1958) Gegen die neuere Energetik [1896]. *Phys. Abh. Vortr. (Braunschweig)*, **1**, 459–465.
56. Planck, M. (1958) Verhalten der Theorien zu einander [1914]. *Phys. Abh. Vortr. (Braunschweig)*, **3**, 103–104.
57. See Hiebert, E. (1978) Walther Nernst (1864–1941). In: *Dictionary of Scientific Biography*, **15**, 432–453; Hiebert, E. (1978) Chemical thermodynamics and the Third Law. In: *Proceedings of the XVth International Congress of the History of Science, Edinburgh, 1978*, pp. 305–313; and Hiebert, E. (1978) Nernst and electrochemistry. In: *Proceedings of the Symposium on Selected Topics in the History of Electrochemistry*, Princeton, NJ, pp. 180–200.
58. Nernst, W. (1907) *Experimental and Theoretical Applications of Thermodynamics to Chemistry*, London, pp. 6–7.

DISCUSSION

ARMIN HERMANN

Lehrstuhl für Geschichte der Naturwissenschaften und Technik, Universität Stuttgart, Stuttgart, Federal Republic of Germany

In the three lectures, on the history of physics, on the history of biochemistry and on the history of physical chemistry, three different approaches to the history of science were presented. Fruton gave a sketch of the internal history, Heilbron dealt mostly with social history, whereas Hiebert spoke about the small group of scientists who founded physical chemistry and about their intellectual interchanges. He showed how important the great men are, in contradiction of a conviction widespread among the younger historians.

Heilbron pointed to a change in the philosophy of physics between 1870 and the 1890s, from seeking truth to a mere description of reality using some principles or theories. I agree: About the turn of the century there was a predominance of positivism or descriptionism. A quotation Hiebert gave is added proof of this statement: Ostwald, replying to a letter from Planck, claimed that he basically was not interested in the question of truth or falsity but rather in what was appropriate or inappropriate.

Apparently both standpoints or philosophies had existed in the development of physics from the time of Galileo. Galileo's aim was twofold: to discover God's plan of the world and to apply science; whereas for Francis Bacon the value of scientific knowledge could be found only in its application. Both philosophies have been present through the centuries, and I agree that, in the struggle between these two philosophies at the end of the nineteenth century, descriptionism had won the upper hand, so that it is possible to speak of '*laissez-faire* physics'. Even so, a man like Planck developed his idealistic philosophy, influenced by Plato; he called his Natural Constant 'the absolute' and founded quantum theory.

Heilbron writes: 'The wide descriptionist consensus of 1900 . . . did not include agreement about the best mode of description.' In practice, also, the descriptionist was bound very strictly to theories, principles and models that were acknowledged in the community, and he felt himself in general not free to change over at will from one picture of nature to another. Mach and Ostwald are examples: Mach declared the atom to be an anthropomorphic construction; and when the reality of single atoms was proved, it was regarded as a severe defeat of positivism. So even with a strong descriptionist preoccupation, a physicist could be, and in general was, strongly in favour of a special theory or picture.

In this connection, one may put a question to Erwin Hiebert: Wilhelm Ostwald was on the one hand an ionist and on the other hand against the atom. How could Ostwald combine these two extremes in his mind?

Heilbron also spoke about the fears and hopes that people associated with the advancement of science, a question that was already discussed following Professor Ziman's paper. In my opinion, Heilbron overemphasized somewhat the *fin-de-siècle* mood of *Kulturpessimismus*, the view that increasing knowledge would *not* lead to a better future. Of course, there were writers such as Max Nordau and Alfred Döblin, and in religious circles men who saw the greatest danger in encouraging science and the scientific spirit; but the great majority of scientists and the great majority of the public believed that science and technology would lead mankind to a better future, to a new age of science, in which not only the material conditions of human life, but also the human character would be improved.

Are there limits to scientific knowledge? In 1872, on the occasion of the fiftieth anniversary of the German Association of Scientists and Physicians, Emil Du Bois-Reymond formulated his famous *Ignoramus-Ignorabimus*. Eight years later he posed seven principal problems, seven *Welträtsel*, which are unsolvable, even by future science. A lot of scientists were strongly opposed: there are no such limits, science will progress without end. Even moral, ethical and social problems can be and will be solved by science. The great majority believed this.

In Heilbron's picture, physics at the end of the century was in trouble; pessimism was widespread in science, and only new discoveries, such as Röntgen's, gave physics new hope. My impression is that there was, rather, a continuous belief in science. I admit, that around 1880 physics was regarded by a lot of physicists as a highly developed science, in which one could expect no particular surprises, since the mechanical world picture and the energy principle had solved all the great problems. Heilbron describes the increase in the number of professional physics students, new institutes and outlay. Whereas students of chemistry were expected to go into industry, the average student of physics wanted to become a school physics teacher. (At least this is true of Germany.) Most of the students, of course, dreamed of an academic career; but even the best students, such as Planck and von Laue, took the state examination for school-teachers in order to be sure of getting a job. Only after the turn of the century did industry offer posts to physicists. So, in addition to the academic physicist, the industrial or technical physicist appeared.

Electrotechnology developed out of physics, and on the one hand the academic physicists were happy that these applications proved the importance of their science. On the other hand, there was a fear, as we learned from Heilbron, that science itself would change, and that the physics in the technical high schools devoted to the applications would become dominant. So they stressed the view that not only would pure science lead to important applications, but that basic innovations would only be possible when science had developed in its own right.

'What we do here is pure science', Adolf Harnack said, referring to the first Kaiser Wilhelm Research Institutes; and he continued: 'But we do not doubt that what we do here, and what is thought here, both great and pure, will also lead to material bliss.' This formula proved to be valid. As we know, new technology has grown out of science; but technology has in turn reacted against science. So science and technology reinforce each other. At the turn of the century this was already so, but at what point in the development, in which year, did scientists become aware of this reciprocal action?

Having concentrated on Heilbron's paper, I can comment only briefly on those of Hiebert and Fruton. Both described the development of new fields, physical chemistry and biochemistry, and (one may add) also theoretical physics, which developed before the First World War. Hiebert's study is based on an overwhelming richness of sources, but I am sure that even in this case the use of the Nobel archives would be of great value and would enable him to make his already very vivid biography even more interesting.

All three new disciplines suffered from neglect, and I want to raise and comment on the following question: What conditions must be fulfilled in order that promoters can establish their new field in the academic world? In my opinion:

(1) It is necessary to have a particular group of scientific problems, some scientific success in solving riddles, and a certain interest in the scientific community or neighbouring disciplines for solving the riddles.
(2) Some stimulating representatives, who are able to impress the scientific community, are also necessary. Physical chemistry had Ostwald, van't Hoff, Arrhenius and Nernst. The awarding of the Nobel prizes to van't Hoff, Arrhenius and Ostwald, as Hiebert pointed out, was a great help. Theoretical physics had Boltzmann, Clausius, von Helmholtz and Kirchhoff, and then Planck and Einstein.
(3) The representatives of the new discipline must be able to stimulate young people and enable them to find academic posts.
(4) An understanding and cooperation between the founders and pioneers of the new discipline is necessary, in particular a sort of agreement about the main methods and principles.

III

New paths of scientific inquiry at the turn of the century: Medicine and physiology

THE EMERGENCE OF SCIENTIFIC MEDICINE: A VIEW FROM THE BEDSIDE

STANLEY JOEL REISER

Francis A. Countway Library of Medicine, Harvard Medical School, Boston, Massachusetts, USA

Technology in Medicine

As the nineteenth century drew to a close, physicians bathed in the sunshine of an array of dazzling instruments and tests developed during that period to evaluate disease. Doctors in great numbers retreated from the bedside to stand in line at the laboratory door, awaiting answers to diagnostic queries. Behind it, surrounded by equipment, technical specialists generated scientific wisdom, which they confidently dispensed to the eager clinicians. Science had come to the aid of doctors and patients who needed data about illness – but its arrival would have mixed results.

Turn-of-the-century physicians were heirs to an armoury of instruments that enabled them to probe the hidden reaches of the human frame with unparalleled exactness. They could subject body fluids to intense chemical analyses, examine the microstructure of any tissue, monitor through graph-generating machines the physiological motions of the most important internal organs. With the aid of the X-ray, they could even see through the skin into the body cavity itself.

These instrumental aids were the products of advances in physics, chemistry, biology and physiology, and other scientific disciplines. By using them in patient care, physicians felt connected to the framework of analysis that marked the pursuit of scientific insights. As Rudolf Virchow wrote:

> 'In the seventeenth century, anatomical theatres; in the eighteenth, clinics; in the first half of the nineteenth century, physiological institutes; so now the time has come to call into existence pathological institutes and to make them as accessible as possible to all.'[1]

The growth of bacteriology demonstrated the influence of this new science on medicine. Work during the nineteenth century, culminating in the rigorous demonstration published in 1882 by Robert Koch of the consistent association of a certain bacterium with tuberculosis, caused many physicians to interpret a particular bacterial presence as an absolute, pathognomic sign of illness. Cases reported in the medical literature confirmed prevailing medical opinion that scientific search for bacteria produced better results than physical examination of the patient. The following is a typical example:

> 'A few weeks ago a boy of sixteen years came to the Massachusetts General Hospital, out-patient department, service of Dr Ernst. Up to one week before he had been quite well, when he had an attack of vomiting accompanied by headache. During the week he had had two slight haemorrhages. He had the day he presented himself, a hot, moist skin, some cough with bloody expectoration; severe pain in the right chest; a temperature of 100°, and a pulse of 90.
>
> 'Physical examination showed diminished resonance over the whole right chest, vocal fremitus and resonance both slightly increased, and moist rales, both coarse and fine, were present.
>
> 'The general condition of the patient suggested pneumonia, and directions were accordingly given for treatment. But partly from the element of doubt which presented itself, and partly as an exercise in differential diagnosis with the microscope, Mr Rowen, Dr Ernst's assistant, was asked to stain some of the sputum to see whether the pneumo-coccus or the tubercle bacillus could be found, and numerous bacilli of tuberculosis were found, thus giving a positive diagnosis.'[2]

This approach to diagnosis was also accepted for cholera:

> 'Experience during the last epidemic of cholera at Hamburg and elsewhere is conclusive that there is no diagnostic symptom or pathological lesion of cholera; there is only one thing, and that is the determination of the organism, the comma bacillus of Koch, in the discharges.'[3]

A similar medical consensus existed that a dependable diagnosis of diphtheria was possible only through bacteriological examination. The uncertainty of being able to make this diagnosis on the basis of the disorder's main symptom, sore throat, had, before the discovery of its bacterial cause, led many excellent doctors to treat most cases of sore throat as diphtheria.[4] But when the Klebs-Loffler bacillus was discovered, its presence became recognized as absolutely pathognomic of diphtheria.

Diagnostic Use of the Laboratory

Confidence in this bacteriological evidence, and recognition that physicians had neither the ability nor the time to search for it, led communities and boards of health all over the world to establish laboratories for that purpose.[5] One of the first was opened in 1893 in New York City. At chemists' shops throughout that city physicians were provided with free culture tubes and specific directions for their use, which was simple: The throat of the patient was swabbed, the matter obtained drawn over the surface of the culture medium, and the tube then returned to the shops for collection each evening by health department agents. Diagnoses could be learned the next day, in person or by telephone.[6]

The combined use of special posting places to leave specimens for analyses and the telephone to receive results was an important innovation: it facilitated the use of distant institutions in the routine practice of diagnosis.[7]

Within three months, the New York Board of Health had examined 431 cultures, from which it found 301 active cases of diphtheria. So severe were the consequences of the disease – forced isolation, household quarantine, and the disruption of business and social relations – that family doctors welcomed the prospect of being relieved 'of the risk and responsibility' of making a positive diagnosis.[8] 'Every city and country town should have connected with it a bacteriological laboratory', wrote an enthusiastic physician after reporting on the New York innovation.[9] By 1895, the Laboratory of the Philadelphia Polyclinic was offering the same service.[10] Two years later, the chief of out-patient medicine at the Massachusetts General Hospital, Richard Cabot, condemned the Boston Board of Health for failing to furnish facilities for the diagnosis of typhoid fever similar to those for diphtheria. There was not a city of any size in the United States except Boston, he reported, that was not offering the test as a matter of public health. He argued that municipal laboratories should aid practitioners to use bacteriological procedures of diagnosis.[11] The report of St Thomas' Hospital Laboratory in London for 1898, its first year of existence, reveals that physicians were beginning to use laboratories for more than bacteriological identification. In addition to conducting typhoid and diphtheria tests, blood and sputum samples were examined microscopically; and numerous chemical examinations were performed on tissues and fluids.[12] Thus, the laboratory had assumed responsibility for examining chemical variations in the body during disease. Many doctors believed that the time was at hand when, without ready access to laboratories run by experts, 'no physician can do his patients, himself, or his science justice'.[13]

The increasing constraint placed on the doctor's time by the new techniques of observation was important in fostering these developments. While in the eighteenth century, general external observations of the body and questioning of patients constituted the essence of the diagnostic examination, the contemporary use of a large number of instruments and chemical and bacteriological examinations had turned diagnosis into a process that took hours or days. Although patients profited from it, doctors were severely taxed. It became clear that scientific-minded physicians would need either a clientele limited in number, or else more assistants to help them with investigations. The help could come either in the form of a highly trained scientific aide employed personally by the doctor to collect data and perform laboratory investigations, or from laboratories established by others to serve the doctor's needs.[14]

The large majority of physicians chose the latter alternative. They recognized that the instruments necessary to examine fluid and tissue specimens, and organisms, entailed not only great expense but several kinds of expertise. For example, improved bacteriological technique had led, by 1897, to the discovery of several organisms that are very similar to the one which causes typhoid fever. A greater number of tests were thus required to differentiate them. To prove that an organism was the typhoid bacillus, Losener proposed that eleven biological and chemical tests be performed.[15] For such work an institutional locus seemed preferable.

Thus, by enabling physicians to make correct diagnoses, or to choose the best remedies, the knowledge and assistance of the laboratory became a dominant factor in efforts to practise scientific medicine. Doctors increasingly viewed the most complete and painstaking bedside examination as incapable of giving definitive responses to diagnostic questions. The answers received with confidence were those generated through analytical investigation, which could be conducted competently only in specialized laboratories. As observed in 1897:

> 'The day is not far distant when every large hospital must have its pathological laboratory, as much as its operating room with a trained and well-paid head, busied as much with the solution of problems arising in the living, as with the determination of the cause of death.'[16]

Clinical Research Facilities

Alongside the laboratories for clinical diagnosis were hospital facilities for clinical research. They applied refined physical, chemical and biological methods to finding basic explanations of illness through study of the living patient. Human disease became regarded by many physicians as experiments that Nature made on the organism, experiments that were even better than some conceived in the laboratory.[17] In a 1901 article, 'The human body as an analytical laboratory', a physician extolled the quantitative accuracy of experiments in which the intake and output of food and chemical substances fed to patients were measured. The body, he proclaimed, was a 'perfect temple' for experimental investigation.[18] William Osler shared this opinion:

> 'Disease is an experiment, and the earthly machine is a culture medium, a test tube and a retort – the external agents, the medium and the reaction constituting the factors. . . . Each dose of medicine given is an experiment, as it is impossible to predict in every instance what the results may be.'[19]

Since the final test of every new procedure had to be made on people, Osler viewed the hospital as a biological laboratory, 'a clearinghouse for the scientific traders who are doing business in all parts of the body corporate'.[20]

In establishing the first department of scientific research in 1904, the Board of Trustees of the Massachusetts General Hospital noted that although special research institutions had been created recently, a large part of medical investigation 'must be carried on in close proximity to patients'. They thought it important that benefactors see the hospital as a place that promoted knowledge.[21]

Frederick Shattuck, professor of medicine there, thought of the hospital as 'a laboratory for the relief, cure and study of the experiments wrought by disease on human beings'.[22] He stressed the similarity of objective instruments, reactions and methods in laboratory and hospital procedures: 'The clinician in his daily work . . . [is] animated by the scientific spirit and should pursue scientific methods, i.e., be as scientific as his brother of the laboratory.'[23]

This connection between patient care and research, punctuated by occasional stories about the misuse of human beings in research, made the public in Great Britain and the United States uneasy: 'There is a large class of thoughtful, conscientious well-balanced people to whom the physiological laboratory and the clinical wards of a hospital are a source of misgiving and honest perplexity,' wote one American journal.[24] Disclosures of research by the German clinician Neisser, who inoculated several prostitutes with syphilitic serum in an experiment, was an occasion for widespread criticism of human research.[25] Yet the subtle connection being drawn in the minds of medical students and doctors between laboratory and clinical procedures was more important than sensational disclosures of wrongdoing. The search for pathology through dissection of the body at autopsy and through physical examination of the living patient during the first half of the nineteenth century had centred the search for basic knowledge of illness in hospitals. However, laboratory experimentalists like Claude Bernard initiated in the second half of the century a new concept of research, predicated on the notion that human diseases were best studied in the laboratory by animal experimentation. Research work shifted from hospital to laboratory, the latter having the complicated instruments and experimental objects (animals) necessary to illuminate physiological mechanisms. By the beginning of the twentieth century, the apparatus used in laboratory research had become simplified and more portable, cheaper and more dispersed, so that fine physiological determinations could now be made by physicians on patients. The notion of shifting the locus of scientific advancement back to the hospital was becoming increasingly possible, and appealing to the physician. In the shift, however, the attitude of laboratory research towards its test objects was adopted by some physicians at the hospital, along with its methods and instruments. Hence, viewing patients as scientific objects was one of the prices medicine and patients would pay for the new possibilities raised by clinical research.

These attitudes worried some clinicians. Concern with the science of medicine seemed to divert the attention of doctors from the art of medicine. The term 'art' described a number of behaviours and functions thought critical to practice, among them: the ability to approach patients properly, to inquire about personal secrets with delicacy, to announce prognoses with honesty and sympathy, to make examinations with minimum pain or offensiveness, to inspire confidence. The art of medicine encompassed a concern for the patient as an individual. It stood for an ability to understand the effects of the illness as well as the therapy on the life of the patient. In applying the science of medicine to disease, physicians analysed and dissected patients; in applying the art they looked to the whole, and sought to keep things together. Before the age of medical science, shrewdness and knowledge of human nature were often more important for successful practice than technical knowledge, ignorance of which had compelled the doctor to treat the patient rather than the disease. The arrival of science and technology focused the doctor's efforts on the disease. Thus, the concerns of turn-of-the-century practice continued to be one-sided – but in an opposite direction.[26] With these events, a hierarchy of exactitude was being established in medicine. The analysis of disease through elaborate instrumental agencies seemed more accurate than the examination of the physical signs through the doctor's senses which, in turn, appeared more precise than listening to the patient's narrative of felt symptoms. The technology of practice grew in influence, and in number as well. 'Now,' wrote a doctor in 1907, 'a Hercules could not carry all the appliances, even the so-called transportable ones, which in diagnosis are at our disposal, to say nothing of those found only in clinics and laboratories.'[27]

Search for Exactitude in Clinical Findings

Another critical factor in these developments was the idea, which originated with nineteenth-century German scientific medicine, that a high level of exactness should be sought at all times in stating diagnostic results. Doctors of the period generally viewed laboratory reports as inherently accurate and clinical evidence as filled with uncertainty.[28] Laboratory methods of examination implied the use of exact instrumental techniques, in contrast to less exact diagnostic procedures applied at the bedside.[29] 'One of the ideals dearest to the student of medicine is the ideal of accuracy. I suppose there is no ideal to which any of us pay homage more regularly,'[30] noted Richard Cabot. The general public too demanded 'scientific accuracy' from doctors.[31] The prospect of having exactly-stated clinical data was appealing to clinicians. For several centuries, they had looked with envy upon the ability of scientists in other fields to express evidence precisely: 'In the attempt to stir ourselves out of the slough of vague guesses in which the practice of medicine has wandered so long, we catch at anything that seems to make for greater exactness in every department of our work.'[32] The new technology, which replaced qualitative estimates with quantitative or graphic measurements expressed in standard units, seemed most able to generate precise results. However, many clinicians failed to appreciate that the accuracy desired should be governed by the ends sought. Recording white blood counts in units of less than ten, for example, just as recording blood pressure in units smaller than five, was a useless pursuit of accuracy, and scientific pretension too. Such figures could not be more than roughly approximate, for the number of factors that might influence them could not be measured. Other examples of misdirected exactness were reported in the medical literature, such as that of a well-known Boston practitioner visited by a woman complaining of gastric trouble. The doctor examined the abdomen extensively, took fluid for analysis from the stomach, measured its size and position, examined the urine and blood quantitatively, and finally prescribed therapy. She got no better. The doctor's technological and laboratory examinations were rigorous, but his effort in history-taking slipshod. The cause of the discomfort was pregnancy, a diagnosis that could have been made only through careful questioning of the patient.[33] The physician had selected the wrong place to exercise his knowledge. He was needlessly meticulous in some areas, careless in what turned out to be the most significant avenue of investigation. Maurice Richardson, Professor of Surgery at Harvard Medical School, thought that more errors in diagnosis resulted from incomplete or inaccurate histories than from imperfect physical or laboratory examination, even in cases where thoroughness in the latter was possible.[34] Beneath the demand for expressing clinical data in exact terms probably lay a desire for clearness in reporting data. But the clarity sought could easily be undermined by the misleading appearance of accuracy of quantitative evidence, and the growing clutter of numbers that increasingly filled the pages of medical records.[35]

There were other causes for the capitulation of physicians to laboratory reports. Doctors who had completed medical school before the development of laboratory courses had no experience in laboratory method, could not appreciate the dangers of excessively exact reports, and were unable to be critical of them. They sent a specimen, received a report. To such doctors the scientific methods used were mystical. The opsonic index, Gram stain, agglutination phenomena – all were names lacking associated concepts. Such doctors often neglected to make an adequate physical examination or to take a comprehensive history of their patient's illness. When ordering tests and using instruments, they thought themselves exquisitely modern and the patient fortunate to have the assistance of science in his case. Writers of journal articles reporting uncritically on the value of the new medical science also contributed to this result:

> 'The hardworking doctor lays aside his weekly journal after reading a brilliant contribution on the wonderful results of bacteriological cultures of the blood, and sighs over the uncertainty and perplexity of heart sounds and hyper-resonance and rashes. He feels the unreliability of his conclusions, and the possibility of an overlooked reflex weighs on his mind. He yearns for the certainty of the test tube and the apochromatic vision of the microscope.'[36]

Further, early reports of tests tended to be optimistic; later observations that expressed scepticism were printed more slowly and were less likely to attract the attention of the busy doctor. Frequently, the doctor relied for information on brief conversations with colleagues, incomplete journal abstracts, or on half-understood discussions at a medical society. Many physicians were unaware that the clinical meanings of numerous tests were still unclear, that they should be regarded as only approximately correct, that other clinical evidence should be used in interpreting the significance of a test result:

> 'The laboratory finding must be fitted into the symptom-complex as a cardinal symptom perhaps, but must not be permitted to usurp the position of prime importance unless its inherent nature entitles it to such a position. The fact that its place of birth was the laboratory does not give it such rank that it may look down on the symptom or finding whose origin was the lowly ward bed,'

the American doctor James Herrick noted.[37]

Physicians themselves created laboratory errors by carelessness in collecting, preserving and sending biological specimens for examination. Preparation of blood slides for the diagnosis of malaria was a case in point: for the microscopist to have a reasonable hope of making the diagnosis, blood had to be taken during a particular phase in the disease cycle, spread on a slide so that corpuscles did not bunch up in thick masses, and done before the patient received quinine. But, in the usual sequence of events, the physician treated the patient for a week or so with large doses of quinine, collected a large, thick drop of blood, and expected the microscopist to do the rest. Even more difficult was the preparation of a section of solid tissue for laboratory analysis. Few doctors consulted pathologists about the conditions of preparation before taking the specimen. Usually the doctor took any sort of shred, put it into any convenient solution and sent it to the laboratory, expecting an absolute diagnosis to follow, with suggestions for treatment provided besides. Physicians who wanted a laboratory examination of sputum for tubercle bacilli frequently contaminated the specimen by sending it in jelly or vaseline jars, ink bottles, whiskey flasks and even red handkerchiefs. Doctors were not aware of the influence on results of errors in technique. They usually compounded this problem by not including details of the case with the specimen, on the belief that concealing clinical data helped the laboratory to make unbiased judgements. Physicians also placed restrictions on the type of test procedures to be performed, thus tying the hands of the laboratory analyst who might suggest better means of gaining the desired ends.[38]

Doctors generally failed to appreciate that personal judgement was present as much in the laboratory as in the operating room. A laboratory report from an untrained chemist is as potentially dangerous as a surgical knife in the hands of a new intern. To doctors who lacked actual laboratory experience, the entire process of scientific testing seemed sure and automatic: 'You press the button, we do the rest' was the prevailing medical image of the laboratory.[39] Thus, the possibility of error in technique and interpretation by laboratory personnel was not sufficiently considered by doctors in evaluating their results. Those who understood these problems were appalled at 'how innocently, how literally, how trustfully, physicians accept the verdict of the laboratory as decisive'.[40]

Sources of Error in Laboratory Results

Experts recognized three basic kinds of laboratory errors by the turn of the century: of technique, of attitude and of incorrect reasoning on the basis of accurate results. Errors of technique were frequent and pervasive. At times they stemmed from carelessness – foreign substances might fall into specimens that required microscopic examination, twenty-four-hour urine samples were not collected precisely, or specimens from different patients were confused and substituted.[41] At times, errors were generated by inherent flaws in the techniques themselves: for example, small amounts of nonpathological albumin were precipitated in nearly all urine tests, and frequent variation in culture media produced changes in bacterial growth. There was also a difference in the skilfulness of laboratory personnel, not only in performing a procedure but in judging the end reaction of a given test. Unless workers standardized the various solutions employed, it was hard to know what to call end reactions which, in many analyses then used in clinical chemistry, were not sharply delineated.[42]

A study made in Germany and publicized in England revealed how unobservant people generally were. Students were questioned on their ability to recall a scene they had witnessed eight days before. Disturbing dissimilarities appeared in their recollections: a quarter of the students incorrectly reported what had been done, and a third were mistaken about what had been said. The experimenters were astonished not so much by the incompleteness of the observations as by the large numbers of those who fundamentally changed the reality of the event they had witnessed. This study reflected badly on the trustworthiness of observation in general, and cast doubt upon the historical accounts by patients of their symptoms. These sombre results were darkened further by the scientist and mathematician Karl Pearson, who believed that trained observers were not much more reliable than average people. He thought that at least half of the scientific observations made and data collected were worthless because they had not been safeguarded by strict precautions against error.[43]

Errors of attitude were classified into several kinds. First, failure frankly to admit the inherent limitations of laboratory methods. The best experts were conservative in their opinions, alert to the possibility of error in technique, false interpretation and lack of knowledge about even common laboratory reactions. They were concerned about artefacts and had the courage to admit ignorance. However, most failed to acknowledge these limitations. Second, laboratory staff often failed to recognize the gravity of the duty imposed on them by doctors who believed in their findings. A laboratory report could be the basis for prescribing a dangerous therapy or for dispensing a grave prognosis. Laboratory workers were generally remote from patients and felt little of the personal responsibility for their actions that doctors did.[44] Physicians were partly at fault for this result, since they did not encourage a mutual exchange of data with laboratory staff.

The errors of drawing unwarranted conclusions from accurate results was equally serious. What conclusions were appropriate when albumin was present in urine? Did the presence of sugar in urine constitute diabetes? How should one interpret negative results from a culture that was suspected of containing the diphtheria or tubercle bacillus? It was hoped that no physician would interpret the failure to find any given bacillus as proof of recovery from or absence of the disease. Yet, for example, so many doctors in St Louis believed this to be so in cases of tuberculosis that the city bacteriologist wrote on each of his negative reports: 'The examination of sputum from [patient "X"] does not show the presence of the tubercle bacilli. This does not signify . . . [the absence of] tuberculosis. Should the clinical signs still lead you to suspect tuberculosis, please send other specimens.'[45]

In addition to the problems raised by observer variation and error, the pathological significance of accepted laboratory tests was questioned fundamentally. In 1905, Richard Cabot systematically compared the clinical records and post-mortem findings in all cases of acute and chronic nephritis that came to autopsy at the Massachusetts General Hospital between 1893 and 1904. He examined over 200 records, looking for answers to two basic questions. The first: When no anatomical lesion was found at autopsy, how often was the urine normal? He found frequent discrepancies between clinical diagnoses and autopsy findings; for example, in acute nephritis seventy-five percent of the diagnoses during life were wrong, in subacute nephritis fifty percent were in error, in chronic nephritis eleven percent. His second question: Could these diagnoses have been made correctly if the results of the urine analyses had been interpreted properly? Cabot did not think so. Some kidneys that had showed no lesions on post-mortem examination were associated with as much urinary abnormality during life as those that had. He concluded that chemical and microscopic examination of the urine was often insufficient to make a correct diagnosis, even when evaluated with all available clinical data.[46]

This study not only challenged the validity of laboratory analyses, but also the pathological certitude offered by the anatomical lesion. The Johns Hopkins' pathologist William Welch, in discussing Cabot's findings, observed that pathologists could not accurately predict from an examination of the kidneys after death what the function of those organs had been during life, nor construct clinical histories of kidney diseases from post-mortem appearances. Medicine could not say for certain upon what precise anatomical lesions urinary change depended, notwithstanding the many theories about the subject. There could be marked lesions of the kidney without chemical urinary changes, and vice versa. Anatomical and physiological change did not always proceed *pari passu*. A deceptive appearance of health in the face of hidden destruction of tissue could be maintained – a precarious compensation of the organ masking the danger to life. Hence, tests of organ function now seemed necessary.

To this end, turn-of-the-century physicians were developing tests to evaluate kidney activity. These revolved basically around the induced elimination of a foreign substance given to the patient in known dosage.[47] Some clinicians now began to distinguish functional and anatomical diagnoses, giving increased importance to the former. These doctors wanted to know not only if structural changes had occurred in an organ, but if they interfered with its function, and if so how much. Anatomical diagnosis asked: How does the organ feel or look? Functional diagnosis asked: What can the organ do? Some physicians began to retreat from the prevalent maxim – when you find a lesion, treat it. They became interested not only in the question of whether a lesion existed, but whether it interfered with function. This distinction was particularly important in gynaecology. There were many anatomical changes in the adult female pelvis for every one that really affected function. The chief business of the diagnostician was to determine which lesion to take seriously and which to disregard.[48]

The dangers of misinterpreting the meaning of laboratory tests required corrective action. The argument was introduced that laboratory facts were personal facts, in the same way as evidence gathered by the senses of the physician during physical examination. Laboratory data required subjective interpretation and were not readily transferable to another person without some loss of meaning. All so-called objective methods had subjective components. Even when investigators believed themselves to be purely objective, they were unconsciously and unavoidably being subjective, because their accurate methods represented an outward projection of subjective thoughts.[49] Calls were made to consolidate diagnostic operations and to put an end to the division between laboratory and clinical findings. Physicians should not be forced to choose between them, but should make themselves master of all the facts that bear on a diagnosis, any one of which, in isolation, was almost meaningless. They should also remember the patient. 'All that tends to make us build up our diagnosis at a distance from the patient, and without the constant reminders of every side of his case given us by his actual presence before our eyes, – all such tendencies, I say, are dangerous,' Richard Cabot warned.[50] James Herrick called for more laboratory work for the medical student and practitioner. Laboratory procedures of established value should be no more the exclusive property of a select few – laboratory personnel – than were the basic tools of physical examination, such as the stethoscope. He wanted laboratory workers to be trained to observe disease: the more familiar they were with illness, the better they could understand the importance of their decisions and the magnitude of their responsibilities.[51] Some urged, too, that laboratory personnel be consulted more in helping to choose appropriate test procedures.[52]

The tragedy was that knowledge gained in the laboratory and evidence obtained at the bedside often failed to reinforce one another. 'Too many errors are made', noted one editorial, 'when the physician, lacking in knowledge of the significance of laboratory findings, relies on the reports of the pathologist who lacks in clinical experience, and who bases his returns on the examination of abstract specimens from a concrete patient of whose history he is usually almost completely ignorant.'[53]

Dazzled by the brilliance of its splendid results, stung by the taunt that past practice was unscientific, the turn-of-the-century doctor 'hailed the laboratory as his deliverer from the thralls of empiricism, irrationalism, and mere experience and has looked up to it as the embodiment of science, as a never-failing aid in solving the mysteries of disease'.[54] The laboratory method in the minds of many acquired a transcendent importance; in comparison, the techniques of meticulous history-taking or physical examination both seemed unreliable and old-fashioned. Subsequent developments have shown that doctors who followed those technological pioneers generally failed to draw from alternative sources of diagnostic evidence a picture of illness that conveyed its true complexity, that eschewed the narrow vision which fixation on any single form of data produces. The challenge remains.

Notes

1. Camac, C. N. B. (1900) Hospital and ward clinical laboratories. *J. Am. med. Assoc.*, **35**, 220.
2. Stone, A. K. (1890) Clinical value of the bacillus of tuberculosis. *Boston med. surg. J.*, **123**, 515.
3. Society Proceedings (1893) The diagnosis of asiatic cholera. *J. Am. med. Assoc.*, **20**, 537. See also Proceedings of Societies, Society of Alumni of Bellevue Hospital (1894) The early diagnosis and treatment of asiatic cholera with report of twenty-one cases. *NY med. J.*, **60**, 244.
4. Culbertson, J. C. (1893) Diphtheria: its specific diagnosis. *J. Am. med. Assoc.*, **21**, 698.
5. Anon (1895) The early diagnosis of diphtheria and the physician's duty when confronted with cases of doubtful character, presenting symptoms of diphtheria. *J. Am. med. Assoc.*, **25**, 118.
6. Anon. (1893) Review of Koch's *The Diagnosis of Cholera. NY med. J.*, **58**, 258.
7. Anon. (1893) Miscellany, the examination of cholera discharges. *NY med. J.*, **58**, 223–224. See also Anon. (1897) The clinical value of the Widal test for typhoid fever. *Boston med. surg. J.*, **137**, 478.
8. Anon. (note 5), p. 118.
9. Culbertson (note 4), p. 699.
10. Anon. (1895) Abstract, Diagnosis of diphtheria. *Med surg. Rep.*, **72**, 98. See also Anon. (note 5), p. 381.
11. Anon. (1897) Correspondence, health boards and the Widal test for typhoid fever. *Boston med. surg. J.*, **137**, 276.
12. Jenner, L. L. (1898) Report on the clinical laboratory for 1898. *St Thomas Hosp. Rep.* 27, 305–309.
13. DaCosta, J. M. (1897) *Tendencies in Medicine; President's Address Delivered Before the Association of American Physicians*, Washington DC, Association of American Physicians, p. 3.
14. Musser, J. M. (1894) *A Practical Treatise in Medical Diagnosis*, Philadelphia, Lea Brothers, p. 21. See also Musser, J. M. (1898) *The Essential of the Art of Medicine*, Philadelphia, pp. 12, 14.

15. Richardson, M. W. (1897) On the bacteriological examination of the stools in typhoid fever, and its value in diagnosis. *Boston med. surg. J.*, **137**, 433.
16. Emerson, H. C. (1898) On the value of laboratory research to the clinician. *Boston med. surg. J.*, **139**, 267.
17. Fitz, R. H. (1901) Some surgical tendencies from a medical point of view. *Boston med. surg. J.*, **145**, 695. See also Meltzer, S. J. (1909) The science of clinical medicine. *J. Am. med. Assoc.*, **53**, 509; and Wesselhaeft, W. (1907) On the need of reform in methods of clinical research. *North Am. J. Homeopathy*, **55**, 478.
18. Haig, A. (1901) The human body as an analytical laboratory. *Br. med. J.*, 1078–1082.
19. Osler W. (1907) The evolution of the idea of experiment in medicine. *Trans. Congress Am. Physicians Surg.*, **7**, 17.
20. *Ibid.*, p. 6.
21. Washburn, F. A. (1939) *The Massachusetts General Hospital; Its Development, 1900–1935*, Boston, Houghton-Mifflin, p. 117.
22. Shattuck, F. (1906) The clinics. *Boston med. surg. J.*, **155**, 328.
23. *Ibid.*, p. 328.
24. Anon. (1900) Human vivisection. *Boston med. surg. J.*, **142**, 172. See also Tweedy, J. (1909) Experimental research and medical progress. *Br. med. J.*, 1018.
25. Anon. (1900) Experiments on human beings. *J. Am. med. Assoc.*, **34**, 1358–1359.
26. Holt, L. E. (1907) Medical tendencies and medical ideals. *J. Am. med. Assoc.*, **48**, 846. See also Hemmeter, J. C. (1906) Science and art in medicine: their influence on the development of medical thinking. *Ibid.*, **46**, 244; and Shattuck, F. C. (1900) Specialism in medicine. *Ibid.*, **35**, 725.
27. Emerson, C. P. (1903) The accuracy of certain clinical methods. *Johns Hopkins Hosp. Bull.*, **14**, 9.
28. Herrick, J. B. (1907) The relation of the clinical laboratory to the practitioner of medicine. *Boston med. surg. J.*, **156**, 763. See also, Fyke, B. F. (1904) Diagnosis. *Nashville J. Med. Surg.*, **96**, 431.
29. Cabot, R. C. (1907) The historical development and relative value of laboratory and clinical methods of diagnosis. *Boston med. surg. J.*, **157**, 150. See also Anon. (1900) Editorial: Laboratory investigation and its effect on clinical diagnosis. *J. Am. med. Assoc.*, **35**, 1679; and Elsner, H. L. (1902) On the value to the physician of modern methods of diagnosis. *Boston med. surg. J.*, **146**, 103–104.
30. Cabot, R. C. (1904) The limitations of urinary diagnosis. *Johns Hopkins Hosp. Bull.*, **15**, 174.
31. Wallace, R. (1903) Laboratory diagnosis – its relation to the general practitioner. *Interstate med. J. (St Louis)*, **10**, 148–150.
32. Cabot (note 30), p. 174.
33. Cabot (note 30), p. 174.
34. Richardson, M. H. (1908) On the significance of clinical histories before and after operative demonstration of the real lesion. *Boston med. surg. J.*, **158**, 510.
35. Cabot (note 29) pp. 150–152. See also Cabot, R. C. (1904) The ideal of accuracy in clinical work: its importance, its limitations. *Boston med. surg. J.*, **151**, 557–559.
36. Harris, D. L. (1909) The danger of laboratory diagnosis, with especial reference to the use of the city bacteriological laboratory by the medical profession. *St Louis med. Rev.*, **58**, 117–118.
37. Herrick (note 28), p. 766.
38. Levy, E. C. (1900) Comparative value of laboratory and bed-side diagnoses. *Virginia med. Semi-monthly*, **27**, 421. See also Mayo, A. W. (1906) Position of pathology with regard to clinical diagnosis. *Br. med. J.*, 603.
39. Harris (note 36), p. 118. See also Grant, W. W. (1909) Diagnosis in its relation to the laboratory and the bedside. *J. Am. med. Assoc.*, **52**, 746.
40. Cabot, R. C. (1900) The relation of bacteriology to medicine. *Boston med. surg. J.*, **142**, 428.
41. Janeway, T. C. (1901) Some sources of error in laboratory clinical diagnosis. *Med. News*,

78, 700–706. See also Bishop, L. F. (1905) Symposium on clinical pathology. The relation of clinical pathology to actual practice. *Boston med. surg. J.*, **153**, 689.

42. Emerson (note 27), pp. 11–18.
43. Anon. (1905) Errors in observation. *Br. med. J.*, 547–548.
44. Herrick (note 28), p. 764.
45. Harris (note 36), p. 119.
46. Cabot, R. C. (1905) Clinical examination of the urine: a critical study of the common methods. *J. Am. med. Assoc.*, **44**, 837–842. See also Anon. (1905) The chemical examination of the urine. *Boston med. surg. J.*, **152**, 438–439.
47. Achard, C. (1900) The diagnosis of renal insufficiency. *Boston med. surg. J.*, **143**, 239.
48. Cabot (note 29), p. 152.
49. Hemmeter (note 26), p. 247.
50. Cabot (note 29), p. 151.
51. Poore, G. V. (1900) Science and practice. *Br. med. J.*, 984–988.
52. Levy (note 38), p. 421.
53. Anon. (1907) Editorial. The laboratory in diagnosis. *J. Am. med. Assoc.*, **48**, 525–526. See also Anon. (1902) Bacteriology as an aid to diagnosis. *Boston med. surg. J.*, **147**, 685–686.
54. Herrick (note 28), p. 763.

THE RISE OF PHYSIOLOGY DURING THE NINETEENTH CENTURY

BÖRJE UVNÄS

Department of Pharmacology, Karolinska Institutet, Stockholm, Sweden

Introduction

According to Alfred Nobel's will, published in 1895, the prize to be awarded by the Karolinska Institute in Stockholm should be given 'to the person who shall have made the most important discovery or invention within the domain of physiology or medicine'. Today the phrasing 'physiology *or* medicine' may seem both restrictive and unclear, and likely to cause misinterpretations and conflicts within a prize jury. However, it should be remembered that at the end of the nineteenth century physiology dominated experimental medicine. Biochemistry, pharmacology, toxicology and their modern descendants were still developing within the framework of their mother discipline, physiology. It is true that important (not to say revolutionary) discoveries and inventions had been made during the second half of the nineteenth century in fields which we today call bacteriology, microbiology and immunology. The Nobel prizes awarded during the first decade of their existence witness the extraordinary progress made in these fields. Emil von Behring, of Marburg, Germany, received the prize in 1901 'for his work on serum therapy, especially its application against diphtheria, by which he has opened a new road in the domain of medical science and thereby placed in the hands of the physician a victorious weapon against illness and deaths'. Ronald Ross, University College Liverpool, England, was awarded the prize in 1902 'for his work on malaria by which he has shown how it enters into the organism and thereby has laid the formation for successful research on this disease and methods of combating it'. Robert Koch, Berlin, Germany, received his prize in 1905 'for his investigations and discoveries in relation to tuberculosis'; and Charles Laveran, Pasteur Institute, Paris, in 1907 'in recognition of his work on the role played by the protozoa in causing diseases'.

These early attributions of prizes illustrate the liberal attitude that was taken by the Karolinska Institute in interpreting 'domain of physiology or medicine'. From the beginning it has been taken to mean 'all the theoretical as well as the practical medical sciences'. Here, I shall limit myself to describing the development of physiology and its closely related disciplines during the nineteenth century, especially during the second half of that century — the active time of Alfred Nobel.

Alfred Nobel's interest in physiology

Alfred Nobel's deep interest in physiology is evident from his negotiations with a young Swedish physiologist, Jöns Johansson, during the 1890s. Johansson was at that time a lecturer in physiology at the Karolinska Institute, and later professor of the same subject. Nobel had a well-equipped laboratory in his villa outside Paris and invited the young Swede to work there; Johansson accepted and stayed in Paris for five months. At Nobel's request he carried out in the laboratory at Sevran a certain number of tests connected with blood transfusion, in which Nobel took special interest. Nobel writes in one of his letters to Johansson:

> 'I already had a faint suspicion that once outside the organism the blood begins to change immediately and that was why I wanted to have it transfused as quickly as possible and by the shortest possible route. My belief is that without seriously reducing the vitality of the corpuscles it can be conducted through tubes made of a molten mass of borax and sodium silicate. Tubes of such material ought to be able to prevent coagulation without modifying any more corpuscles than those coming in direct contact with them. It is nevertheless probable that the time element is an important consideration and that perhaps blood changes ten times more during the second than during the first.'

The letter shows not only his personal engagement but also his insight into a current medical problem: how to prevent coagulation. The morphological and chemical composition of the blood and the mechanism of its coagulation were problems of great interest to the physiologists of Nobel's time. The French chemists Dumas and Gavarret (1843) analysed the blood's fibrin and albumin content. The corpuscular elements in the blood were described and their functions studied. The identity of haemoglobin with the blood colouring matter was established by Hoppe-Seyler (1868). Ehrlich stained and characterized the white corpuscles (1874), and Metchnikoff reported on their phagocytotic properties (1886); in 1908 they jointly received the Nobel prize 'in recognition of their work on immunity'. In this way, blood was degraded from being a living organ for heat production and the spread of vital forces throughout the body to a transport medium for the distribution of its corpuscular elements and solutes.

Coagulation of the blood was studied by Buchanan in Germany, who extracted the fibrin ferment (1845), and by Hammarsten in Sweden, who showed that fibrinogen is converted into fibrin during coagulation and that calcium is essential for this process. Lister noted at the same time that coagulation of the blood within the blood vessels is dependent on their injury: removal of calcium from the blood either by precipitation or complex formation prevents the coagulation. Nobel was therefore on the right track in his attempt to prevent coagulation by manipulating the chemical composition of the transfusion tubing. It is not until very recently however, that this goal has been reached, by siliconizing the tubes used for blood transfusion. One thing Nobel could not know was that the solution to the transfusion problem lay not only in preventing blood coagulation but had to await the discovery of the human blood groups. Karl Landsteiner received the Nobel prize of 1930 for this discovery, which quite unexpectedly explained the previously recognized incompatibility reactions during blood transfusion.

Additional evidence of Nobel's interest in experimental medical research was furnished by the Russian physiologist I. P. Pavlov, winner of the Nobel Prize in Physiology or Medicine – actually the first 'classical' physiologist to receive the prize – in 1904. In his Nobel lecture of that year he related that about ten years earlier he and his colleague, M. Nencki, Professor of Medical Chemistry at St Petersburg, had received from Nobel a considerable sum for the benefit of their respective laboratories. In the letter accompanying the gift, the donor had described his deep interest in physiological experiments and had discussed the problems of ageing and dying.

Fundamental Discoveries in Physics and Chemistry, and Technical Innovations

The development of the natural sciences, physics and chemistry, formed a good foundation for the growth of physiology. Attempts to apply the laws of physics and chemistry to physiological processes during the eighteenth century had been relatively unsuccessful, but the nineteenth century saw a trail of fundamental discoveries. The Italian physicist Avogadro postulated that at identical temperature and volume different gases contain identical numbers of molecules (1811). A gram molecule of gas was later determined to contain 6×10^{23} molecules. The English physicist and chemist Dalton developed the law of partial pressures of gases, and, to explain 'the multiple proportions' of gases, he postulated his famous atom theory (1808). The importance of colloids and cristalloids in osmotic pressure was investigated by Graham (Scots) in 1861. van't Hoff (Dutch) demonstrated the physicochemical factors involved in osmotic pressure (1885), which are fundamental to physiological processes. Svante Arrhenius (Swedish) laid the basis for current concepts of electrolytic dissociation and of osmotic pressure (1887), fundamental to functional mechanisms. van't Hoff received the Nobel Prize in Chemistry in 1901 and Arrhenius that for 1903.

Many new instruments and techniques were added to the scientific arsenal. The galvanometer and the capillary electrometer allowed accurate measurements of electrical processes in muscles and nerves. The mercury manometer and the kymograph, both inventions from Carl Ludwig's laboratory in the middle of the century, were to become two of the standard instruments in physiological laboratories all over the world for more than one hundred years (and still are in many places). Increased knowledge about the composition of blood and the importance of the cristalloids in osmotic pressure led to the introduction of artificial salt solutions and perfusion techniques for maintaining organs and tissues, which were developed in particular in Ludwig's laboratory at the end of the 1860s. Induction currents were discovered by Faraday (1831), and the induction coil was introduced for electrical stimulation of nerves and muscles by Du Bois-Reymond – a pupil of Johannes Müller. The injection needle was invented by Rind in 1845, and anaesthesia came to Europe at about the same time. The microscope was improved: achromatic lenses were made by Lister. With the introduction of the microtome, with which tissues could be cut into very thin sections, the ground was laid for an explosive development in histology. Instead of being associated with macroscopic anatomy, physiology now became allied with microscopic anatomy (histology). A classic example is the work of Johann Purkinje (a professor at Prague and Breslau), who, besides his many contributions to physiology and pharmacology, described certain large cells in the cerebellum and certain fibres that conduct electrical impulses within heart muscle, both of which today bear his name. Aseptic techniques introduced by Lister were an important advance.

The animal experiment is by no means an invention of the nineteenth century, but it was during that time that it became an essential instrument in the study of body functions, due to the unexpected possibilities that were offered with the discovery of anaesthetics and hypnotics. One must admire those who had the courage and astuteness to conduct animal experiments before that time: one example was Regner de Graaf (1641–1673) who had already succeeded in cannulating the pancreatic duct and observing in a living dog the flow of juice from the duct into a collecting bag. Heidenhain, and later his pupil Pavlov, developed techniques for preparing chronic stomach pouches for the study of gastric secretion. Typically, at that time, chronic experiments could be carried out only on those dogs that happened to have survived the operative procedures. The experiments must have been both embarrassing and technically difficult: I have been told that for his operations Pavlov needed four assistants just to hold the dog in place on the operating table. The situation changed totally with the appearance in the 1840s of anaesthetics and hypnotics. Previously, alcohol and morphine (Sertürner, 1817) had been used to reduce pain and induce sleep; but the volatile anaesthetics, ether and chloroform, allowed operative procedures under full unconsciousness. The volatile anaesthetics were replaced by chloral hydrate at the end of the last century and by barbiturates in the beginning of our century, which had the advantage that they could be given by mouth and by intravenous injection, respectively.

All these today seemingly simple innovations revolutionized experimental physiology; they exchanged the previous more descriptive zooscopy and experimental vivisection for quantitative recording of body functions and paved the way for the successful introduction of physical and chemical principles into physiological research.

Physiology as a Recognized Science

Independent chairs in physiology

By the end of the century, physiology had become the leading experimental medical science on the European continent as well as in Britain and the United States. Special chairs in physiology were established at the leading universities of the world. The first German chairs in physiology were those at Freiburg in 1821 (Schultze) and at Breslau in 1824 (Purkinje). The first chair in physiology in France was established in 1854 at the Sorbonne (Longet), in Britain in 1874 at University College in London (Sanderson) and in Sweden in 1874 at the Karolinska Institute in Stockholm (Lovén).

In the same way that physiology branched from anatomy and histology, pharmacology and toxicology are branches of physiology. During the nineteenth century many of the leading physiologists were also interested in pharmacology, or rather experimental toxicology. It started in France. François Magendie and his successor Claude Bernard are well known for their studies on deadly arrow poisons. Especially famous is Claude Bernard's analysis of the mode of action of curare: it was found to kill by paralysing the neuromuscular system, including the respiratory muscles, by a peripheral blocking action on neuromuscular transmission. The leadership in pharmacology was then taken over by the Germans: Sertürner, a pharmacist from Hamelin, isolated morphine from opium; Rudolf Buchheim in Dorpat and later Oswald Schmiedeberg in Strasbourg became the founders of modern German pharmacology.

Physiological laboratories

Laboratory facilities in the middle of the nineteenth century were very primitive, both as to space and equipment. Existing descriptions all give the same picture. When Claude Bernard in 1841 became 'préparateur' at the Collège de France with the famous François Magendie he was allotted 'accommodation for research in the form of a dark, unwholesome room in the basement called a laboratory'. Many years later (1868), as a consequence of the great personal impression Bernard had made on the Emperor Napoleon III, two well-equipped laboratories were established, one at the Musée and the other at the Sorbonne. Johannes Müller's laboratories in Berlin at about the same time were located in an old house behind the 'Garnison Kirche', and the rooms were dark and unventilated; in these laboratories, which were like nasty-smelling caves, 200 medical students prepared their corpses. The collaborators J. Henle and T. Schwann had to work in their private rooms, among clothes and specimens, living and dead frogs, books and food. Rudolf Buchheim, considered to be one of the founders of experimental pharmacology, when he took over a chair of 'materia medica, diethetics and history and encyclopedia of medicine' at the University of Dorpat (Estonia), arranged a private laboratory for himself and his students in the cellar of the house in which he lived. This induced the authorities to create a university laboratory in pharmacology (1860), probably the first of its kind in the world.

The first independent physiological institutes in Germany were built by Schultze in Freiburg (1821) and by Purkinje in Breslau (1839). However, spacious and well-equipped physiological laboratories with adequate staffs were first seen during the second half of the last century, established by prominent physiologists like Claude Bernard in France (Paris), Carl Ludwig in Germany (Leipzig) and Ivan Pavlov in Russia (St Petersburg). In Stockholm, a new and adequately equipped laboratory was built at the Karolinska Institute in 1885 (Lovén).

Physiological journals

Other signs of the increasing reputation of physiology were the formation of scientific societies and the publication of textbooks and journals especially designed for physiological subjects. Magendie started the *Journal de la Physiologie expérimentale* in 1821. Müller's *Archiv für Physiologie* came in 1834 and his famous *Handbuch der Physiologie des Menschen* in 1834–1840. *Pflüger's Archiv für Physiologie* stems from 1848. The *British Journal of Physiology* first appeared in 1878, the *American Journal of Physiology* in 1898 and the *Skandinavisches Archiv für Physiologie* (now *Acta Physiologica Scandinavica*) in 1891.

Development of Physiology

French physiology

As mentioned above, modern experimental physiology has its roots in France. Indoctrinated by old religious conceptions, obsessed by philosophical speculations and lacking sufficient knowledge of the laws of nature, the physiologists of the eighteenth century were unable to give rational explanations for the bodily functions. These functions were still, at the beginning of the last century, assumed to depend on and to be regulated by vital forces that did not obey the laws of physics and chemistry. Marie Bichat (1771–1802), Professor of Anatomy and Pathology in Lyon, was an eager vitalist and he strongly influenced the coming generations of physiologists both in France and Germany. His pupil, François Magendie (1783–1855), became one of the boldest fighters against vitalism. Magendie was a child of the French revolution, open-minded and critical, and entertained 'distrust of theory and firm faith in experiment'.

According to Magendie, 'vitalism was used only to conceal the ignorance of its adherents'. Magendie's contributions to physiology are numerous; among his most famous observations are those on decerebrate rigidity and on the separate functions of the dorsal and ventral spinal roots, the former being related to sensory and the latter to motor phenomena. Magendie's experiment is an example of the vivisectional type of approach characteristic of French physiology at that time: with a cataract knife he sectioned the dorsal roots of the spinal medulla in a living dog and found that the sensibility in the denervated area disappeared. Similar unilateral sections of the ventral roots – experimentally, a very difficult task at that time – deprived the same half of the body of its motor functions. During the careers of Achille Longet (1811–1871), who held the first French chair in physiology (1854) at the Collège de France, and Claude Bernard (1813–1878), also a pupil of Magendie and professor of physiology at the Collège de France (1855), French physiology rapidly metamorphosized into experimental biology.

Claude Bernard, the most famous French physiologist of all, has many important discoveries to his credit. He discovered the formation of sugar from glycogen in the liver; he observed that a high amount of blood sugar resulted in sugar in the urine; he made fundamental contributions to our knowledge of the function of the salivary glands, the stomach and the pancreas; he discovered the vasomotor nerves, i.e. the nerves which regulate the width of the blood vessels and thereby the peripheral blood flow; he studied the function of the blood, the effect of blood loss and the role of the blood gases. His studies on heat production in the body illustrate the fact that old ideas die hard: since ancient times, the heart had been assumed to produce the body heat, which was then distributed via the blood. Lavoisier, who discovered the role of oxygen in combustion at the end of the eighteenth century, assumed that body heat was produced in the lungs. Claude Bernard measured the temperature directly in the heart and found a higher temperature in the right than in the left side. Since the blood in the right half of the heart originates from the tissues, he concluded that body heat was produced in all of the tissues of the body. He gained most fame for his thesis of the 'milieu intérieur', i.e. that mammals are capable of keeping a constant interior milieu because of the physical and chemical properties of the blood and the cells.

These few examples of the very broad scientific production of Claude Bernard serve to illustrate the fact that at that time a talented and diligent physiologist could master the whole subject and could make new discoveries almost every day. Typical of the enthusiasm and optimism of physiologists at that time is the declaration of Charles Richet, another pupil of Claude Bernard's: 'Quelle plus grande douceur que d'arriver chaque matin a son laboratoire et de se dire: "C'est peut-être aujourd'hui que je vais faire une grande découverte".' Richet, who became Professor of Physiology at the Sorbonne, received the Nobel prize in 1913 'in recognition of his work on anaphylaxis'.

With Longet and Bernard, French physiology reached its peak. German physiology then took the lead in Europe. But before we leave France for Germany, it may be mentioned that high-altitude physiology was pre-eminent in France during the second half of the last century. Paul Bert, a pupil of Claude Bernard's, in his famous book *La Pression barométrique*, deals particularly with mountain sickness. He was the first to show that this sickness is due to low partial pressure of oxygen in the atmosphere and the resulting anoxaemia, and not to the low barometric pressure itself, as was previously believed. In 1875, three French scientists, Croce-Spinelle, Sivel and Tissandier, went up in a balloon – the Zenith. A height of over 26 000 feet (around 8000 metres), with a barometric pressure of 270 mm of mercury, was reached. When Tissandier regained consciousness, the balloon was falling rapidly, but his two companions were dead. Oxygen containers were carried in the ascent, but Tissandier records that his arms became powerless and he was thus unable to raise the mouthpiece to his lips. High-altitude and respiration physiology became a British speciality at the turn of the century, but the French pioneers must be mentioned. Evidently, Alfred Nobel became interested in such adventures, for he offered financial support to the Swedish engineer S. A. André, who later perished with his two companions in an attempt to reach the North Pole in a balloon named 'The Eagle'.

German physiology

The development of German physiology differs in several respects from that in France. Under the influence of the romantic movement (romanticism) and philosophers like Fichte, the brothers Schlegel, Schelling and others, German physiology became very speculative and philosophical during the first decades of the nineteenth century. The development of experimental physiology was retarded, and, as will be shown, vitalistic ideas had their defenders until the turn of the century.

Johannes Müller (1801–1858) is called the founder of modern German physiology. He was strongly influenced by the romantic movement, as is evident in his declaration: 'Die Physiologie ist keine Wissenschaft, wenn nicht durch innige Verbindung mit der Philosophie.' Müller was the discoverer of the specific nerve effects, e.g. that stimulation of the optic nerve always results in the perception of light, stimulation of the auditory nerve in the perception of sound, etc. He was never enthusiastic about the application of physics and chemistry to physiology, but preferred speculations about spiritual influences and vital forces, *Lebenskräfte,* Müller was Professor of Physiology in Berlin (from 1833).

Carl Ludwig (1816–1895) represents the glorious days of German physiology. He held a chair in anatomy and physiology in Marburg in the 1840s, and later in Zürich. After ten years in Vienna he was appointed Professor of Physiology at Leipzig. Here he built up his famous institute, which for thirty years became the Mecca for young physiologists of that time: to spend one or a few years with Carl Ludwig became a must for any ambitious physiologist. Carl Ludwig's list of young guest scientists includes over 200 names from all over the world; many of his pupils became prominent physiologists in England, America, Scandinavia and Russia. In this way, his influence on the international development of physiology was enormous. Ludwig was unparalleled as a teacher: he not only helped his pupils in planning their experiments, but he often carried them out himself and wrote scientific papers without allowing his name to be put on them. Ludwig was a very hard worker and demanded a great deal of his pupils, a fact illustrated by the following anecdote. Ludwig received a Baltic baron, de Cyon, a rich young man who had been to several of the famous physiology laboratories of that time. He was an intelligent but rather lazy fellow, more interested in wine and women than in hard laboratory work. So one day Professor Ludwig reproached him for not using his time properly and urged him to do some physiological experiments. de Cyon replied: 'Jawohl, Herr Professor, ich will morgen früh aufsteigen und eine Entdeckung machen', and he performed a very successful experiment in which he discovered the depressor nerve – called in the old physiology textbooks 'de Cyon's nerve' – which is involved in the reflex regulation of heart activity and blood pressure.

Ludwig introduced the experimental causal-analytical and quantitative physicochemical principles which characterize modern physiology. During his fifty-five years of active science, Ludwig, like his contemporary Bernard, made fundamental observations in almost all fields of physiology. To give a few examples, he studied the nervous control of blood pressure, the heart and the peripheral blood circulation, and for these studies he constructed the kymograph and the mercury manometer, indispensable instruments for later physiologists; and he studied kidney function, metabolism and blood gases. This diversification of his research programme was another reason for Ludwig's enormous influence on the development of international physiology. The prestige of physiology grew; and experimental laboratories were built at all German universities during the second half of the century and all over the world, as Ludwig's enthusiastic pupils returned home and developed their own scientific activities.

Carl Ludwig's textbook of physiology, in two thick volumes, was the first real attempt to introduce chemical and physical principles into physiology. Together with his contemporaries, the three brothers Weber – Ernst Heinrich, Wilhelm Eduard and Eduard Friedrich – all in Leipzig, Emil Du Bois-Reymond and Herman von Helmholtz in Berlin, Carl Ludwig represented the mathematical-physical viewpoint which for years to come retained a strong position in German physiology.

Physiological chemistry

During the second half of the last century German physiology also became chemically oriented. However, the development of the chemical branch of physiology, 'physiological chemistry' as it was then called, was hampered for a long time by vitalistic ideas, which survived in part until the end of the century, when the great discoveries in organic and enzyme chemistry finally killed them. Among the prominent pioneers of modern biochemistry, special mention must be given to Jöns Jakob Berzelius, Friedrich Wöhler and Justus von Liebig.

Wöhler was the man to deliver the first fatal blow to the vital force concept. Chemistry of those days held the view that man could never create, could never synthesize the complex sugars, starches and proteins of the vegetable and animal kingdoms. These had been created by vital forces, a mysterious form of energy which permeated living matter but never inanimate material. The chemist was therefore circumscribed by self-imposed boundaries, and his productiveness was frustrated by this erroneous concept. Wöhler (1800–1882) had already proposed in 1824 that benzoic acid ingested in food is excreted as hippuric acid in the urine, a fundamental discovery in metabolic chemistry. After training in analytical chemistry with the great master chemist of that time, Berzelius in Stockholm, he went back to Berlin and soon reported (1828) the synthesis of urea, the first organic substance to be made outside of the body. By this remarkable achievement, Wöhler not only bettered his master Berzelius, who himself adhered to the vitalist school, but also inaugurated the era of carbon chemistry, which would result in the rise of the powerful organic-chemical industry in Germany at the end of the last century. The revolution that occurred in the physiologists' way of thinking is well illustrated by von Liebig's own words: 'Die schönste und erhabenste Aufgabe des menschlichen Geistes, die Erforschung der Gesetze des Lebens kann nicht gedacht werden ohne eine genaue Kenntnis der chemischen Kräfte.'

von Liebig (1803–1873), educated scientifically by Gay-Lussac in Paris, recognized the great lead that French chemistry had over the strongly speculative German chemistry. He devoted his life to analysing the metabolic processes of the body. In his books *Die Chemie in ihrer Anwendung auf Agrikulturchemie und Physiologie* and *Die Tierchemie oder die organische Chemie in ihrer Anwendung auf Physiologie und Pathologie* (1840–1842) he described the circulation of materials in Nature, the assimilation of organic substances by plants, their metabolism by animals and man and the consequences for agriculture.

By the end of the century, our knowledge of the chemical composition of the bodily organs and fluids had increased tremendously. With the discoveries of enzyme chemistry, the vitalistic influence on chemistry was finally destroyed. Buchner showed in 1897 that the fermentation of alcohol was independent of living microorganisms but depended on an enzyme – which he called 'zymase' – present in the fermenting bacteria. Otto Warburg isolated the enzyme necessary for oxidative processes in the cell, the so-called 'respiratory enzyme' and received the Nobel Prize in Physiology or Medicine in 1931 'for his discovery of the nature and mode of action of the respiratory enzyme'. Otto Meyerhof studied the quantitative relationship between the contraction, the production of heat and carbohydrate metabolism in muscle and was awarded the Nobel Prize in Physiology or Medicine in 1922 'for his discovery of the fixed relationship between the consumption of oxygen and the metabolism of lactic acid in the muscle', jointly with Archibald Hill 'for his discovery relating to the production of heat in the muscle'. Organic synthetic chemistry had such great figures as Emil and Hans Fischer. The chemistry of hormones and vitamins had been established.

It is clear that the leading nation in physiology and chemistry at the turn of the century was Germany, and it remained in that position until the First World War, after which the English-speaking world (first the British and then the Americans) took over.

It might be mentioned that Scandinavian physiology developed under the strong influence of the Ludwig school. Professors in physiology, like Fritjof Holmgren, Christian Lovén, Olof Hammarsten, Christian Bohr and Robert Tigerstedt, were all pupils of Carl Ludwig. Swedish physiology thereby got an early and favourable start and has since enjoyed a good international reputation.

Epilogue

Animal experimentation held a central position in physiological research during the first decades of the twentieth century, but after that the 'harvest time' was over for 'classical physiology'. Great advances were then made within its daughter disciplines, biochemistry, pharmacology and, recently, toxicology. Thanks to the development of new techniques in analytical and synthetic chemistry, research today has adopted more and more the character of molecular biology, as evidenced by the work of several of the people to whom Nobel prizes were awarded in the last decades. The scientific front lines in experimental medical research have enlarged concomitantly to embrace rapidly expanding new fields like cell biology, immunology and genetics, with rapid development to molecular levels.

PAUL EHRLICH: HIS IDEAS AND HIS LEGACY

BERNHARD WITKOP

National Institute of Health, Bethesda, Maryland, USA

The Nineteenth Century. Darwin, Marx, Freud, Einstein

The nineteenth century ended not with a whimper but with a bang, namely in 1918 with the end of the First World War.

In the history of mankind, this era is perhaps the most significant in man's development, which had at the time been newly interpreted by Charles Darwin (1809–1882). More innovative ideas were put forward in this period than in the preceding two millennia.

When Alfred Nobel (1833–1896) died, there was little indication that the dogma of Karl Marx (1818–1883), who had died thirteen years earlier, would convert one-third of the world's people to Communism within less than 100 years. At that time, Freud (1856–1939) had just started on a parlous journey into the world of the subconscious. But the greatest member of this remarkable triumvirate, Albert Einstein (born 14 March 1879, exactly twenty-five years after Ehrlich), was then struggling in Aarau and Zürich to learn the fundamentals of mathematics and physics. In less than ten years, he published the most important paper of 1905 on special relativity. Einstein's understanding friend and teacher Hermann Minkowski (1865–1909) summed up the novel discovery: 'Henceforth space by itself and time by itself are doomed to fade away into mere shadows, and only a kind of union for the two will preserve an independent reality.'[1] And ten years later the fundamental principle of general relativity was formulated: 'Space tells matter how to move and matter tells space how to curve.' What a neat way to sum up the mystery of gravitation!

Darwin, Marx, Freud and Einstein, pioneers in four different dimensions, were anxious to bring order into man's evolutionary, social, psychic and cosmic problems and to prove that the old yardsticks for these four fundamentals of our existence were no longer absolute but relative and changeable depending on the location and situation of the observer. In that respect they followed, knowingly or subconsciously, their great preceptor Spinoza, who said: 'Nothing in the universe is contingent, but all things are conditioned to exist and operate in a particular manner by the necessity of divine nature.'[2]

Modern man, haunted by apocalyptic visions, needs enlightenment, consolation and encouragement. We find these in Einstein's thoughts:

> 'Out yonder lies this huge world, which exists independently of us human beings and which stands before us like a great eternal riddle. . . .'[3]
>
> 'All of our endeavors are based on the belief that existence should have a completely harmonious structure. Today we have less ground than ever before for allowing ourselves to be forced away from this wonderful belief.'[4]

The physicist-philosopher created new unity by bringing Spinoza's ordered geometry closer to the 'prestabilized harmony' of Leibniz.

Social and Cellular Hierarchies: Marx, Virchow, Ehrlich

The progress of science is based on new methods: the miracles of the macrocosm and microcosm became accessible through better optical instruments, telescopes and microscopes. When Nobel was thirteen years old (1846), Le Verrier (re)discovered the planet Neptune (after Galileo, 1612[5]); when he was five years old the cell was discovered in plants, by Mathias Jakob Schleiden (1804–1881) in 1838, and one year later in animals, by Theodor Schwann (1810–1882) in 1839. When Nobel was twenty-five years old (1858), Rudolf Virchow (1821–1902) formulated the principles of cellular pathology. In doing so, he paralleled Karl Marx, because he established a social order or a democratic physiology of local cellular unions serving a complex animal or human organism. This organismic view was new because it no longer considered blood to be merely a fluid body but showed it to be a functional composite of vast masses of independent and heterogeneous live cells controlled and governed by forces not yet understood. In his own words:

> 'The body is a sum of vital units of which each one possesses full and complete life, a kind of social institution in which individual entities depend on each other. Therefore it is only meet that these entities owe allegiance and fealty to the third estate, the conscious organism as a whole.'

Again, it is enlightening to observe the conceptual overlap between Marx' social order and Virchow's cellular hierarchy, i.e. society as an organism and the organism as a society of cells. Likewise, the accomplishments of Marx in the social sciences have been compared by Engels to those of Darwin in biology.

And there is the connection with physics: Ludwig Boltzmann (1844–1906), who introduced statistical theory into mechanics, was keenly conscious of the kindred basis of Darwin's theory of evolution. In modest self-effacement he suggested that the nineteenth century be named the 'Century of Darwin'.

In the same way as the atom slowly revealed itself as a microcosm of many unexpected particles of wondrous and unprecedented properties, the cell and its study, cytology, opened up entirely new vistas. It is at this point that we should take note of a young student of medicine, twenty-one years younger than Nobel, Paul Ehrlich, who at the age of eighteen followed an *idée fixe*. He was prompted by the examples and advice of his cousin Karl Weigert, pathologist and anatomist in Breslau; and at the University of Strasbourg (24 September 1872 to 10 June 1874), his teacher in anatomy, Wilhelm Waldeyer, was asked by his teacher in chemistry, Adolf von Baeyer: 'I never see our "Sorgenkind" Paul Ehrlich: What is he doing?' Waldeyer gave the laconic response: 'Ehrlich färbt am längsten!', a pun on the proverb: 'honesty lasts longest', in this case: 'Ehrlich stains the longest'.

We pause at this juncture to let the late Jacques Monod comment on an important point: we are aware of the

> 'duty which more forcibly than ever thrusts itself upon scientists to apprehend their discipline within the larger framework of modern culture, with a view of enriching the latter not only with technical findings but also with what they may feel to be humanly significant ideas arising from their area of special concern. The very ingenuousness of a fresh look at things (and science possesses an ever youthful eye) may sometimes shed new light upon old problems.'[6]

It is 'the fresh look at things' that characterizes Paul Ehrlich's bold foray into the morphology of the cell.

From Histological Staining to New Drugs

When we say that young Ehrlich had a 'fixation', we are not far from a fundamental truth, as it were the *Leitmotif* of his research: *Corpora non agunt nisi fixata*, i.e. for compounds to have a selective pharmacological effect, they must bind to special cells or tissue.

These ideas are expressed in his long-lost doctoral dissertation to the University of Leipzig (1877) entitled *Contributions to the Theory and Practice of Histological Staining.*[7] Since his entire life was devoted to an amplification and demonstration of these seminal ideas, we may rank this dissertation in the same class as that of van der Waals (Nobel prize of 1910) or Rudolf Mössbauer (Nobel prize of 1961), namely as a direct bridge to the Nobel prize. To Ehrlich, the easiest demonstration of his principle of selective binding seemed to be the histochemical staining of cells with the large number of synthetic dyes that had just become available.

William H. Perkin (1838–1907) had accidentally prepared mauveine, the first aniline dye, in 1856 while boldly attempting to find a synthetic substitute for quinine, the natural antimalarial drug of choice. Of the thousands of synthetic dyes that were subsequently synthesized, Paul Ehrlich picked methylene blue, a phenothiazine dye (Figure 1). He started with the selective staining of bacteria:[8] methylene blue preferentially stained microorganisms and cell nuclei. At this point he must have had a vision of the fundamental concept of chemotherapy, and he coined the terms 'organotropic' and 'parasitotropic'.

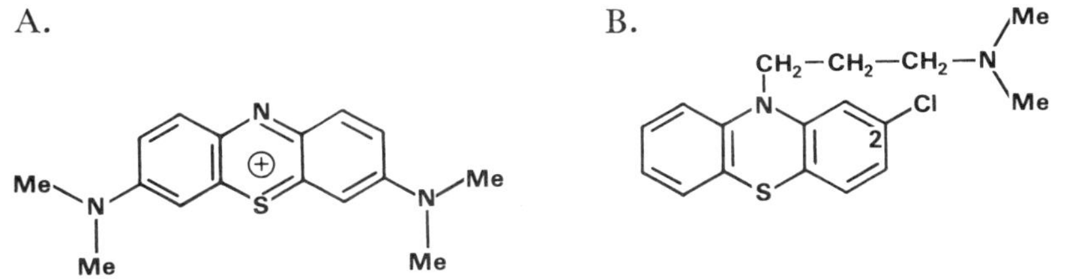

FIGURE 1. Neurotropic staining by methylene blue led to therapeutic applications based on Ehrlich's ingenious concept of a specific receptor. A, Methylene blue: vital staining of nervous tissue (1886); treatment of malaria (1891); analgesic action (1890); suppression of allergic responses, competitor for acetylcholine (1962); inducer of interferon (1973). B, Chlorpromazine: phenothiazine neuroleptics such as this block α-adrenergic receptors in the pontomedullary region (1953–1980).

On 24 March 1882, Robert Koch (1843–1910) announced the discovery of the tubercle bacillus before the Berlin Physiological Society. Koch's staining method was complicated and did not yield nice pictures. Ehrlich, twenty-eight years old, was present on this memorable day and the following day showed Robert Koch, then thirty-nine years old, his elegant way of staining tubercle bacilli, which is still used today.[9] In fact, Ehrlich had already looked at a histological preparation of spleen stained with a mixture of aniline and fuchsin in aqueous solution and nitric acid. In his laboratory journal he noted 'strange red crystals?'. After Koch's discovery, there was no longer need for the question mark: Ehrlich had probably seen the tubercle bacilli before Koch officially discovered them. Similar examples have been recorded by Felix Pinkus (1868–1947), Ehrlich's maternal cousin.[10] 'Every day I see something new under the microscope that I am unable to interpret,' was Ehrlich's comment. But the progression from *idein* (see) to *eidolon* (image) to *idea* (flash of recognition) was realized many times. That is Ehrlich's greatness!

In 1905, Robert Koch received the Nobel prize for the discovery of the tubercle bacillus and for his tuberculin, later found to be unsuccessful. Ehrlich certainly played a part in Koch's achievement.

In 1886 he reported the first results. He called his novel discovery 'Vitalfärbung', or vital staining: only living tissue, metabolically active and fully supplied with respiratory oxygen, was stained. Of the large body of cells, methylene blue selectively stained certain nerve fibres – sensory, olfactory and in the medulla oblongata. Ehrlich's own description indicates his aesthetic enjoyment:

> 'If a frog is injected with a small amount of methylene blue and the tip of its tongue is excised, the most subtle dark-blue ramifications of the nervous system of the tongue become visible against a colourless background in a picture of impressive beauty.'[11]

This was a methodological breakthrough, because up to that time nerve endings could only be made visible by the precarious gold-staining method of Cohnheim. A little-known fact: before Freud started mapping the subconscious he delighted in observing the differentiation of nerve cells and became convinced of the evolutionary process of the nervous system, although his name is not mentioned among the pioneers of neuron theory, such as Wilhelm, and Ramon y Cajal (1852–1934; Nobel prize of 1906), or in Waldeyer's famous monograph of 1881. Ehrlich was not satisfied with this remarkable discovery; two questions remained: (1) Why does methylene blue stain nervous tissue? (2) Why are nerves stained selectively?

He probably started the first investigation of the relation between structure and activity. He modified methylene blue, a ditertiary base, to thionine, with two primary amines; he observed the same staining effects with methylene azure, the sulphone of methylene blue; when he removed the sulphur function entirely and used (toxic) Bindschedler's Green (dimethylphenylene green), he proved that the selective staining depended on the presence of the sulphur group and of the intact phenothiazine ring.

With regard to the affinity of nerves to methylene blue, he assumed that basic groups in the nerve protein and the presence of oxygen in the living tissue were the chemical requirements for selective staining.[12]

The 'fresh look at things' described by Monod is embodied by Ehrlich's shelving of dead histochemical preparations in favour of living tissue and, even more important, the elevation of a descriptive histological problem to an exact chemical level. Thus, Paul Ehrlich, as a chemist, gained the vantage point necessary to attack complex problems in physiology, pharmacology, immunology and clinical therapy.

An intriguing physiological observation was made by G. N. Durdufi,[13] a Russian student of Paul Ehrlich's: when frogs were perfused with methylene blue in physiological saline, nervous degeneration after denervation could be localized and the interdependence of nerve fibres was revealed. For Paul Ehrlich, the logical next step was the translation of the neurotropic staining effect into a therapeutic application.[14] At the time this was a highly unusual and innovative idea; it also came very close to a discovery, made long after Ehrlich's death, that was so fundamental and far-reaching that it made it possible to send home more than half of the population of mental patients who were hopelessly pining away in psychiatric wards and asylums.

To quote Ehrlich again: 'Certain dyes show an affinity to certain organs. This biological fact should serve as a stimulus for therapeutic application in the sense of a localized organ therapy.' Methylene blue of medicinal purity was now administered to patients at the prison hospital in Berlin-Moabit in oral doses of up to 1 g *per diem* or injected as 2% solutions containing 80 mg. Its analgesic effect was indeed surprising: The pain associated with many kinds of neuritis, rheumatic affections of muscle, joint or tendon or even angiospastic migraine disappeared within two hours after administration. Ehrlich carefully noticed, however, that it had no effect on juvenile neurasthenia or on patients with psychic emotions. We shall return to this observation.

If Perkin inadvertently made the first aniline dye by trying to synthesize the antimalarial drug quinine, Ehrlich closed the cycle by demonstrating an antimalarial effect of his favourite dye methylene blue,[15] again because he observed its affinity to plasmodia.

Ehrlich's studies of methylene blue deserve so much prominence and attention because they are a case in point for the momentous beginning of a trail in research that – later and with other molecules – led him to found modern chemotherapy in the form of his highly successful arsphenamine ('Salvarsan'). 'Der Chemotherapeut muss chemisch zielen lernen!' (The chemotherapist has to learn how to take his aim in a chemical way.) His own admonition may have guided Ehrlich in developing his methylene blue to one of the most important drugs of modern time, chlorpromazine, also a phenothiazine derivative but without colour.

Most classical drugs of the nineteenth century were discovered or developed at universities by professors. It was industrial research in France during the Second World War that resuscitated Ehrlich's interest in derivatives of methylene blue, which were found to have not only antimalarial but also antiallergic effects. Then, by accident, some of these phenothiazines, especially one called chlorpromazine, were discovered to prolong the effect of anaesthetics. Chlorpromazine became an aid in the treatment of surgical shock. Finally, it was tested on a mentally disturbed patient – and medical history was made. The tortuous road thus started with malaria, via synthetic dyes, led back to malaria, then to allergy, to surgery, to schizophrenic and mental disorders![16]

The Receptor Concept: the Beginning of Molecular Immunology

In order to understand the complex mechanism of action of the neuroleptic chlorpromazine, we need a concept that – again – was created by Paul Ehrlich. If we credit Marx with a scheme for the equidistribution of wealth, Freud with guidelines for our dreams and Einstein with curved space and the relativity of time, we may, in the same vein, add Paul Ehrlich as the creator of the concept of the receptor. Modern (clinical) biochemistry, molecular biology, endocrinology and quantitative pharmacology are unthinkable without Ehrlich's idea, which, expressed in its simplest form, states: A receptor is the combining group of the protoplasmic molecules to which a foreign group or molecule, when introduced to or into the cell, attaches itself.[17]

Ehrlich borrowed this concept from his studies on immunity, which he started in 1891 with investigations on the toxalbumins, abrin and ricin.[18]

The earliest description of the immune response was given by Thucydides when the 'plague' struck Athens in 430 BC:

> 'All speculation as to its origins and its causes, if causes can be found adequate to produce so great a disturbance, I leave to other writers . . . for myself, I shall simply set down its nature and explain the symptoms by which it may be recognized by the student, if it should ever break out again. This I can the better do, as I had the disease myself, and watched its operation in the case of others. . . .
>
> 'Yet it was with those that had recovered from the disease that the sick and the dying found most compassion. These knew what it was from experience, and had now no fear for themselves; for the same man was never attacked twice – never at least fatally. And such persons not only received the congratulations of others, but themselves also, in the elation of the moment, half entertained the vain hope that they were for the future safe from any disease whatsoever. . . .'

In characterizing the hope of protection against all disease a vain one, Thucydides clearly appreciated the specificity of the protection afforded by the immune system.[19]

Sir Almroth Wright (1861–1947), a great admirer of Ehrlich's and a protagonist of practical immunization for everybody, expressed this concept even more succinctly in Item 1 of his immunological 'creed': 'No one recovers from an acute or chronic bacterial disease unless it is by the production of protective substances in his organism.'[20] This is, of course, a statement of fact that is not even correct in view of the effects of sulpha drugs and antibiotics. A historical explanation was provided by Ehrlich's theory of side chains,[21] his way of picturing receptors (Figure 2).

In addition to having the gifts of imagination and intuition, Ehrlich was blessed with an eidetic mind. In the platonic tradition, he thought in symbols and pictures, pictures so strong that in the end they shaped reality. Perhaps he was influenced by the 'Philosophie des Als-Ob' of Hans Vaihinger (1852–1933), which was related to the English and American schools of pragmatism. Vaihinger's 'as-if' or fictions are scientific assumptions whose improbability or impossibility is known but which are still useful as auxiliary concepts. Ehrlich was aware that his pictorial side chains no longer had any connection with precise chemical constitutional terms. He once remarked: 'Die dummen Leute denken, ich stellte mir die Sache wirklich so vor!' (The stupid people think that I really visualize it that way!) Thus his concepts eventually represented the graduation from fiction to hypothesis to theory to fact and were probably beyond what he dared to hope.[22]

To Ehrlich, every compound that interacted with the cell or the protoplasm possessed one side chain or 'haptophore' that could attach to the cellular or humoral receptor; these had to fit like a lock and a key. The other side chain was free for specific effects. Poisons, such as the plant proteins abrin and ricin, were to him welcome models of bacterial toxins. The interaction between toxin and host thus became to Ehrlich a purely chemical process. He clearly

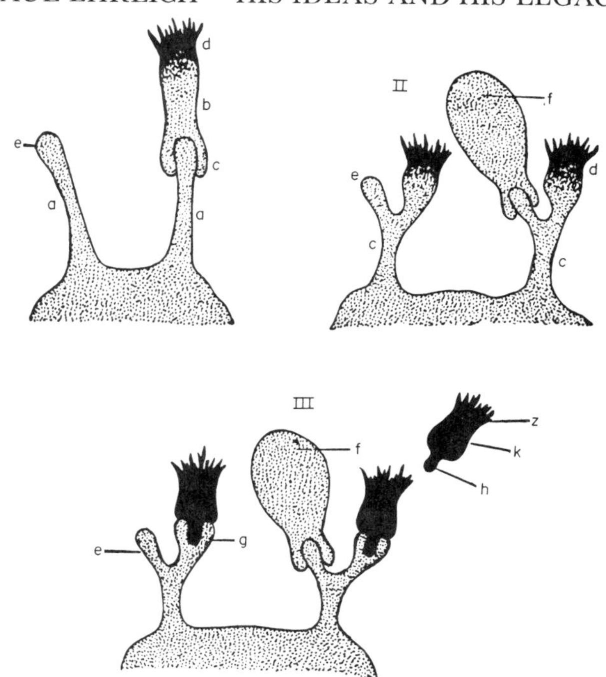

FIGURE 2. Schematic representation of receptors according to Ehrlich. I, Receptors of the first order: *a* is the receptor, with *e* its haptophore group; *b* is a toxin or enzyme, with *c* its haptophore group and *d* its toxophore group. II, Receptors of the second order: *c* is the receptor, with *e* its haptophore group and *d* its ergophore group; *f* is a food molecule or antigen. III, Receptors of the third order: *e* is the antigen-combining haptophore group, and *g* is the complement-combining haptophore group of the receptor (amboceptor); *k* is complement, with its haptophore group *h* and its ergophore group *z*. [From Ehrlich, P. (1901) *Schlussbetrachtungen; Erkrankungen des Blutes; Nothnagel's Specielle Pathologie und Therapie*, Vol. VIII, Vienna.]

saw the difference between these toxins, which caused an abrupt appearance of immunity after one week, and drugs like cocaine, arsenic or morphine, which produced slow tolerance (*Gewöhnung*) over one week in a linear time curve. The former toxins he called 'antigens', because they induced the immune system to produce 'receptor substances' or antibodies in such abundance that after the invading toxins had been neutralized by forming a complex, so many of these free, defensive molecules were left that a new infection was impossible. The immunity was specific and lasted, depending on the toxin, for weeks, years or a lifetime. This was the beginning of serum therapy, pioneered by Ehrlich's friend and contemporary Emil von Behring (1854–1917), recipient of the first Nobel prize in medicine in 1901, an achievement in which Ehrlich played a major part.

It was Ehrlich who pointed out to von Behring how to make serotherapy against diphtheria effective: repeated injections of toxin, in increasing amounts, led to sufficient concentration of antitoxin units,[23] whose potency he was able to determine.[24] In Ehrlich's words (1899): 'It was I who helped him [von Behring] to achieve his success with diphtheria.' Likewise, von Behring in Copenhagen (1901) confided to Carl Salomonsen: 'What would have become of my projects, had it not been for Ehrlich who developed them!'

Theodor Langhans, professor of pathology in Berne, in a letter written on 12 January 1901 to the Nobel Committee, suggested Ehrlich and von Behring as the most deserving candidates for the Nobel prize in medicine or physiology. Ehrlich's candidacy came up every year after that until 1908, when he shared the prize with Ilya Metchnikoff, whose name also appears for the first time in 1901, in a letter dated 29 December 1900, from Edouard Nocard, Paris. Ehrlich was suggested for a second prize, in chemistry, by Emil Theodor Kocher (Nobel prize of 1909) in a letter dated 20 October 1910. Two more nominations were submitted in 1912 and 1913, this time for a second prize in medicine or physiology.[25]

Ehrlich's ideas, concerning the miracle of the immune response, were so farsighted and premature that it required the last twenty years of progress in molecular biology and immunology to vindicate them and to show their ingenuity.

It is instructive to read Ehrlich's Nobel lecture, which was given without any slides.[26] No tiresome experimental or clinical detail is presented, only general observations and his basic ideas on receptors, 'nutriceptors', chemoreceptors, 'arsenoceptors', his grandiose theme to which subsequent generations added many intriguing variations. Where Virchow left off, Ehrlich carried on, with the creation of a prophetic vision in which he credited the cell with functions that posterity is still trying to localize, isolate and define.[27]

The first scientific theory in the West started with Aristotle: The intellect (*nous*) approaches the objects of its investigation, which are opaque; it encircles them; and all of a sudden, by virtue of the 'noetic light', the object's inner principles become visible, intelligible. Thinking is the ability to grasp the essential amid the confusing multitude of details. This description fits Ehrlich's inductive approach to the immune problem. In the experimental domain it required an ingenious simplification: diphtheria toxin was too unstable – it deteriorated on standing to a toxoid, no longer lethal, but still capable of inducing the formation of antibodies effective against active toxin. He switched to simpler, more stable toxins, such as snake venoms, abrin (the poison of the abrus bean) and ricin (that of the castor bean). However, most of these toxalbumins, which we now call 'lectins', were mixtures and therefore produced different antibodies.[28] In spite of these handicaps, Ehrlich's mind was able to abstract from 'the confusing multitude of details' four fundamental principles: (1) The principle of *immunological specificity*, that each pure toxin produces only one kind of antibody. (2) The principle of *complementarity*, that the antigenic determinant (Ehrlich's 'haptophore') fits exactly into the combining site (Ehrlich's side-chain receptor) of the antibody. (3) The *heterogeneity* of antitoxins, which complicated Ehrlich's attempts to standardize therapeutic sera by biological methods. He nevertheless succeeded, and even Pappenheimer's method of quantitative flocculation, introduced in 1937,[29] has not replaced Ehrlich's original methods.[30] (4) The principle of *selection* versus *instruction.* This principle forms the core of immunology. When Landsteiner demonstrated the seemingly limitless number of antigenic specificities, he (and Pauling and Haurowitz) postulated a theory in which the antigen instructs the antibody to form the proper combining site by means of a template. Ehrlich, from the very beginning, held a different view, which was later confirmed by Lederberg, Jerne and MacFarlane Burnett (Nobel prize of 1960): Antibody diversity is selective from the start. This is understandable in modern terms of *de novo* protein synthesis resulting from transcription and translation of a nucleotide sequence in the requisite deoxynucleic acid.[30]

Selection and not *instruction*, i.e. the Darwinian and not the Lamarckian concept (transmissibility of acquired traits), forms the basis for understanding the role of evolution in the immune system. But Paul Ehrlich went beyond Darwin and Lamarck in his pioneering concept of *Immunity With Special Reference to Cell Life*, the original title of his publication in the *Proceedings of the Royal Society*.[30] The phenomenon is still a miracle: an embryo not yet born will be able to make antibodies against a chemical not yet synthesized, i.e. it will be able to mount a specific immune response against an exogenous or xenobiotic invader of the body. Should one think that the immune response brings about a genetic change which permits the adaptation (in the Lamarckian sense) in the next generation without the involvement of (Darwinian) selection?

Ehrlich's theory of antibody generation has been called 'frighteningly up to date'.[31] Its reincarnation came in 1954 when Niels K. Jerne, then Director of the Paul Ehrlich Institute in Frankfurt, unknowingly reformulated it in more modern terms. His insight came from a reading of the following passage from Soren Kierkegaard's *Philosophical Bits or a Bit of Philosophy*:

> 'Can the truth [the capability to synthesize an antibody] be learned: If so, it must be assumed not to pre-exist; to be learned, it must be acquired. We are thus confronted with the difficulty to which Socrates calls attention in Meno (Socrates, 375 BC), namely that it makes as little sense to search for what one does not know as to search for what one knows; what one knows one cannot search for, since one knows it already, and what one

> does not know one cannot search for, since one does not even know what to search for. Socrates resolves this difficulty by postulating that learning is nothing but recollection. The truth [the capability to synthesize an antibody] cannot be brought in, but was already inherent.'

By replacing the word 'truth' by the words in brackets, the statement can be read to present the logical basis of selective theories of antibody formation. Or, in the parlance of molecular biology: synthetic potentialities cannot be imposed upon nucleic acid, but must pre-exist.

Eighty years of side-chain theory

As we celebrate the eightieth birthday of the Nobel prize, we are aware that Ehrlich's momentous *Side Chain and Receptor Theory* had its eightieth anniversary just two years ago, in 1979. As Kabat expressed it on that occasion:

> 'One might perhaps suspect that Ehrlich's important work in developing histological staining reactions for examining cells, which showed the relatively uniform appearance of small lymphocytes, constituted a difficulty in getting others to accept the concept that each of these smaller "look-alike" lymphocytes was programmed to react to a different antigenic determinant and had a different specific receptor on its surface.'[32]

Let us stray for a moment from epistemology and immunology to Paul Ehrlich the man. Ehrlich was asked: 'How do you, when you have a series of inoculations to give to a horse, judge whether the animal is in a condition to receive the next of the series?' 'There is,' said Ehrlich, 'only one possible way to find out if the horse is again in a condition to receive an inoculation, and that is to put a man in charge of the horse who really becomes fond of him. He will be able to tell you whether you should give the inoculation or to withhold it.' Sir Almroth Wright considered this attitude important enough to comment as follows:

> 'This emotional attachment of the man for the subject is the only way in which one is able to judge of the value of the information tendered on such matters as when and where to fish, or to do any gardening or agricultural operation, or read the prospects of the weather, or look for birds' nests or butterflies. You must get hold of a man who has a native taste for these things.'

Unexpectedly, an element of Zen wisdom is introduced into Paul Ehrlich's style of research and thinking.

The Complex Complement

With Julius Morgenroth (1871–1924) and Hans Sachs, at the turn of the century, Ehrlich attacked the most difficult problem in immunology, the problem of complement.[33] The lysis of erythrocytes is mediated by two factors, namely, complement and a specific antibody against the erythrocytes. The ease with which cytolysis of erythrocytes can be observed has led to their widespread use for the detection and assay of complement.[34] Extending these observations, Ehrlich concluded that cytotoxin immunity is based on two components, a specific 'amboceptor' which links a non-specific, lytic 'complement' to the body of the cell antigen.

Jules Bordet (1870–1961; Nobel Prize in Medicine or Physiology, 1919) and his coworkers in Belgium had also studied this phenomenon and given the names 'substance sensibilatrice' and (after Buchner) 'alexine' instead of complement.

Ehrlich analysed the combination of bacteria or blood cells with antibody or complement in terms of molecular combining groups on those molecules and corresponding receptors on the cells. He was the first to recognize that antibody molecules have two different molecular combining groups, one with an affinity for the appropriate immunodeterminant on the cell surface, and the other with an affinity for complement. Furthermore, he advanced the possibility that complement may be an enzyme, and pointed out the significance of the specifically directed enzymatic action that is inherent in the antibody-complement system. However, Ehrlich's keen insight was far ahead of the technical capabilities of the early 1900s. The biochemical tools for elucidating the composition and mode of action of the complement system became available only in the second half of the twentieth century.

Now we recognize complement as a delicately-poised homeostatic system (with twenty components) of enzyme precursors in normal serum, which is involved in inflammation, anaphylaxis, phagocytosis, the destruction of invading microorganism, the historical test of August von Wassermann (1866–1925) for the sero-diagnosis of syphilis and even the control of tumour cells.[35] In 1907 Ehrlich prophetically postulated immunological surveillance of cancer and suggested the control of cancer by immunological methods.[36]

A schematic presentation of the molecular events (Figure 3) that lead to ana-

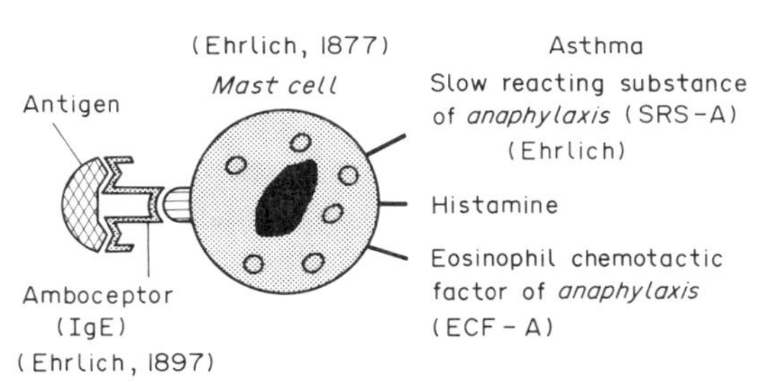

FIGURE 3. Mast cells, complement, amboceptor and anaphylaxis, four of Ehrlich's pioneering discoveries, were not fully understood in their interdependent functions until 1979 (cf. Samuelson, note 37).

phylactic shock and hypersensitivity[37] illustrate both the advances made after Ehrlich's time as well as his share in it: he discovered mast cells by a special staining technique; he postulated the 'amboceptor'; and, finally, he discovered the phenomenon of anaphylactic shock. Julius Axelrod suspects that gamma-E globulin receptor-histamine-mediated release from mast cells may involve a subtle change of lipid conformation triggered by methylation.[38]

Ehrlich's pictures came to life with the full knowledge of structure and function of antibodies.[39] As antibody binds antigen, it may facilitate a pivoting movement of the heavy and light chains around a 'hinge', thus exposing one or two new sites that bind complement. How Ehrlich would have been delighted by these new and dynamic aspects of his side-chain theory.

He would have been even more impressed by the three-dimensional structure of an intact human immunoglobin.[40] What a vindication to see Ehrlich's postulate of binding sites for antigen and complement represented in such a precise way! But he was lucky that he lived in the nineteenth century, when words and ideas had not been replaced by curves and graphs: today, his intuition and imagination with regard to reality and receptors might have been brushed aside as 'subjective', which is jargon for 'unreliable'. Nowadays experience must be translated into numerical form to be convincing. This is one of the striking differences in sensibility in our time and in Ehrlich's time.

This view of science, as a game to be played by rules that extend to the arts and to aesthetics, shows Ehrlich as *Homo ludens* to be equally creative as *Homo sapiens*.[41,42] In the memories of his grandchildren, grandfather Ehrlich was always ready to play and to extemporize little games which he enjoyed as much as they, because in his approach to people and problems he also had a trait best characterized by the word 'spielerisch' ('playful' is an inadequate translation). He liked to pose before the delighted children (and before the world) as a *Zauberer* (magician: magic bullet!) or miracle worker (*Wundertäter*, thaumaturgist) with various tricks involving his favourite dyes. Metaphorically and idiomatically: 'Ehrlich sind viele Erfolge spielend zuteil geworden!' (To Ehrlich many a success came as if he was playing or [an intranslatable double-entendre] in an easy way.)

A letter sent to Ehrlich, an ardent reader and admirer, by Conan Doyle testifies to his hobbies and relaxation. Sherlock Holmes probably appealed to him because he displays the features of the scientist serving society while impressively enacting the very epitome of aloofness. That Holmes was an amateur relates him to other amateurs, such as Boyle, Darwin and Gregor Mendel. Amateurism, or *Liebhaberei*, characterizes most pioneers of the nineteenth century and Ehrlich's first staining 'games' – *Homo ludens*, indeed! It was up to the twentieth century to develop the term and image of the serious professional, the scientist, as a social type, as easily recognizable in films or the mass media as a postman or train conductor.

Haemolysins and Membranes of Red Blood Cells

Intuition and 'unconscious automatic induction' interweave. As early as 1891, Ehrlich devised a method for distinguishing blood samples from different animals. He used arachnolysin, a spider venom, and phrynolysin, a toad venom,[43] to differentiate the blood of rabbits, rats, ox, mice and men, on the one hand, and guinea-pigs, horses, sheep and dogs on the other. Again, no effect was seen without binding; thus, binding must be governed by special receptors on erythrocytes. Red-cell membranes show considerable differences in glycolipids depending on the species.[44] Ehrlich's experiments with impure animal venoms were later expanded by studies of cytolytic bacterial toxins on cell membranes.[45] The earliest records of the existence of oxygen-labile cytotoxins are Ehrlich's studies on the haemolytic effects of tetanolysin.[46]

The biochemical and morphological alteration of membranes by venoms and bacterial toxins is today as active[47] an area as it started to be in the pioneering days of Ehrlich – with the difference that, while the methods have reached a surprising level of refinement and sophistication, the basic ideas have not changed to the same degree.

Drugs and Drug Receptors

Ehrlich suggested in 1898 that the distribution of a drug in an organism is the link between its chemical structure and its pharmacological activity.[48] Even earlier, in 1894, with Alfred Einhorn, he had established relationships between anaesthetic action and receptor groups on cocaine and derivatives,[49] at a time when the structure of cocaine had not yet been elucidated by Willstätter (1898). Bioactivity starts with the descriptive pharmaceutical phase, proceeds to the pharmacokinetic and finally to the pharmacodynamic phases.[50] Ehrlich carried his pictures from immunology to pharmacology and postulated that a drug 'fits' its biological receptor,[51] again as a key fits a lock, a concept that has recently led to (computer-assisted) three-dimensional molecular modelling for the design of drugs.[52]

We should have no illusions, and must admit that hardly anything is known about the structure and shape of drug receptors. In some fortunate cases, for instance with antibiotics, we know the exact mechanism which then involves a specific enzymatic process.[53]

To Ehrlich, the most important problem of medicine and the challenge facing the drug designer were comparable: the former involves cause and seat of disease, or *de sedibus et causis morborum*, as it was phrased by Giovanni Battista Morgagni (1682–1771), the founder of pathology, when he was eighty years old; correspondingly, the therapeutics of the future will have to know *de sedibus et causis pharmacorum*.[11] Rational design of drugs is still a difficult and risky task; we do not know enough about relationships of chemical structure and biological activity to be successful.[54] As the example of penicillin shows, there is, besides the classical macromolecular drug-receptor concept, a possible interaction that involves metabolic, degradative or synthetic processes, i.e. the macromolecule becomes an enzyme. When drug interaction is coincidental, though still specific, we now speak of a 'binding site'.

All the more must we admire the monumental effort and perseverance that led Paul Ehrlich to develop Salvarsan (arsphenamine), the 'magic bullet' and cure for syphilis. Here we recognize the transition of scholar to inventor,[55] which moves the scene from science to technology,[56,57] whose difficult relationships[58] became painfully apparent to Ehrlich after Salvarsan had become a world success. His life was shortened by this experience.

The Start of Chemotherapy

In the *Festband*, in honour of Ehrlich's sixtieth birthday, we are introduced to all the illustrious students, colleagues and collaborators who were involved in Ehrlich's favourite programme of his latter years, viz. chemotherapy. These are Albert Neisser (1855–1916), the first to occupy a full chair in dermatology; Julius Morgenroth (1871–1924); Sachahiro Hata (1872–1938); Louis Benda (1873–1945);[59] Kyoshi Shiga (1870–1957); Alfred Bertheim (1879–1914); Paul Karrer (1889–1971); Hugo Bauer (1874–1968);[60] Heinrich Bechhold (1866–1937); Jacob Benario (1886–1916); and many others.

In New York at the Rockefeller University, a special room houses Ehrlich's legacy: more than 45 000 handwritten pages of manuscripts, laboratory journals and letters. On page 149, dated 7 September 1909, of a slim booklet entitled: *Präparate, 1906–1912*, one can read about Hata's experiments with preparation number 606 (made in May 1909 by A. Bertheim): *salzsaures Aminoarsenophenol* and the well-known structure (Figure 4).[61]

Ehrlich explained the action of Salvarsan on the basis of his receptor theory and postulated an 'arsenoceptor' in the cell wall of trypanosomes.[62] Ehrlich's

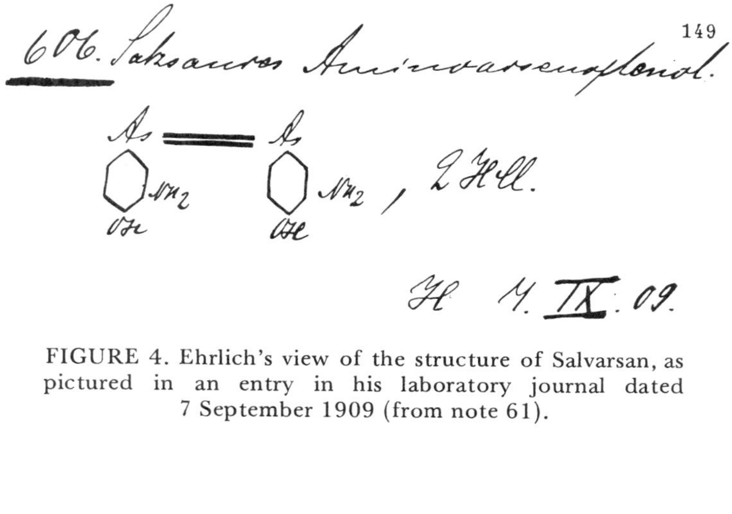

FIGURE 4. Ehrlich's view of the structure of Salvarsan, as pictured in an entry in his laboratory journal dated 7 September 1909 (from note 61).

hypothesis that reducing sulphydryl groups were interacting with arsphenamine, possibly through *ortho*-quinoid intermediates (*para*-arsenoquinone-*ortho*-imine), was taken up by my colleagues at the National Institutes of Health, Carl Voegtlin (1879–1960)[63] and Sanford Rosenthal, who studied both glutathione[64] and sulphydryl groups of cell-wall proteins.[65,66] In 1943, Sir Henry Dale commented on the mechanism of action of arsenicals and sulphanilamides in biochemical terms:

> 'But now in place of vaguely conceived chemoreceptors, labels for observed but unexplained affinities, it begins to be possible to think in terms of interference with activities of vital enzyme systems, or supplanting substrate molecules on the specific surfaces of enzymes.[67]

Ehrlich in his pictorial way compared Salvarsan to a poisoned arrow, the benzene nucleus of the molecule corresponding to the shaft, the *ortho*-aminophenol group, the 'haptophore', to the tip, and the trivalent arsenic radical, the 'toxophore', to the poison on the tip.[68] Such chemical detail relies heavily on the traditional structure of Salvarsan. However, Paul Walden had already asked the question: 'Is there no *cis*- and *trans*-isomerism analogous to *cis*- and *trans*-azobenzene?'[69] The answer is 'No', because there are no arsenic compounds of that structure.

The roentgen-ray analysis of crystalline arsenobenzene[70] was carried out after doubts about its structure had been raised,[71] and demonstrated a trimer with six atoms of arsenic arranged in a puckered, chair-like ring comparable to that of cyclohexane, with the six aromatic substituents in an all-*trans* arrangement. Salvarsan, which is a complicated mixture whose composition depends on the mode of preparation, contains such a six-membered ring structure in addition to five-membered rings and two types of linear polymers, such as $HO(AsROAsR)_nOH$ in which n varies from 3 to 400. Ehrlich's magic bullet (*Zauberkugel*) has now become magic buckshot (*Zauberschrot*)!

As we look into the seeds of time we are now able to see which grain did grow and which did not. Ehrlich's seminal ideas grew into a large 'tree of knowledge', and new disciplines with many ramifications are still to come.

Epilogue

In his sensitive homage on Ehrlich's centennial, Theodor Heuss[72] found moving words. He thought it worth mentioning that Ehrlich had instructed his collaborators not to publish 'preliminary communications' before complete and thorough experimentation. By contrast, our modern predicament: the frequent births of new scientific journals, the excess and 'entropy of information' (a term created thirty years ago by Claude E. Shannon), the fragmentation of literature, the lack of tradition, the illusion of Cartesian progress – could be summed up no better than in the following quotation from Robert Burton:

'Heretofore learning was graced by judicious scholars, but now noble sciences are vilified by base and illiterate scribblers, that either write for vain-glory, need, to get money, or as parasites to flatter and collogue with some great men; they put out trifles, rubbish and trash. Amongst so many thousand authors you shall scarce find one, by reading of whom you shall be any whit better, but rather much worse.'[73]

Burton (1577–1640), the witty English clergyman, would have approved of Ehrlich the man, his style and his substance.

Acknowledgements

I am indebted to Paul Ehrlich's grandchildren, Hans Wolfgang and Günther Schwerin, as well as Susanne von Schueching, for the experience of many conversations with the personalities and destinies of their grandfather and grandmother as focal figures; to James Hirsch, Dean, Rockefeller University, for hospitality and helpful criticism; to Professor H. Schipperges, Institut für Geschichte der Medizin, Heidelberg, for background material; and to Professor Otto Westphal, Max Planck Institut für Immunobiologie, Freiburg, for stimulating discussions.

Part of this paper was prepared during the tenure of an Alexander von Humboldt Senior US Scientist Award at the University of Hamburg during April to September 1979.

Notes

1. Reid, C. (1970) *Hilbert*, Berlin, Springer, p. 12. Minkowski's successor in Göttingen was Edmund Landau (1877–1938), the son-in-law of Paul Ehrlich.
2. Spinoza, B., *The Ethics*, Proposition XXIX. 'The ranking of the most important persons in history' is a matter of individual outlook and preference, as the book with the same title (1978) by Michael H. Hart shows (New York, Hart Publishing Co.). Compare: Fassmann, K., ed. (1976) *Leben und Leistung der sechshundert bedeutendsten Persönlichkeiten unserer Welt*, Zurich, Kindler. In that compendium: Seidler, E., *Paul Ehrlich*, Vol. IX, pp. 224–235.
3. Einstein, A. (1949) *Einstein: Philosopher-Scientist*. In Schilpp, P. A., ed., *Autobiographical Notes*, Evanston, Ill., Library of Living Philosophers, p. 4.
4. Einstein, A. (1934) *Essays in Science*, New York, Philosophical Library, p. 114.
5. Drake, S. & Kowal, C. T. (1980) Galileo's sighting of Neptune, *Sci. Am.*, **243**, 74–81.
6. Stent, G. S. (1978) *Paradoxes of Progress*, San Francisco, W. H. Freeman & Co., p. 122.
7. Michaelis, L. (1919) Zur Erinnerung an Paul Ehrlich, Seine wiedergefundene Doktor Dissertation. *Naturwissenschaften*, 7 (11).
8. Ehrlich, P. (1881) Über das Methylenblau und seine klinisch-bakterioskopische Verwertung. *Z. klin. Med.*, 2, 710.
9. Ehrlich, P. (1882) Über die Färbung der Tuberkelbazillen. *Dtsch. med. Wochenschr.*, **8** (19), 269–270; Ehrlich, P. (1882) Diskussion zur Färbung des Tuberkelbacillus. *Dtsch. med. Wochenschr.*, **8**(26), 365; Koch, R. (1883) Kritische Besprechung der gegen die Bedeutung der Tuberkelbacillen gerichteten Publicationen. *Dtsch. med. Wochenschr.*, 9, 137–141.

10. Pinkus, F. (1915) Paul Ehrlich's Wirken. *Med. Klin.*, **11** (**40**), 1116–1117; Pinkus, F. (1915) Paul Ehrlich's Wirken. *Med. Klin.*, **11** (**41**), 1143–1145.
11. Ehrlich, P. (1960) *Ansprache bei Einweihung des Georg-Speyer-Haus*. In: Himmelwert, F., ed., *The Collected Papers of Paul Ehrlich*, London, Pergamon Press, pp. 42–52 (53–63 in English).
12. Ehrlich, P. (1893) *Verteilung und Wirkung chemischer Körper*, Leipzig, Georg Thieme, pp. 39–51.
13. Durdufi, G. N. (1887) *Arch. Slaves biol.*, **III**(3).
14. Ehrlich, P. & Leppmann, A. (1890) Über schmerzstillende Wirkung des Methylenblaus. *Dtsch. med. Wochenschr.*, **16** (23), 493–494. More recently, methylene blue has been found to be a competitor of acetylcholine [Cook, R. P. (1962) The antagonism of acetylcholine by methylene blue. *J. Physiol.*, **62**, 160–165] and an inducer of interferon [Diederich, J., Lodemann, E. & Wacker, E. (1973) Basic dyes as inducers of interferon-like activity in mice. *Arch. ges. Virusforsch.*, **40**, 82–85].
15. Erhlich, P. & Guttmann, P. (1891) Über die Wirkung des Methylenblaus bei Malaria. *Ber. klin. Wochenschr.*, **28**(**39**), 953–956.
16. Racker, E. (1979) *Science and the Cure of Diseases, Letters to Members of Congress*, Princeton, NJ, Princeton University Press.
17. Pepeu, G., Kuhar, M. J. & Enna, S. J. (1980) *Receptors for Neurotransmitters and Peptide Hormones, Advances in Biochemical Psychopharmacology*, Vol. 21, New York, Raven Press. Another series of many future volumes has just been started: O'Brien, R. D. (1979) *The Receptors: A Comprehensive Treatise*, Vol. I, New York & London, Plenum Press.
18. Ehrlich, P. (1891) Experimentelle Untersuchungen über Immunität. I. Über Ricin. *Dtsch. med. Wochenschr.*, **17** (32), 976–979; Ehrlich, P. (1891) Experimentelle Untersuchungen über Immunität. II. Über Abrin. *Dtsch. med. Wochenschr.*, **17** (44), 1218–1219.
19. Cf. Stryer, L. (1975) *Immunoglobulins*. In: *Biochemistry*, San Francisco, W. H. Freeman & Co. A more recent description [Hoyle, F. & Wickramasinghe, C. (1980) *Diseases from Space*, New York, Harper & Row] enlarges on some of Thucydides' details: 'An astonishingly virulent and remarkably localized disease struck the city of Athens in the summer of 430 BC, a disease with symptoms unlike anything known to modern medicine', a febrile disease, with vomiting and a variety of upper respiratory symptoms accompanied by severe headaches, convulsions, extreme thirst, sleeplessness and paresis, according to P. W. Medawar, probably not the 'plague' but a severe viral infection with encephalitic complications, certainly not, as Hoyle believes, a disease from outer space.
20. Colebrook, L. (1954) *Almroth Wright, Provocative Doctor and Thinker*, London, William Heinemann Medical Books.
21. Aschoff, L. (1902) Ehrlich's Seitenkettentheorie und ihre Anwendung auf die künstlichen Immunisierungsprozesse. *Z. Allgemeine Physiol.*, **1**, 69–248.
22. Koch, R. (1924) *Das Als-Ob im Ärztlichen Denken*, München & Berlin, Gebr. Paetel.
23. Schick, B. (1954) Ehrlich and problems of immunity. Paul Ehrlich Centennial. *Ann. N. Y. Acad. Sci.*, **59**, 183–197.
24. Boehnke, K. E. (1914) *Die Serumprüfung und ihre theoretischen Grundlagen*. In: *Paul Ehrlich, eine Darstellung seines wissenschaftlichen Wirkens*, Jena, Gustav Fischer, pp. 292–331.
25. I am much indebted for the help and cooperation of Professors Bo Holmstedt and Jan Lindsten and of Dr Wilhelm Odelberg who searched the records of the Karolinska Institute and of the Royal Academy of Sciences, respectively; see also Liljestrand, G. (1962) In: *Nobel Foundation, Nobel, The Man and his Prizes*, Amsterdam, Elsevier, p. 213. The letter of Professor T. Langhans is reprinted in *Naturwiss. Rundsch.*, 34, 361–379 (1981).
26. Ehrlich, P., Über Partialfunktionen der Zelle, Nobel Vortrag gehalten am 11. Dezember 1908 in Stockholm. Cf. Int. *Arch. Allergy appl. Immunol.* (1954) (Paul Ehrlich Centenary Issue), **5**, 67–86.
27. Cf. Westphal, O., Westphal, U. & Brede, H. D. (1979) Das Erbe Paul Ehrlichs. *Med. unserer Zeit.*, 3, 107–116.

28. Cf. Kabat, E. A., Heidelberger, M. & Bezer, A. E. (1947) A study of the purification and properties of ricin. *J. biol. Chem.*, **168**, 629–639.
29. Pappenheimer, A. M. & Robinson, E. S. (1937) A quantitative study of the Ramon diphtheria flocculation reaction. *J. Immunol.*, 32, 291–300.
30. Ehrlich, P. (1900) On immunity with special reference to cell life. *Proc. R. Soc. (London), Ser. B*, **66**, 424–448; cf. Gottlieb, P. D. (1980) Immunoglobin genes. *Mol. Immunol.*, **17**, 1423–1435; Rubin, L. P. (1980) Styles in scientific speculation: Paul Ehrlich and Svante Arrhenius on immunochemistry. *J. Hist. Med.*, 35, 397–425.
31. Golub, E. S. (1981) *The Cellular Basis of the Immune Response. An Approach to Immunobiology*, Sunderland, Mass., Sinaver Associates, Inc., pp. 5–9.
32. Cf. Kabat, E. A. (1980) *Structural and genetic insights into antibody complementarity.* In: Cohen, E. P. & Köhler, H., eds, *Membranes, Receptors, and the Immune Response, 80 Years after Ehrlich's Side Chain Theory*, New York, Alan R. Liss, pp. 1–46.
33. Ehrlich, P. (1910) *Studies in Immunity*, New York, John Wiley & Sons. This collective volume contains more than twenty contributions on the haemolysins and complement.
34. Mayer, M. M. (1978) Complement, past and present. *Harvey Lectures*, **72**, 139–193; cf. Alper, C. A. (1979) *Snakes and the complement system. B. Early complement research.* In: Lee, C.-Y., ed., *Handbook of Experimental Pharmacology*, **52**, 863–880.
35. Mayer, M. M., Michaels, D. W., Ramm, L. E., Shin, M. L., Whitlow, M. B. & Willoughby, J. B. (1981) Membrane damage by complement. *Crit. Rev. Immunol.* (in press).
36. Cf. Old, L. J. (1977) Cancer immunology. *Sci. Am.*, 236, 62–79. Ehrlich was fascinated by the possible use of diphtheriatoxin bound to antitumour antibodies as 'magic bullets' against malignant disease. Recent advances with monoclonal antibodies should lead to the construction of new hybrids, which could be used as pharmacological tools that would specifically 'home in' on new targets, such as cancer cells.
37. Cf. Samuelson, B. (1980) The leukotrienes: a new group of biologically active compounds including SRS-A. *Trends pharmacol. Sci.*, **1**, 227–230.
38. Hirata, F. & Axelrod, J. (1980) Phospholipid methylation and biological signal transmission. *Science*, **209**, 1082–1090.
39. Edelman, G. M. (1970) The structure and function of antibodies. *Sci. Am.*, **223**, 34–42; Edelman, G. M. (1969) The antibody problem. *Ann. Rev. Biochem.*, **38**, 415–466.
40. Silverton, E. W., Navia, M. A. & Davies, D. R. (1977) Three-dimensional structure of an intact human immunoglobin. *Proc. natl Acad. Sci. USA*, **74**, 5140–5144.
41. Eigen, M. & Winkler, R. (1975) *Das Spiel*, München & Zürich, R. Piper; Huizinga, J. (1938) *Homo Ludens*, Amsterdam.
42. Staudinger, H. J. (1977) Wissenschaft und Spiel. *Phil. nat.*, **16**, 273–292.
43. Sachs, H. (1914) *Tierische Toxine, Paul Ehrlich, eine Darstellung seines wissenschaftlichen Wirkens, Festschrift zum 60. Geburtstag*, Jena, Gustav Fischer, pp. 209–232.
44. Jamieson, G. A. & Greenwalt, T. J., eds (1971) *Red Cell Membrane, Structure and Function*, Philadelphia, PA, J. B. Lippincott, p. 187.
45. Alouf, J. E. (1977) *Cell membranes and cytolytic bacterial toxins, the specificity and action of animal, bacterial and plant toxins.* In: *Receptors and Recognition*, Series B, Vol. 1, London, Chapman & Hall.
46. Ehrlich, P. (1898) *Klin. Wochenschr.*, 35, 273; cf. Bernheimer, A. W. (1976) *Sulfhydryl-activated toxins.* In: *Mechanisms in Bacterial Toxicology*, New York, John Wiley & Sons, pp. 86–97.
47. Bernheimer, A. W. (1976) *Mechanisms in Bacterial Toxicology*, New York, John Wiley & Sons.
48. Parascandola, J. (1974) The controversy over structure-activity relationships in the early twentieth century. *Pharmacy Hist.*, **16**, 54–63.
49. Ehrlich, P. & Einhorn, A. (1894) Über die physiologische Wirkung der Verbindungen der Cocainreihe. *Ber. Dtsch. chem. Ges.*, **27**, 1870–1873.
50. Ariëns, E. J. (1979) Receptors: from fiction to fact. *Trends pharmacol. Sci.*, **1**, 11–15.
51. Parascandola, J. & Jasensky, R. (1974) Origins of the receptor theory of drug action. *Bull. Hist. Med.*, **48**, 199–220.
52. Gund, P., Andose, J. D., Rhodes, J. B. & Smith, G. M. (1980) *Science*, **208**, 1425–

1431; Humblet, C. & Marshall, G. R. (1980) Pharmacophore identification and receptor mapping. *Ann. Rep. med. Chem.*, **15**, 267–276.

53. Strominger, J. L., Blumberg, P. M., Suginaka, H., Umbreit, J. & Wickus, G. G. (1971) How penicillin kills bacteria: progress and problems. *Proc. R. Soc. (London), Ser. B*, **179**, 369–383.
54. Burger, A. (1979) Relationship of chemical structure and biological activity in drug design. *Trends Pharmacol. Res.*, **1**, 62–64.
55. Julius Morgenroth speaks of 'Erfinder erster Ordnung' (inventor of first order) in the sense of Max Eyth (1836–1906), the German writer and engineer (inventor of the steam plough).
56. Cf. Buzzati-Traverso, A. (1977) *The scientific enterprise today and tomorrow.* In: *Science and Technology*, Paris, UNESCO, p. 25.
57. Anon. (1979) *A five year outlook.* In: *Science and Technology*, San Francisco, W. H. Freeman & Co.
58. *Ibid.*, p. 412.
59. Morgenroth, J. (1914) *Chemotherapeutische Studien*, Jena, Gustav Fischer pp. 541–582.
60. Bauer, H. (1954) Paul Ehrlich's influence on chemistry and biochemistry, Paul Ehrlich centennial. *Ann. N. Y. Acad. Sci.*, **59**, 150–167. This review shows the logical extension of Ehrlich's seminal work, for instance the transition from the dyes trypan blue and afridol violet to Bernhard Heymann's (1861–1933) [cf. Lommel, W. (1933) Bernhard Heymann. *Ber. Dtsch. chem. Ges.*, **66A**, 65–67] colourless Germanin (Bayer 205), the end product of 2000 compounds, which in 1941 was superseded by pentamidin [Ashley, J. N. *et al.* (1942) *J. chem. Soc.*, **103**]. The antimalarials, pamaquine (Plasmochin), quinacrine (Atebrin) and paludrine, the bactericidal Rivanol and trypaflavin and finally prontosil, the sulphonamides and the antibiotics are all on the trail started by Ehrlich.
61. Cf. Bäumler, E. (1979) *Paul Ehrlich, Forscher für das Leben*, Frankfurt, Societäts, Fig. 50.
62. Ehrlich, P. (1913) Chemotherapeutics: scientific principles, methods and results. *Lancet*, **ii**, 445–451.
63. Voegtlin, C. (1925) Arsphenamine and related arsenicals. *Physiol. Rev.*, **5**, 63–94.
64. Rosenthal, S. & Voegtlin, C. (1930) Biological and chemical studies of the relationship between arsenic and glutathione. *J. Pharmacol. exp. Ther.*, **39**, 347–367.
65. Rosenthal, S. (1932) Action of arsenic upon the fixed sulfhydryl groups of proteins. *Publ. Health Rep.*, **47**, 241–256.
66. Parascandola, J. (1977) Carl Voegtlin and the 'arsenic receptor' in chemotherapy. *J. Hist. Med.* **32**, 151–171. This detailed and critical historical review contains most of the leading references on chemotherapy.
67. Dale, H. (1943) Modes of drug action: a general discussion. *Trans. Faraday Soc.*, **39**, 319–322.
68. The strong (organotropic) binding of arsenicals to model membranes of erythrocytes was demonstrated by Hogan, R. B. & Eagle, H. (1944) *J. Pharmacol. exp. Ther.*, **80**, 93; the parasitotropic part of the mechanism of action still needs to be elucidated. [Cf. Burger, A. (1979) In: Wolff, M. E., ed., *Medicinal Chemistry*, 4th ed., Part II, New York, John Wiley & Sons.]
69. Walden, P. (1941) *Geschichte der Organischen Chemie seit 1880,* Berlin, Julius Springer, p. 871.
70. Hedberg, K., Hughes, E. W. & Waser, J. (1961) The structure of arsenobenzene. *Acta crystallogr.*, **14**, 369–374.
71. Rasmussen, S. E. & Danielsen, J. (1960) The structure of arsenobenzene. *Acta chem. scand.*, **14**, 1862; see also Kraft, M. Y., Borodina, G. M., Strelt'sova, I. N. & Struchkov, Y. T. (1960) *Dokl. Akad. Nauk SSSR,* **131**, 1074–1076; and the latest review: Smith, L. R. & Mills, J. L. (1975) Cyclopolyarsines. *J. organometall. Chem.*, **84**, 1–15.
72. Heuss, T. (1954) Ansprache zur Hundertjahrfeier der Geburtstage von Paul Ehrlich und Emil von Behring. *Behringwerk-Mitteilungen*, **29**, 25–28.
73. Burton, R. (1951) *The Anatomy of Melancholy* [1621], New York, Tudor Publishing Co., p. 18. Prophetically, Burton here anticipates a feeling of malaise about the short-

comings of the literary life of his time, which was, ironically, Shakespeare's time. Many intellectuals of more recent times have resumed these laments, especially the historian Henry Adams (1838–1918), the grandson and great-grandson of American presidents, in his application of the entropy concept not only to energy, in a physical sense, but to the sciences, literature and the arts in a metaphorical sense [cf. Mumford, L. (1973) *Apology to Henry Adams*. In: *Interpretations and Forecasts: 1922–1972*, Ch. 32, New York, Harcourt, Brace, Jovanovich, Inc.]. I wonder in this context whether the invention of a mechanism for the systematic publication of fragments of scientific work, a key event in the history of modern science, was such a blessing. Modern man feels insecure because the elusive meaning of progress or the constitutionally guaranteed 'pursuit of happiness' must be defined not only in scientific but also social, ethical and philosophical terms.

THE RISE OF TROPICAL MEDICINE: MILESTONES OF DISCOVERY AND APPLICATION

LEONARD JAN BRUCE-CHWATT

Wellcome Museum of Medical Science, London, UK

Let's the learn'd begin
Th'enquiry where disease could enter in,
How those malignant atoms forced their way,
What in the faultless frame they found to make their prey.'

John Dryden (1631–1700)

Si l'intelligence de l'homme a mise à sa disposition des moyens qui lui permettraient . . . de supprimer, dans un délai plus ou moins long, certaines maladies infectieuses, ce sont des raisons purement humaines qui l'empèchent . . . d'arriver a ce résultat. Les difficultés matérielles et sociales qui s'opposent aux efforts des hommes pour l'extinction des maladies infectieuses ne doivent pas diminuer . . . la valeur des progrès déjà réalisés. Certes, l'oeuvre à entreprendre est immense . . . mais les victoires nouvelles sont de tous les jours.

Charles Nicolle: *Destin des maladies infectieuses* (1933)

Introduction

Few if any chapters of medical history have seen such an explosive rise and development as that of tropical medicine, which was born and grew during the years 1830–1900. There is no need to emphasize that the origin of tropical medicine was closely related to the history of European exploration and expansion, pioneered in the fifteenth and sixteenth centuries by the Portuguese and the Spaniards. There followed the English and the Dutch in their search for the passage to Cathay; and, in laying the foundations of their vast Empires, the French extending their possessions in North America, in Africa and in the West Indies, and the Germans in their drive for new colonial territories.

The seas that had been the great barrier between the Old and the New World became in the eighteenth and nineteenth centuries the highways that linked the continents. Human migrations took place on a large scale, and the impact of the slave trade on the depopulation of Africa and on the implantation of new ethnic groups and their diseases in distant lands was tremendous.[1]

The European colonial expansion was based on a mixture of intellectual curiosity, courage, cruelty, folly, idealism, faith, greed and many other human virtues or vices in the service of trade, religion, adventure or military conquest.

This period of history cannot and must not be judged without fully understanding its roots in the contemporary political, economic, social and moral climate.

In the history of Europe, the first half of the nineteenth century saw many sweeping social, technical and political changes. Following on the lead of Britain in 1807, the slave trade was gradually abolished, so that by about the 1880s most of the European and American governments had prohibited the infamous traffic. But at the same time the phenomenal growth of European industry caused inhumane and insanitary conditions of work and led to the exploitation of men, women and children in factories or mines, and to ill health, starvation and crime. The industrial jungles of London, Birmingham and Manchester were not exaggerated in the melodramatic descriptions of Dickens. Keating's recent book gives a sober but no less disturbing account of that period.[2]

And yet there rose slowly a new spirit of understanding and compassion and a call for greater social justice. Johann Peter Frank (1745–1821) in Germany and Edwin Chadwick (1800–1890) in England were the early exponents of the relationship between poverty, environment and illness. Chadwick's nationwide survey of the sanitary conditions of the labouring population, published in 1842, was a powerful indictment of the unacceptable face of the new capitalism. The social impact of this industrial growth was stressed by the arrival of cholera, which invaded Europe from India in 1817 and spread in successive pandemics: it has been estimated that during four subsequent epidemics of this disease no fewer than 250 000 people died in Britain alone. Spurred by John Snow, who showed in 1849 that those Londoners who used water from the Broad Street pump suffered a mortality eight times higher than that of other inhabitants, the government began to connect ill-health with the pollution of the environment.

But the new social and political trends had as yet little effect on the major epidemics. Epidemic typhus decimated the armies of Napoleon on their retreat from Moscow; and its outbreaks in eastern Europe were notorious. Periodic outbreaks of smallpox occurred in many sea-ports and caused great fear lest it should spread more widely. Plague epidemics swept through several continents and claimed many victims. Haiti owes its early independence to yellow fever: Bonaparte's expeditionary force, sent to the island in 1801, lost 23 000 out of its 30 000 troops and was obliged to return to Europe.

It was only in 1851 that the first, and very ineffectual, International Sanitary Conference took place in Paris, in an attempt to establish an agreement on quarantine regulations. Nine more of these conferences followed between 1859 and 1897; and slowly some sort of understanding of the principles of quarantine began to emerge, although their main goal was to protect Europe and North America from exotic diseases.[3]

The early involvement of the medical profession with the illnesses of tropical countries was related mainly to the preservation of the health of ships' crews, of armies or of isolated military garrisons; this was later extended to the protection of civilian officials in the colonies and of selected overseas trading communities. The Indian (then Bengal) Medical Service had already been established in 1763. Balfour[4] names many of these early pioneers.

Curative and preventive medicine for improving the conditions of economically valuable native labour forces came sooner in some colonial territories than in others. Its introduction was usually determined only by the profit motive, and the appalling working conditions in some tropical plantations in the Americas, in Africa and elsewhere are too well known to need any emphasis.

In this gloomy picture the contribution of Christian missions provided a brighter light. There are few reliable records of the early provision of medical aid to the overseas territories of European powers, although a few Dutch and Danish physicians were working in West Africa at the end of the eighteenth century; but missionary hospitals already existed in Mozambique and Angola in the sixteenth and seventeenth centuries. David Livingstone's travels have captured the imagination of many, and in the nineteenth century medical work became an important missionary activity. At the same time, the attitude of the colonial powers began to change. Grants in aid to charitable medical institutions in the tropics became established; and soon hospital medical services were organized in the colonial territories and provided for, although usually on a less than adequate scale.

Most of the colonial territories were very unhealthy for newcomers from Europe, whether soldiers, traders, missionaries or explorers. The climatic conditions, the environmental factors of food, water and sanitation, the unremitting struggle against hostile forms of life (and particularly insects), the difficulties of day-to-day existence, and also the loosening of moral principles, all contributed to an extremely high mortality among Europeans in colonial territories. Thus, the six major British expeditions to West Africa between 1805 and 1841 had an average mortality of nearly fifty percent.[5] During the period 1881–1887, the annual death rate of European civilians employed by the governments of the Gold Coast and the colony of Lagos in Nigeria averaged between 5.4 and 7.6 percent. Statistical data for the comparable period in East Africa are not available, but even as late as 1910 the annual death rate of European officials was about 1.5 percent.[6] Many other similar examples could be quoted.

However, the level of disease and death among the indigenous inhabitants of the tropics was also very high, both before and after the colonial conquests. There is a school of thought, especially in the United States, that primitive societies were havens of Arcadian peace and health before the coming of the European and that all the epidemics, famines and wars were a consequence of colonization.[7] There is some truth in this simplistic view, because of a cardinal epidemiological law that infectious diseases spread with greater speed in large non-immune groups; and no doubt wider contacts between isolated tribal societies and the outside world, through travel, trade and conquest, must have increased the incidences of some communicable diseases. There are few valid records concerning this particular point, but the available data show that many native populations had not been spared by severe epidemics in precolonial times.[8] Thus, in the Indian province of the Punjab during the cyclical epidemics of malaria in the early nineteenth century, the number of people who died annually of various fevers varied between 2.5 and 3 million. The successive outbreaks of yellow fever in the Caribbean islands left tens of thousands of native victims; and the same could be said of African epidemics of smallpox, typhus and yaws, or of plague in China, long before the colonial period.

This then is the background to the rise of tropical medicine, the main substance of which deals with infectious and parasitic diseases. The roots of modern bacteriology date from 1837 when Schwann and Caignard-Latour showed the the processes of fermentation and putrefaction were due to a '*contagium vivum*' in the shape of a yeast; in the same year, Donné found bacteria and spirochaetes in secretions from the genital tracts of men and women. In 1850, Rayer and Davaine saw rod-shaped organisms in the blood of animals which had died of anthrax and soon proved that these bacteria could transmit the infection. The doctrine of contagion *versus* spontaneous generation was finally and convincingly proved true in 1860 by Pasteur. The devastating epidemic of cholera in Europe in 1865–1866 spurred the search for bacterial causes of disease and culminated with Koch's discovery of *Vibrio cholerae*. Soon Koch's methods for culturing microorganisms opened the remarkable period of 1876–1880, when bacteriology became an established experimental science. The discovery of filterable viruses by Ivanovsky and others in 1892–1896 led to the confirmation of the presence of toxins and eventually to the foundations of immunology by Metchnikoff, although the first bacterial vaccines had already been introduced by Pasteur and the benefits of serotherapy by von Behring.

The link between tropical medicine and the sciences of protozoology and helminthology, however, was not only much closer but also much earlier. To see bacteria one needed a microscope, while larger parasitic bodies, such as tapeworms or roundworms, could be seen with the naked eye. No wonder then that the major part of tropical medicine deals with diseases due to protozoa and helminths. Some knowledge of intestinal worms voided by man and animals dates from far back in antiquity and the Middle Ages.[9] Francesco Redi (1626–1697), the father of parasitology, described and illustrated tapeworms from dogs and cats, and liver fluke and roundworm. Nicolas Andry, in his book *De la Génération des Vers dans le Corps de l'Homme*, published in 1718, produced a series of remarkable though often fantastic engravings of tapeworms and other helminths. There was further development of parasitology by Pallas in Russia, by Göze, Bloch and Rudolphi, and others followed during the late eighteenth and early nineteenth century.

Ten Milestones of Tropical Medicine in the Nineteenth Century

In retracing the early phase of modern tropical medicine, one could select a fair number of major discoveries considered to be milestones on the road travelled in the nineteenth century. Only ten of these signposts have been chosen here, quite arbitrarily, because they are splendid examples either of a rational formulation of a hypothesis (later confirmed by an experiment) or the result of the pursuit of a chance observation, or formed by an accumulation of apparently unrelated data which were gradually fitted into the jigsaw puzzle. Such discoveries, due to chance, coincidence or serendipity, are rarely, if ever, the work of a single genius: each is based on the cumulative experience of several men devoted to the pursuit of knowledge.[10]

(1) *Steenstrup and the liver fluke*

The first milestone in the history of parasitology during the lifetime of Alfred Nobel was the work of a Danish naturalist, Johann Japetus Steenstrup, born in 1813 in North Jutland. Steenstrup studied zoology and geology in Copenhagen, and his palaeobotanical work on peat bogs was of considerable value. But his short paper in 1842 on the alternation of generations in the development of the animalcules known as cercariae was of fundamental importance. Although some knowledge of the encystment of the cercariae in the bodies of snails of the genus *Planorbis* or *Limnaeus* already existed, Steenstrup showed that when cercariae that have been encysted in snails for some time are extracted from the cyst, they have the form of flukes of the genus then known as *Distoma*, which he found in the liver of some specimens of snails. He soon found that these free-swimming cercariae were liberated from an oddly shaped body in the viscera of the snail. He then assumed that this body, which he called the 'parent-nurse', is formed in the snail from an egg produced by a full-grown liver fluke and concluded that 'all trematode animals are developed in this way'. This conclusion elucidated the life cycle of one species of liver fluke and confirmed the seminal concept of 'alternation of generations'.

Steenstrup's work was soon expanded by Dujardin, Davaine and Cobbold and formed the basis for the definitive elucidation of the life cycle of other blood and liver flukes.

(2) *The human blood flukes*

The next milestone in parasitology in the nineteenth century was the work of Theodor Bilharz. Born in 1825 at Sigmaringen in Germany, he studied medicine in Freiburg and Tubingen. In 1850 he joined his teacher, Griesinger, in Egypt, where a 'bloody urine disease', already known to ancient Egyptians, had been prevalent for centuries. In 1851, Bilharz discovered in the blood vessels of a patient who had died of complications of 'bloody urine' some worms that he named *Distoma haematobium*. The prevalence of these worms was very high in Egyptians at autopsy, and particularly in men. In 1857, Cobbold found similar worms in the body of an ape in the London zoo; he realized that this was a new species and gave it the generic name *Bilharzia.* When some patients from South Africa were seen in London with a complaint of blood in the urine and the voiding of characteristic eggs, it became clear that the disease was widespread in Africa. In 1862, Bilharz went on an expedition to Abyssinia, where he died of typhus at the age of 37.

Although the diagnosis of the disease and its complications then became better known, the way in which the parasite was transmitted remained a mystery, in spite of a few more or less inspired guesses by Looss and, especially (in the 1880s), by Sonsino, who was persuaded that water snails were the vectors of the disease. In the meantime the picture became obscured by the discovery of two related intestinal diseases, one in Japan and the other in the West Indies; the infected patients were voiding bilharzial eggs in the faeces, and the tale-telling eggs were of a different shape from the Egyptian variety. It took nearly twenty years to unravel this conundrum, although the problem engendered even more interest when many British soldiers became infected and ill during the Boer War. Eventually, in 1918, the mystery was solved in Egypt by Leiper, who confirmed that there were two species of blood fluke, one *haematobium* and the other *mansoni* (in addition to the third, Japanese species, described by Katsurada in 1904), and that the infection is transmitted by cercariae which penetrate the skin of people who bathe or wade in water in which there are snails that have been infected by eggs in the urine or faeces of infected people.

The sad comment to this much abbreviated story is that the names of the parasite (*Bilharzia*) and the disease (bilharziasis), after that of the man who first discovered it, were changed to *Schistosoma* and schistosomiasis because of a trifling chronological priority that commands the rules of zoological nomenclature.

(3) *The hookworm's progress*

The clinical picture of a disease, already seen in ancient Egypt and characterized by anaemia, dropsy, wasting and intestinal disturbances, was described in the Ebers' papyrus 3500 years ago. It was also known to ancient Chinese physicians, to Hippocrates and to the Romans. It was particularly common among negro slaves in the West Indies and in South America and was called 'oppilaçao' or 'cançaco' (fainting or constant fatigue) by Brazilian physicians. It also had many other names, such as the Egyptian 'chlorosis', 'mal-coeur', 'tuntun', 'water itch', etc., which were related to some of its symptoms.

The intestinal worm that causes this disease was first discovered in the gut of a dying patient in 1843 by Angelo Dubini, a Milanese physician teaching in Padua. The parasite received the name *Ankylostoma duodenale.* This finding was soon confirmed in Egypt by Griesinger and Bilharz. In Brazil, Wucherer drew attention to the clinical symptoms of severe anaemia; and in Italy Grassi showed in 1878 how to diagnose the disease (ankylostomiasis) from the presence of the worm's eggs in the stools. This proved to be of particular importance in 1880, during work on the St Gotthard tunnel between Italy and Switzerland, when a severe outbreak of ankylosomiasis among the miners interfered with completion of the tunnel. The problem also became a serious one in German and Cornish mines. The early assumption that the infection spread from man to man by faecal contamination was only partly true: in 1898, Looss in Germany accidentally infected himself by letting a culture of *Ankylostoma* larvae contaminate his hand, and a few weeks later found that he was excreting hookworm eggs in his stools. This experiment was repeated in 1902 by C. A. Bentley in Assam, and it thus became established that there are two routes of infection: by the ingestion of hookworm eggs and by penetration of larvae through unbroken skin.

Although knowledge of the life cycle of the hookworm pointed to the method of preventing infection and led the Rockefeller Foundation in 1913 to attempt eradication of this disease of poverty, the results of sanitary and medical measures have been only moderately successful, in spite of the discovery of active and reliable drugs. This widespread disease, which is related to unsatisfactory socio-economic conditions, remains a serious health problem in the tropics.

(4) *Worms and elephant-men*

Elephantiasis, or the monstrous condition in which a leg, a scrotum or, more rarely, a different part of the body is permanently swollen and thickened so that it resembles a foot of an elephant, was known to the great Arab and Persian physicians Avenzoar and Avicenna. One of the symptoms of the infection is 'milky urine', due to the presence of chyle from the lymphatic duct. In 1863, Demarquay reported the presence of tiny worms in such urine; but it was Wucherer in 1866 who saw several similar cases in Brazil where he practised. In 1874, Thomas Servis published a full account of these filarial worms in the blood of patients in India.

The explanation of the way in which filarial worms are transmitted, however, did not come until 1877, the year that some scientists mark as the beginning of medical entomology. Although Bancroft in Australia and Lewis in India found adult worms in the bodies of people with microfilariae in the blood, it was Patrick Manson, working in China, who discovered in 1878 the presence of developmental forms of such worms in mosquitoes and thus deduced from experimental studies that filariasis (which in the meantime had been given the adjective Bancroftian) is transmitted by the bites of mosquitoes. A further important observation by Manson was that microfilariae appear in the peripheral blood of infected persons mainly at night. Manson's life and work earned him the eponym 'Father of Tropical Medicine'.[11]

The confirmation by Grassi and by Low in 1899 of developmental forms of filariae in the body of mosquitoes was of monumental value, since it extended to other filarial infections Manson's elucidation of the role of blood-sucking insects in tropical pathology.

(5) *Guinea worm becomes* Dracunculus

It is debatable but not unlikely that the ancient tribes of Israel suffered from the affliction that the Bible describes as 'fiery serpents' (*Numbers,* 21:6). Egyptian, Greek, Roman, Persian and Arabic writers knew of the symptoms, and the early Dutch physicians described them in the East Indies. The disease was particularly common in Africa, the Middle East, the West Indies and parts of South America. A series of remarkable engravings showing the traditional method of extracting the worm was published in 1674 by Welsch (or Velschius), a German physician. Several surgeons of the British Indian Army commented in the early 1800s on the disability that peculiar leg ulcers caused among the troops, and on the fact that long worms emerged from these ulcers when the leg was immersed in water. The worm, which was given various names, such as 'Dracunculus' (little serpent) and *Filaria medinensis* (Medina worm), but was commonly known as 'guinea worm', was described by Cobbold, who observed in 1864 that when the female worm emerges from the ulcer it liberates a large number of tiny embryos whose natural habitat is water. The mode of infection was unknown until 1870, when a Russian explorer and traveller, Aleksei Pavlovitch Fedchenko, made a remarkable discovery in the course of a visit to Turkestan. Influenced by an earlier observation of Leuckart, that the embryos of a fish parasite develop in a minute crustacean (*Cyclops*), Fedchenko described his experiment in which the embryos of guinea worms made their way into the body cavity of *Cyclops* and there underwent further development. He concluded that *Cyclops* was the intermediate host of the human *Dracunculus* but could not prove that the disease occurs after swallowing water containing the infected *Cyclops.* This fact was confirmed by Leiper in West Africa twenty-seven years after Fedchenko's discovery, and suggested a simple way of controlling the disease. However, guinea-worm disease (dracunculiasis) is still a common infection among people in the tropics and remains one of the major health problems in rural areas of the Third World. Its prevalence is closely related to environmental conditions, and in particular the type of water supply.

(6) *Sleeping sickness*

The story of the discovery of the cause of African trypanosomiasis, or sleeping sickness, goes back to 1803, when Winterbottom, a naval surgeon, observed the symptoms of somnolence and enlarged glands in negroes from West Africa. A number of French physicians noted the presence of this fatal affliction, or 'maladie de sommeil', in slaves shipped to the Caribbean plantations. Corkscrew-like, microscopic parasites occurring in the blood of various animals and named 'trypanosomes' were observed by several scientists and especially by Gruby in Paris in 1843. (Gruby soon tired of laboratory science and became a successful clinician: among his patients were George Sand, Chopin, Liszt and Dumas.) For about forty years, trypanosomes were regarded as simple curiosities, but in 1880 Evans, a British veterinary officer serving in the Punjab, discovered the presence of these parasites in the blood of horses, mules and camels that had died of a febrile disease known as 'surra'. It was suspected but not proved that this disease could be transmitted by the bites of stable flies.

During the next decades little attention was paid to trypanosomes; however, many African travellers, and Livingstone in particular, were reporting the ravages of a disease in domestic animals and people, which the local population called a 'fly disease' and attributed to a biting fly called the 'tsetse'. This fly was identified in 1852 by Arnaud, a French entomologist, as being of the genus *Glossina.* In 1897, Surgeon-major David Bruce was investigating a serious and spreading disease of cattle and horses in South Africa called 'nagana'; he found trypanosomal parasites in the blood of diseased animals, and he infected dogs with this blood.

Up until the end of the nineteenth century there was no evidence that human sleeping sickness was in any way connected with 'nagana'. However, there was no doubt that the disease was spreading fast, as a result of the opening up of new territories, easier travel and the extension of trade. It is possible that Henry Stanley's expedition in 1887 for the relief of Emin Pasha contributed to the extension of sleeping sickness from the Congo to East Africa.[12]

Forde and Dutton, in 1891, first reported the presence of a trypanosome in a European patient with sleeping sickness, who had contracted the disease in the Gambia. Later, in 1902, a Commission of the Royal Society was sent to Uganda, where an extensive epidemic was reported. Unfortunately it would be impossible to give a detailed account of the work of this Commission, of the indefatigable travels on foot by some of its members, of the many wrong guesses made by them, of the personal and interteam jealousies and accusations. The mystery of the way in which African trypanosomiasis is transmitted was not solved until 1903 when Bruce published his report to the Sleeping Sickness Commission, pointing out Castellani's discovery of parasites in the cerebrospinal fluid of sick patients. The report concluded that the disease is indeed transmitted by the bite of infected *Glossinae.*

Many and difficult questions related to the identification of sources of virulent trypanosomes, to the various species of tsetse flies and to their habits, to the diagnosis, treatment, prevention and control of this widespread African disease, occupied hundreds of scientists and physicians for the following fifty years. Some of those questions have been answered, while many others are still awaiting solution. It is significant and rather humiliating that human and animal trypanosomiasis still remains a serious medical, sanitary and economic problem in tropical Africa.[13]

At the end of the nineteenth century, the trypanosomiasis that affects human species appeared to be confined to Africa. However, during the first decade of the twentieth century a Brazilian, Carlos Chagas, described a disease with quite different symptoms that is caused by a related parasite, found in South America and transmitted by bites of reduviid bugs. The importance of American trypanosomiasis, justly known as 'Chagas' disease', has been recognized fully only recently. Its prevalence is related to bad housing conditions, and it can still not be treated satisfactorily.

(7) *Dumdum fever and oriental sore*

Kala-azar, 'black disease' or 'dumdum fever', had been known for centuries in large areas of southern Asia, in the Middle East and especially in India. In 1825 there occurred in Jessore a severe epidemic of this 'black disease', with symptoms of progressive emaciation, gross enlargement of the liver and spleen and a darkening of the skin. The disease spread rapidly, and in Burdwan in India the outbreak of 1862 killed three-quarters of a million people. Several investigators, and those not among the least experienced, thought that the disease was a form of malaria or a severe hookworm infection. For forty years no progress of any kind was made, although cyclical outbreaks of this puzzling disease were observed in China and even in some Mediterranean islands. The first clue appeared only in 1900 when Leishman, an army pathologist, found in the spleen of a British soldier who had died of dumdum fever a large number of small, round bodies similar to degenerated trypanosomes, but obviously of a different nature. Major Leishman published his finding in May of 1900, and a few weeks later in Madras, Captain Donovan made the same discovery in several patients at autopsy. Laveran and Mesnil, believing that the parasite was new to science, named it *Piroplasma donovani*, but they soon realized that their generic distinction was in error. Ronald Ross suggested a new genus named after Leishman with the specific name of *donovani*. Thus was born *Leishmania donovani*, one of the most tantalizing and versatile pathogens in tropical medicine.

The wide distribution of closely related parasites of this genus and their many side-effects became obvious when more knowledge was acquired about another human disease in Africa and Asia, known as 'oriental sore', 'Aleppo boil', or 'bouton de Biskra' and many other names. A similar but more serious disease known as 'espundia' occurs in Central and South America. This infection takes the form of a localized, chronic skin ulcer with occasional invasive character, widely distributed in many tropical countries.

These ulcers were well known in the Middle East and greatly feared by young women because of the tendency to appear on the face and thus affect the chances of a good marriage. Although the presence in the ulcer of unusual round cells was noticed in India by Cunningham in 1885, it was then generally believed that 'oriental sore' was due to the use of hard or salty water for washing and drinking. The real precedence for the discovery of the causative agent belongs to a Russian military doctor, P. F. Borovsky, who in 1898, when serving in Turkestan, discovered and correctly described the parasite, later named *Leishmania tropica.*

Thus, at the end of the nineteenth century, two groups of quite different human diseases were found to be caused by a pair of genera of *Leishmania* parasites that are virtually indistinguishable one from another; but the way in which they are transmitted remained a mystery for another twenty years. It took an enormous amount of effort on the part of many scientists to incriminate a sandfly of the genus *Phlebotomus* as the main vector. But even today a number of problems related to these diseases remain unsolved, and their prevention and treatment are only moderately successful.

(8) *Texas fever*

Texas fever, a disease that causes severe and fatal anaemia in cattle, invaded the United States from the West Indies at the beginning of the eighteenth century, if not before, and caused fearful losses. A major epizootic occurred in 1796 and again in the 1850s, when cattle were trailed from Texas to the west of the country, giving rise to the name of the disease. A particular feature of the disease was that cattle from the northern parts of the United States fell ill when in contact with cattle from the south or even when allowed to graze in summer on pastures previously used by southern cattle. In 1889, Salmon, the chief of the Bureau of Animal Industry, requested Theobald Smith, a young and promising pathologist, to work on the worrying problem, which every year caused enormous financial losses and threatened the large meat industry of the country, in spite of a prohibition to transport the cattle north of the 37th parallel.

Recognizing that destruction of the red blood cells was the most striking symptom of the disease, Smith and his coworker Kilbourne soon found small, pear-shaped bodies in these cells, particularly in those of blood from spleen, liver and kidneys. They succeeded in infecting healthy cattle with such blood and gave the new parasite the name *Pyrosoma (Piroplasma) bigemina.* But that was not enough to explain the mode of infection and to seek a way of preventing it. Some cattle owners believed that ticks were involved in this disease, so Smith began his four-year study of the ticks found on cattle. Experiments started in 1889 at a station near Washington; and, although the answers were clear within a year, Smith continued for three more years, to be quite sure of his conclusions. These, submitted in 1892, proved: (1) that the parasite infection was indeed transmitted by a tick called *Boophilus bovis*; (2) that the infection of the adult tick was transmitted to the next generation of ticks and thus maintained the infectivity of the vector population; and (3) that proper preventive measures could be devised. Many regard the report of Smith and Kilbourne as a scientific classic, demonstrating for the first time the transmission of a parasitic disease by an arthropod.

Full details of the development of the parasite in the tick were finally explained forty years later (by Dennis in 1931). But, once again, the 'fickle finger of Fate' amended the name of the parasite (*Pyrosoma* or *Piroplasma*) discovered by Smith. It appeared that Babeş in Rumania had described the parasite in 1888, and, although his work was of minor practical importance, the generic name of the agent of Texas fever has now been changed to *Babesia.* It has recently been found that *Babesia* may infect human beings and cause a serious and sometimes fatal disease.

(9) *Yellow fever*

No other tropical disease has a history more dramatic than that of yellow fever. From the beginning of the European exploration of Africa and of the New World, this infection, known as 'black vomit' and various other names, was deadly to travellers and settlers in the new overseas territories. Its course was often fatal, its nature was cryptic and its menace real, not only overseas but also in many sea-ports of Europe and North America. The early story of this disease was told by Carter and by Harold Scott.[14]

During the two-and-a-half centuries from 1645 to 1900, there were 153 major outbreaks of yellow fever in the West Indies and the Americas. Many guesses were made as to the cause of the disease and its means of transmission, and it would be tedious to quote them all. However, in 1848 Dr Josiah Knott of Mobile, Alabama, and, in 1853, Louis Daniel Beauperthuy, a Venezuelan, reasoned that the disease (then often confused with malaria) was due to a vegetable or animal 'poison' introduced into the human body by bites of mosquitoes. In 1881, Carlos Finlay of Havana suggested in a series of papers that mosquitoes carried the infection from man to man. These papers were largely ignored, and Finlay was regarded as a harmless crank, since the official view was that yellow fever was a 'semi-malarious disease' arising from dirty, damp soil and ships' bilge-water.

Nearly twenty years later, American soldiers in Cuba and elsewhere were still dying in large numbers from yellow fever, although they lived in well-maintained quarters, and although the city of Havana had become, under General Gorgas, the cleanest city in the Caribbean. It was obvious that yellow fever was not due to filthy conditions. A Yellow Fever Commission, under Major Walter Reed, was appointed to study the disease. Within two years, Reed and his colleagues, Carrol, Lazear and Agramonte, had succeeded in demonstrating the truth of tentative suggestions of Carlos Finlay incriminating the mosquito *Aëdes aegypti.* The heroic story of this demonstration, which involved the infection of American volunteers and which caused the death of Lazear and severe disease in Carrol, has often been told and deserves much more than this brief mention.

The final conclusion of the American Yellow Fever Commission in Cuba, dated 1901, was: 'While the mode of propagation of yellow fever has now been determined the specific cause of the disease remains to be discovered.' It was not until the 1930s that the idea that the yellow fever agent was a virus came to be confirmed. In the meantime, the history of the search for the true cause of the infection assumed the character of a Greek tragedy with the deaths from this disease of two distinguished workers, Stokes and Noguchi in West Africa. However, the next twenty years brought a series of discoveries, initiated by Max Theiler, which gave us one of the most remarkable vaccines ever produced by science and provided the means for excellent control of the disease. But that, as Kipling says, is another story!

(10) *Laveran, Ross and malaria*

A febrile disease, which often showed a strange periodicity in its attacks, which caused anaemia, with palpable enlargement of the spleen, and which was not infrequently fatal, was known to Hippocrates and to many Greek, Roman and Arab physicians. For centuries, there was no known cure for this 'marsh fever' or ague; but in about 1640 a new drug, known as 'Peruvian bark' or 'cinchona powder', was brought to Europe from the New World. In the eighteenth century the Italian name 'mal'aria', or spoiled air, was introduced into Europe, indicating the belief that some miasma was causing the disease. Already by the middle of the nineteenth century, many morbid anatomists, and Meckel among the first, had noted the presence of a brown pigment in the blood and organs of persons who had died of malaria.

This pigment was the starting-point of the major discovery by Charles Laveran, a French army doctor serving in Algeria. On 6 November 1880, Laveran, examining a drop of fresh blood from a young soldier, noticed that besides the normal erythrocytes there were in the blood some pigmented, spherical or crescentic corpuscles which extruded motile filaments. Laveran saw these bodies on an unstained slide using an old army microscope, with a relatively low magnification of about 400: One can only admire his eyesight and his powers of observation! Further studies confirmed Laveran's conviction that what he had seen were the parasites responsible for malaria infection. He named them *Oscillaria malariae*, and the full account of his discovery was published in 1881. But Laveran's conviction of the significance of this finding was not shared by other scientists in Italy and elsewhere. Eventually, in 1885, Golgi confirmed Laveran's observation; he described two species of malaria parasites and explained their development in relation to the periodicity of febrile attacks.

Although several physicians, including Laveran, surmised that mosquitoes abundant in swampy areas might in some way be connected with the presence of malaria, no one was able to confirm this hypothesis. The elucidation of the mystery of the transmission of the parasite by a mosquito was due to the insight, patience and some luck of another military doctor serving in India.

His name was Ronald Ross. The story of his discovery has been told many times before. He is one of the most romantic figures in the saga of tropical medicine, and only a brief and inadequate mention of his work can be given here. Supported and stimulated by Patrick Manson, but interrupted by various exigencies of his military service, Ross pursued the hypothesis that mosquitoes were somehow involved in the transmission of Laveran's parasites. He worked for nearly two years, dissecting different species of mosquito, with no result. He was near complete exhaustion, but in August 1897 'the Angel of Fate' (as he wrote) came to his aid. A few 'dapple winged' mosquitoes which had previously been fed on a malaria patient and which were dissected a few days later showed in their stomachs several circular, pigmented cysts which Ross intuitively recognized as a form of malaria parasite. He had still a long way to go in confirming his discovery, but the first and most important step had been taken. When he was posted to Calcutta, Ross was able to extend his studies, working on malaria in birds. Within a year, he had traced the mode of infection with avian parasites (then known as *Proteosoma*) to the bite of the female mosquito. Although Ross did not at that stage work on human malaria, he had no doubt that the mechanism of transmission was the same.

The credit for confirming that the *Plasmodium* of human malaria passed through the same developmental stages as the *Proteosoma* of birds must go to Italian scientists, and particularly to Bignami and Grassi, who at the end of 1898 described the full development of malaria parasites in the stomach wall of *Anopheles*, the invasion of their salivary glands by the infective forms, and, thus, the role of *Anopheles* in the transmission. The bitter personal feud between Grassi and Ross, their mutual accusations of plagiarism, deceit and dishonesty are one of the less edifying aspects of this period. A spectacular demonstration of the feasibility of malaria prevention by mechanical protection from mosquito bites was arranged by Manson and the Italians in the Roman Campagna in 1900. It closed this historical chapter of tropical medicine; however, Paul Ehrlich's work at the same time opened the way to new and remarkable advances in chemotherapy of the disease and to the possibilities of malaria control by mosquito abatement in the twentieth century.

By 1970, the whole of the European continent, most of North America, all of Australia and large areas of other parts of the world had been freed of endemic malaria.[15] Yet the surge of new discoveries and the tremendous progress in our fight against malaria have not led to its expected conquest in the tropics. In 1980 – one hundred years after Laveran's discovery – this disease of the developing world is still the 'million murdering death' of Ronald Ross's dramatic metaphor.

What Now? The Time for Decision

Garnham[16] pointed out that most of the tropical diseases of man are in fact zoonoses, or infections naturally transmitted between animals and the human species. The interchange of pathogens, whether viruses, bacteria, protozoa or helminths, between animals and humans evolved during the early age of our own species; and these relationships have a dynamic ebb and flow, influenced by both our internal response to the parasitic invasion and the external environment. Charles Nicolle's grandiose scenario of the origin, growth and disappearance of infectious diseases is as true today as it must have been during prehistoric times.[17]

But man's activity, combined with the growing numbers of the human species, has today assumed the momentum of a geological force; and its impact on the changing pattern of some diseases presents a fascinating picture. The few milestones of the early phase of tropical medicine that correspond chronologically with the life span of Alfred Nobel preceded the tremendous growth and flowering of our knowledge of tropical diseases in the past eighty years. One would be justified in asking: How much have all these scientific achievements changed the health aspects in what is known today as the 'Third World'? The answer is harsh and grim.

The available data, from the periodic reports of the World Health Organization (WHO),[18] indicate that:

- The burden of ill-health of the three-quarters of mankind who inhabit the developing tropical countries is enormous. Helminthic diseases of the intestine affect about 650 million people, of whom at least 300 million suffer from hookworm (ankylostomiasis), with its well-known sequelae of anaemia and general debility.
- Those suffering from urinary or intestinal schistosomiasis number not less than 200 million, although some authors give a higher figure. It is a tragic paradox that efforts to increase food production by extending irrigation also provide better conditions for reproduction of the snail-host of schistosomiasis and increase the prevalence and severity of the disease.
- Malaria is still endemic in areas inhabited by 1600 million people; the attempted eradication programme was successful mainly in Europe, including the USSR, Australia, most of North America, some parts of South America and a few countries of the Far East. In southern and south-eastern Asia the success was temporary and the present massive resurgence of malaria, with at least twenty million cases per annum, causes much concern; in the endemic areas of tropical Africa, about one million children die of malaria every year.[19]
- Filariases of various types are prevalent in most of the tropics, and the disability or blindness that they cause affect not less than 300 million men, women and children.
- About fifteeen million people suffer from leprosy; while African sleeping sickness, together with South American Chagas' disease (both due to trypanosomal parasites), count some ten million victims.

- The recent seventh pandemic of cholera spread westwards from Indonesia invading twenty-three countries and causing probably some five million cases.
- High on the list of fatal communicable diseases of the tropics are five childhood infections: measles, whooping cough, diphtheria, tetanus and poliomyelitis. The annual death toll from these infections is about five million. Measles alone can be a devastating illness in tropical countries, where it is superimposed on parasitism, anaemia and malnutrition in populations living under unfavourable environmental conditions; the annual number of children dying from this infection is close to one million.
- Some 500 million people in developing countries suffer from malnutrition; those in greatest danger are infants and young children in whom the diarrhoeal diseases are extremely common. Of 122 million babies born each year, about ten percent will die before reaching their first birthday and another four percent before they are five years old.

In this dismal catalogue of major communicable diseases there is one bright page: The old scourge of smallpox has now been eradicated, and this great achievement of the WHO shows what can be done through improved technology and coordinated effort.

The developing countries are struggling with all these health and nutritional problems. They require technological and financial resources, but these are difficult to apply in conditions of rapid population increase, political instability and other adverse factors. Only one-fifth of the population in developing countries have access to clean water, and a large proportion of their ill-health is water-related, either directly, through contamination with bacteria that cause diarrhoeas and dysenteries, or indirectly, by proliferating insects and snails — vectors of viral and parasitic infections.

This, then, is the background to what Dudley Stamp called 'the geography of life and death' in the Third World.

No doubt social and economic strides alone can bring about significant improvements in health, but control of major communicable disease speeds up and supports the general developmental progress. In the Third World, as elsewhere, the healthier people are, the more they can contribute to social and economic advance, which in turn releases new resources for better health of the community.

For the past thirty years, the WHO has been responsible for coordinating and assisting health programmes on a global scale. Recently, the previous emphasis of the WHO on technical approaches to health problems veered to cover wider social goals. The outcome of this trend, due largely to pressure from developing countries, which now form the majority of WHO member states, was a new policy, reflected in the symbolic vision of 'Health for All in the Year 2000' and adopted at the World Conference on Primary Health Care held in 1978 at Alma Ata, the capital of Soviet Kazakhstan.[20]

The concept of primary health care invoked at the Alma Ata conference reflects a number of trends which have been obvious during the past decade, such as the disappointment with and flight from the high technology of present hospital medicine, the greater participation of the community in health activities, the belief that health is one of the basic human rights and, finally and unmistakably, the growing political force of the Third World.

Behind the crudity of the utopian promise, which raises too many expectations, there is an important concept of primary health care, which aims at providing preventive and curative services as widely as possible. Although these services will vary from country to country, they should include proper nutrition, an adequate supply of safe water, basic sanitation, maternal and child care, family planning and immunization against the major infectious diseases. To achieve all this a clear national policy is needed to provide sufficient human, material, technical and financial resources as well as international collaboration.

Experience shows that whatever the theoretical, computerized planning predicts, the practical possibility of eradicating diseases other than smallpox is not yet attainable. There is no room for complacency, although significant advances have been made in immunology and in the development of vaccines and drugs. In diarrhoeal diseases, including cholera, there is the promising development of new antisecretory drugs and of potential vaccines. Progress is also expected in the development of vaccines against typhoid fever, meningococcal meningitis, bacterial pneumonias, respiratory viruses and hepatitis. While our range of chemotherapeutic drugs against bacterial and parasitic diseases has become impressive, the efficacy of systemic antiviral agents has until now been viewed with scepticism. However, today's experience with interferon and with amantadine has given hopes for antiviral therapy in many infectious diseases.

Yet a note of caution is needed. While vaccines and chemotherapy are highly cost-effective tools of public health action, administering them to whole communities or groups at risk poses big problems. No immediate, dramatic results must be looked for, since control of many communicable diseases cannot go ahead of the provision of primary health care services or before improvements in nutrition, environmental conditions and socio-economic levels.[21]

However momentous and justified the moving spirit behind the 'Health for All' idea, a number of obstacles to the fulfilment of this simplistic, messianic dream must be recognized: Are the technical and human resources available and manageable? Are the costs of all this within our means? At the present time, most of the developing countries have between one and two dollars per head per year to spend on health, compared with an average of one hundred dollars in Europe and at least five times as much in the United States. The estimated operating cost of comprehensive primary health care in developing countries would be not less than six dollars per head per annum.[22] The probable overall cost of primary health care is in the order of twenty billion dollars – a rough estimate, since it is virtually impossible to predict the recurring cost of primary health care in all developing countries. But it may be revealing to point out that the world's *annual* military expenditure now exceeds the aggregate sum of 500 000 million dollars. This absurd and obscene disparity reflects the sickness of our present runaway world.

It is in this climate of acute anxiety that the Brandt Commission's report on the North-South dialogue appeared in 1980 under the subtitle: *A Programme for Survival*.[23] This remarkably bold document covers a broad area of problems facing mankind. It outlines the issue of excessive population pressure in the developing countries, the intractable nature of the poverty of some 800 million people, the deterioration of the environment, the iniquities of international trade, the energy crisis, expenditure on increasingly powerful military weapons and the decline of the present system of international aid. The report recommends some sweeping changes, such as international taxes on arms, on minerals, on oil and on other geological resources. It proposes to double the current annual twenty billion dollars of official assistance to developing countries, besides other drastic plans. It gives only a modicum of attention to the problem of population pressure, since it accepts the theme 'development reduces the birth rates' approved by the Bucharest Congress of 1974, even if it calls (rather discreetly) for a better balance between people and resources. Reactions to the Brandt Report have been mixed, varying from laudatory to hostile. It is difficult to see how the recommended drastic actions will materialize in the present situation of recession, inflation, hoarding of petrodollars, high unemployment and renewed tensions between West and East. The next step will be the 1981 summit conference in Mexico to decide how to face the grim future of our overcrowded and mismanaged world.

The *Sixth Report on the World Health Situation*[18] seems to exult in the projection that 'barring unforeseen developments', by the year 2000 the life expectancy of people in developed and developing countries will have increased by five and ten years, respectively, while the infant mortality in the developing world will have fallen to about one-quarter of the present figure.

Although the report is at pains to say that 'two decades of fearsome struggle lie ahead', it is guardedly optimistic. One wonders how many would share this rosy view. The instability of most of the developed countries is now acutely felt; their industrial base, depending as it does on cheap energy, is shaky; their economic and military security in view of the threat of nuclear war is uncertain; their political institutions and their moral stance are equally precarious. But there is little evidence that developing countries are choosing a different path, since they are part of the present political and economic systems. A different, more rational and cooperative system based on a greater respect for the capacity of our globe to carry its ever-increasing burden is not in sight.

Acknowledgements

For this outline of the rise of our biological knowledge, I have borrowed freely from Kean, Mott and Russell's *Classic Investigations in Tropical Medicine and Parasitology*,[24] a collection of original scientific studies excellently translated into English. Other aspects of the growth of parasitology during the nineteenth century were based on Foster's *History of Parasitology*[9] and on Sir Harold Scott's admirable *History of Tropical Medicine*.[14]

Notes

1. Hoeppli, A. (1969) *Parasitic Diseases in Africa and the Western Hemisphere*, Basel, Verlag für Rechts und Gesellschaft.
2. Keating, P. (1976) *Into Unknown England*, Manchester, Manchester University Press.
3. Howard-Jones, N. (1975) *Scientific Bases of International Sanitary Conferences*, WHO, Geneva.
4. Balfour, A. (1925) Some British and American pioneers of tropical medicine. *Trans. R. Soc. trop. Med. Hyg.*, 19, 189–219.
5. Bruce-Chwatt, L. J. & Bruce-Chwatt, J. M. (1980) *Malaria and yellow fever*. In: Sabben-Clare, E. E. *et al.*, eds, *Health in Tropical Africa during the Colonial Period*, Oxford, Clarendon Press.
6. Kuczynski, R. R. (1948) *Demographic Survey of the British Colonial Empire*, 2 vols., Oxford, Oxford University Press.
7. Hartwig, G. W. & Patterson, K. D. (1978) *Disease in African History*, Durham, North Carolina, Duke University Press.
8. McNeill, W. H. (1977) *Plagues and Peoples*, Oxford, Blackwell.
9. Foster, W. D. (1965) *A History of Parasitology*, Edinburgh, London, Livingstone.
10. Beveridge, W. I. B. (1980) *Seeds of Discovery*, London, Heinemann Educational Books.
11. Royal Society of Tropical Medicine and Hygiene (1978) *Symposium Proceedings, Medical Entomology Centenary*, London.
12. Duggan, A. J. (1971) *Tropical medicine*. In: *British Contributions to Medical Science, Woodward-Wellcome Symposium*, London, Wellcome Institute of the History of Medicine.
13. Duggan, A. J. (1980) *Sleeping sickness epidemics*. In: Sabben-Clare, E. E. *et al.*, eds, *Health in Tropical Africa during the Colonial Period*, Oxford, Clarendon Press.
14. Scott, H. H. (1939) *History of Tropical Medicine*, London, Edward Arnold.
15. Bruce-Chwatt, L. J. & de Zulueta, J. (1980) *The Rise and Fall of Malaria in Europe*, Oxford, Oxford University Press.
16. Garnham, P. C. C. (1971) *Progress in Parasitology*, London, Athlone Press.
17. Nicolle, C. (1961) *Le Destin des maladies infectieuses* [1933], Geneva, Alliance Culturelle du Livre.
18. World Health Organization (1980) *Sixth Report on the World Health Situation*, Geneva.
19. Bruce-Chwatt, L. J. (1979) Man against malaria: Conquest or defeat? *Trans. R. Soc. trop. Med. Hyg.*, 73, 605–615.
20. World Health Organization (1980) Contribution of health to the New International Economic Order. *WHO Chronicle*, 34, 274–278.
21. Zahra, A. (1980) WHO's communicable diseases programme. *World Health*, November.
22. Golladay, F. L. & Liese, B. (1980) *Paying for primary health care*. In: Wood, C. & Rue, Y., eds, *Health Policies in Developing Countries*, London, Academic Press and Royal Society of Medicine.
23. Brandt, W., Chairman (1980) *North-South: A Programme for Survival. Report of the Independent Commission on International Development Issues*, London, Pan Books.
24. Kean, B. H., Mott, K. E. & Russell, A. J. (1978) *Tropical Medicine and Parasitology – Classic Investigations*, Vols I and II, Ithaca & London, Cornell University Press.

DISCUSSION*

CHARLES LICHTENTHAELER

Institute for History of Medicine, University of Hamburg, Hamburg, Federal Republic of Germany, and University of Lausanne, Lausanne, Switzerland

My paper may not be at all pleasing to the naïve optimist, but it may perhaps hold the attention of those who know that for the diseases of medicine itself, diagnosis must precede therapy and that certain ailments are incurable.

To the Memory of Sten Lindroth

Introductory Remarks

Eager as I am to plunge straight into my subject, I should not like to do so without first thanking you for having invited me here. I am not only honoured but also particularly encouraged by your kind invitation, for I have not always been regarded with extreme benevolence by all my colleagues in my discipline.

I have been asked to introduce the discussion of our medical meeting by reviving for you the great trends in medicine at the time of Alfred Nobel (1833–1896), i.e. towards the end of the last century. In order to give fuller relief to my portrayal of this period I shall begin with a few words on the medicine of our own times. This, I trust, will enable you, by contrast, to gain a better grasp of the medicine of the late nineteenth century, a medicine which was still relevant and familiar to my own teachers (the period of my studies lasted from 1933 to 1939) but which gradually sank into the past until it had become 'historical' to the new generations in the years following the Second World War. I shall restrict myself to providing certain indications, which you will be able to fill out in your own minds from your respective experiences. And as I shall be obliged, owing to lack of time, to make frequent generalizations, I am well aware that all my statements will be open to contradiction in one way or another. But in spite of this, let us attempt to find our direction.

The success of modern medicine

Never had medicine made such tangible progress in so few years as it has between 1945 and the present day. The most impressive advances have been those relating to clinical diagnosis and to medicinal and surgical therapeutics. It would be pointless to cite examples, for we have all seen such discoveries appear

*Translated from the French by Alan Duff.

on the horizon, many of which would have been unthinkable at the time of our university studies. There are others which have caused less of a stir, yet which have nonetheless changed the shape of the world: take, for instance, the complete disappearance of great epidemic diseases, of vitamin-deficiency diseases such as scurvy and rickets, and of cretinism, the massive reduction of infant mortality, and the conversion of the traditional mental asylums into open psychiatric wards. Nor are the social advances any less spectacular. We are all insured against disease and accident. In our hospitals, the worker, surrounded by swarms of specialists, is better treated than were kings at the time of Alfred Nobel; think only of the cancer of the throat which afflicted the German Emperor Frederick III (1831–1888) and of its political consequences.

The Defects of Modern Medicine: Paradox and Crisis

There is always a reverse side to the coin. Let us turn swiftly to the defects of medicine today, and to its paradox. At the very moment when medicine should be more admired than ever on account of its prodigious successes, it is in fact the opposite that is occurring: medicine is being called into question! There is a double crisis – both scientific and professional. Let us begin by recalling some of the unwelcome and unforeseen consequences of modern medical progress: the overpopulation of the globe and the agedness of its inhabitants, the prolongation of the lives of biologically reduced beings, the harmful side-effects of new medicines, and the increasingly frequent professional mistakes made by poorly trained doctors who have been left behind by the dizzying acceleration in the rhythm of new discoveries. In our consumer civilization, personal hygiene has also deteriorated: obesity and scoliosis are becoming endemic, so too are alcoholism and nicotine and drug addiction.

Lastly, some of the imperfections of modern medicine are due precisely to its astounding advances. Diagnoses, remedies and medical journals are now available in their thousands. A gigantism has set in which is forcing medical science and medical practice to break down into a myriad of specialities and sub-specialities. We are losing sight of the whole, as many of our patients are discovering to their own cost. Let us keep as close a track as possible of the strengths and weaknesses of medicine today, for it is indeed both very effective and very powerless. It succeeds admirably when it has the means and is able to implement them – the curing of serious infections by means of antibiotics or the operatory correction of cardiac defects, that is elating!, but it is utterly at a loss when these means fail and even more so when they are missing. When this happens, the doctors immediately seek shelter behind their diagnostic apparatus and the (positive or negative) results of their laboratory tests; they shuttle their patients from specialist to specialist, ending up with the psychiatrist; and yet it is not always the patient who is disturbed but rather our medical practice itself, being too much enslaved by its habits to be able to think and act.

Our doctors' task is further complicated by several additional factors. Insurance schemes – that necessary evil – and the intrusive state intervene between the medical practitioner and the patient, thus altering their relations. The 'peculiar interchange' between doctor and patient, of which the writer and doctor Georges Duhamel spoke, has practically ceased to flourish except with renowned specialists – and with charlatans! After 1945, the classical faculties of medicine, all imbued with humanism, became transformed into polytechnical colleges for engineer-doctors; and for far too long the authors of this dubious reform, which I have spent my life fighting against, remained blinded to its dangers. In our disturbed age, more than half of the patients suffer from psychosomatic afflictions and speak a language other than that of the doctor-technicians who bustle around them.

The result of all this is that, a century after Alfred Nobel lived, our medicine is creating miracles . . . but the surgeries of the 'heretic' doctors and bonesetters of all descriptions, qualified or not, are becoming no emptier. 'The doctor' is certainly still the guiding figure in our unbalanced age, but 'doctors' are becoming the focus of increasingly vehement polemics. One cannot but conclude that in our sphere not everything has happened as it should.

The Common Cause: 'Modern Medical Naturalism' and its Underlying Cartesian and Baconian Philosophy

Modern medical development, then, is characterized by both striking successes and stunning failures, each being highlighted by the other. It would be a mistake to concentrate exclusively on one or the other; one must consider both the successes and the failures alike and seek their common cause, for in my opinion they do have one. And by so doing we shall also be preparing ourselves for an understanding of the medicine of Alfred Nobel's time, a medicine so different from our own in more than one respect.

In my *History of Medicine*,[1] I spoke of the 'experimental revolution' of the physiologist, pharmacologist and Paris hospital doctor, François Magendie (1783–1855), during the first decades of the last century. The future teacher of Claude Bernard (1813–1878) rejected Hippocrates, Galen, the numerous 'medical systems' that flourished in Paris around 1800 and even the pathological anatomy which at the time, with Corvisart, Bayle and Laennec, was enhancing the fame of the Paris Medical Faculty – and this is a point to which we shall be returning. He wished to place medicine on quite a different basis, by introducing it to the schools of the natural sciences, physics and chemistry, which at that time were passing through one of the most creative phases in their history. Medicine thus assumed a new scientific form, by conceiving itself in turn as a natural science; the scientific movement which originated here I have called 'modern medical naturalism'.[2] Now, as we all know, modern physics and chemistry are essentially experimental sciences. By following in their wake, physiology and medicine in their renewed guise aspired to the same condition. This was stated by Claude Bernard in 1865 in the very first pages of his celebrated *Introduction to the Study of Experimental Medicine*:

> 'It is thus clear . . . that medicine is moving towards its definitive scientific path . . . gradually abandoning the realm of the systems in order to assume an increasingly analytical form and so gradually enter into the method of investigation common to the experimental sciences.'[3]

Now, a full century later, we are enjoying the triumphs of these pioneering efforts: man has walked on the moon, and our doctors are using the artificial lung.

It must likewise be recognized that our present setbacks are also closely related to this 'experimentalization' of modern medicine, which over the past hundred years has been taken to increasingly extreme lengths. More precisely, the inadequacies of our official medicine stem from the limitations to the application of its experimental method. These limitations are of at least three orders: (1) Claude Bernard's method, in spite of itself, is eclectic. It cannot encompass reality, natural or morbid, in any form other than that of experimental facts, immediate causal relations (Claude Bernard's 'proximate causes') and natural laws. Yet it is quite evident that it is only one aspect of reality as it appears to us both in the outer world and at the patient's bedside. (2) Even what the experimental method does grasp of reality must be regarded in a certain way, to the exclusion of other ways. This is an analytical method which breaks down the real into its elements, yet is unable to regroup them. Our theories and doctrines are, certainly, based on experimental data, but they are not themselves experimental in the true sense. Let us mention in addition the natural propensity of the laboratory worker to quantify everything. Taking his lead from Galileo, he aspires to measure the measurable and also to render measurable whatever is resistant to measurement. He dreams, as a mathematician, of converting everything, even appearances, into figures. (3) Lastly, it must be said, no matter how axiomatic it may be, that by claiming to be strictly scientific the experimental method drives out of the observer's field all that is incompatible with science as conceived by this method. The physicist and the biologist are undeterred by this, except in extreme cases when they are forced by science to go further than they wished (as with the atomic bomb or genetic manipulation). The doctor, on the other hand, finds that the whole traditional and venerated domain of what he regarded as the medical 'art' is thus ousted, excluded. The experimenter will argue that this is the very precondition of scientific progress, and one which negates progress achieved according to this principle: *experimentali modo*! For science purports to be 'absolutely impersonal; it is the truth envisaged in its ideal aspect'. I quote Claude Bernard himself,[4] who said – after Victor Hugo – that 'art is me, science is us'.[5] But medical practice is not to be confused with the scientific search for new knowledge.

Thus, the successes and failures of modern experimental medicine can be explained in a simple and convincing way. Our doctors succeed when experimental reasoning can be applied unhesitatingly, when they are confronted with 'clear and distinct' morbid cases along Cartesian lines, and when they have sufficient of the means needed to treat them. They remove a swollen appendix or arrest an attack of paroxystic tachycardia by means of an injection. And all alike are satisfied – patient, doctor and entourage. But the complications set in as soon as the disorders are difficult to grasp, when they are multiple or due to multiple causes, when they affect the patient's personality, his inner life, and his social life as well; in short, as soon as the clinical situation diverges too greatly from the experimental conditions. The doctor then no longer knows what to do; these cases bother him; and the patient – who had been expecting to be taken in hand by his doctor – no longer feels supported, but rather abandoned to himself, to his present sufferings, his anguish and his fear of the morrow.[6] He is often all the more disillusioned if he too has allowed himself to be carried away by the Promethean *hybris* of modern medical progressivism, imagining science and technology to be all-powerful and believing that everything had become 'reparable'. Nor should we expect too much of the psychologists and sociologists, for their disciplines have been shaped by the model of the natural and experimental sciences. They too have nothing but doctrines and procedures to offer, and they are likewise unable to get beyond science and scientific techniques.

Modern medical naturalism – with its metaphysics which is scientistic, positivist, progressivist and realist, not to say materialist, but at all events mechanist – has been prospering for 150 years under the auspices of three seventeenth-century masters, Galileo, Descartes and Francis Bacon.[7] In this medicine, the doctor – *res cogitans et agens* – *scientia potestas est* (F. Bacon)[8] – is reflecting and operating upon his patients, regarded as *res extensae* in a Newtonian space and time. In our hospitals and medical consultation rooms, which are conceived in a highly 'functional' fashion, a double asepsis is being carried out: to eliminate not only all kinds of microbes but also anything that might disturb the Bernardian cult of the medicine-science. The patients yield to this so long as there is promise of their being helped; but they begin to shudder when the experimental framework proves to be too restricting for them and turns into a straitjacket. The official medicine of our day occasionally seems to me like a fairy-tale castle set in a ruined park. The patient is saved if the castle drawbridge is lowered to let him in: once inside, he will be cured of even the most serious ailments! But if he has to remain outside the castle gates, his odyssey from one quack to another will often be long and cruel. Thus we are brought up against the inherent ambiguity of our modern medicine. On the one hand, it makes us witnesses of its stunning achievements, but on the other hand, it introduces us to international congresses of overtly self-confident 'heretics'.

To conclude, the imperfections of our medicine today also spring, basically, from medical naturalism. The successes and failures of our university medicine have therefore a common origin: they are the light and the shade of the same scientific current; and they both originate from the same single cause, which is revealed to the modern observer in two distinct aspects. The first concerns only medical science; the second, medicine in its entirety. (1) Modern medical naturalism demands that medical science be strictly experimental: 'Experimental medicine or scientific medicine', said Claude Bernard.[9] But the experimental method, as conceived by the physicists and chemists at the beginning of the last century, is exclusively analytical, as I mentioned a short while ago. Now, the doctor cannot make do with a method which breaks down all morbid cases into *membra disjecta*: he must, in addition and at all cost, be provided with a method of reasoning capable of coordinating the fragments. In 1948, in my first Hippocratic study,[10] I showed that towards the year 400 BC, the School of Cos devised such a method which was both different and complementary to experimental reasoning and could in addition be used as it was by the modern clinician, i.e. independently of the Hippocratic pathology, now obsolete for us, but within which it was originally formed. Since then, more than thirty years have passed, but no doctor has attended to these observations, and I am not speaking of historians. All are still prisoners of the new Cartesian and Bernardian 'scholasticism' which I had already mentioned for the first time in 1951/1952.[11] (2) On the other hand, while it may be natural for a physicist or chemist to consider his discipline as a science, and only as such, the doctor cannot adopt the same attitude towards medicine. The unrestricted incorporation of medicine into a science reduces and 'curtails' it in an intolerable way. It is exceptional for patients to appear before us as simple *res extensae* (this may occur, when they regard their own illness as if it were a car breakdown); and for the patient the doctor must always be more than a *res cogitans et agens*, even if this irritates him! What we are confronted with, however, are twin confusions: (a) between scientific medicine and experimental medicine, (b) between medicine as a whole and medical science. There can be no doubt that owing to this double confusion medicine has experienced a pragmatic upsurge unparalleled over a thousand years of progress. But now we are also experiencing the inconveniences of this upsurge, and in consequence are being forced at last to see things as they are: the medicine we teach and practise today is in reality nothing but the gigantic and disproportionate hypertrophy of a single line of research which for more than a century has been confused with what I have called modern medical naturalism. Here, a historical 'law' is being verified once again: All great advances are unilateral; progress is always made at the expense of something.[12] Modern medical progress, like all violent upward movements in history, is borne forward by an autonomous, tyrannical and irresistible dynamism, the dangers of which we have felt ever since the moment when the unfavourable side-effects became so flagrant that they could no longer be ignored. Thus it is that, in spite of the unique inventions and discoveries, a wind of unease is disturbing our medicine and that today it is increasingly – and better – typified by crisis than by progress.

I have not yet spoken of the worst danger by which we are at present threatened: the breaking up of our medicine, its splitting into two forms of medical practice of more than unequal value. On the one hand, a medicine at an extremely high scientific and technical level, Cartesian and Bernardian to the supreme degree, benefiting from all that has been won by successive industrial revolutions, interventionist and efficacious, a medicine of engineers, of specialists, of apparatuses, of computers, preferably concentrated in the large hospitals; and on the other hand, a sort of medical pigsty, in which the best must rub shoulders with the worst, and whose representatives spring from the most heteroclite backgrounds: university-trained specialists and generalists, qualified 'heretics' or freebooters propagating the doctrines and methods of a thousand disparate sects – charlatans, in fact. Two medicines, then; the one of high standing, the other of extremely uneven worth – as in the scholastic Middle Ages and at the beginning of the modern times, between 1200 and 1800, when there were on the one hand the university 'doctors' and on the other the trade guilds of the surgeons and barbers, and the bonesetters as well. As far as I know, this medical compartmentalization has never been regarded as beneficial. Yet in our times, too, the patients switch from Descartes to the diviner and to yoga! Is this a further victory for modern medicine? No, it is a disaster, and one which can scarcely be remedied, for the engineer-doctors are caught up in their experimentalist, analytical and 'quantificatory' ideology, while extra-hospital medicine is becoming a hundred-headed monster incapable of dialogue. The ditch is being dug, but the bridgemakers are nowhere to be found. We should not paint the devil on the wall, but these separatist tendencies are becoming manifestly clearer before our eyes; and, as a result, all are suffering – doctors, patients and society.

Provisional Conclusion: Modern Medical Progress has Cost Us Dear

Let us conclude this examination of modern medical naturalism. By transforming our classical faculties of medicine into polytechnical colleges for engineer-doctors, we have gained plenty and lost much. We have gained a powerful mastery over detail in our 'specific' operations. But if we step back far enough to view our profession in its largest sense, we realize with dismay that the medicine of the polytechnic no longer coincides – far from it – with the full field of medicine as it is practised today. At the time of Alfred Nobel, there were in addition to the faculty doctors no more than a few homeopaths, naturists and the inevitable quacks; today, by contrast, our profession is splitting up in a most disquieting way.

In my opinion, we took things too far after the Second World War in banning medical 'art' from medicine, under the pressure of biochemical progress and the miraculous Anglo-Saxon and Scandinavian therapeutic advances. For centuries, this 'art' had, as it were, enveloped medical science, allowing it full freedom of action, but hurrying to its aid whenever science ran up against its own limitations at the patient's bedside. The traditional practitioner, with his background in the humanities, was capable then of engaging in a heart-to-heart talk with the patient, a dialogue which would bring him aid. This medical wisdom, nourished by the Greek, Roman and Christian classics, took the place of an ailing medical science.

Today, by contrast, we know and are able to do incomparably more than our humanist predecessors; but when we too come to the end of our means we are caught short and tend to abandon our patients to the narrow, sectarian views of the psychotherapists and to the speculations of the 'heretics'. This is not, of course, to suggest that all their opinions are mistaken; and indeed the patient's faith can shift mountains. Generally speaking, however, our scientific 'asepsis' leads ultimately to nothingness. The laboratory has killed the man in us; the new philosophy – mechanist, activist, but rough and unfinished – has left huge gaps in medical practice, gaps which will become even more serious once the older generations of doctors have departed from the scene. For a long time I have been saying that in the modern race for medical progress the individual personalities of the doctor and the patient have been forgotten.[13] It is not the pioneers – Magendie, Claude Bernard, and in Germany the pupils of Johannes Müller – who are responsible: what they wanted was a more thorough and effective medical science; their successors, on the other hand – those 'specialized barbarians' foreseen by José Ortega y Gasset[14] – lacked the necessary breadth of vision. And when science reaches the point of self-worship, decadence is close at hand. What we need is a medical science and a medical 'art', and unless we are able to re-establish the rights of this art we shall be plunged into crisis.

All this, moreover, serves to explain the present difficulties of medical ethics, for the experimental method as such remains silent as to the limits of its own application. It also serves to explain – and I return to the issue – why public opinion always places 'the doctor' above other professions, in that it is simply projecting on to him its aspirations towards a less unsettled world; while 'doctors' are subjected to mounting criticism. A thousand contradictory reasons are advanced to explain this paradox; in what was said above I endeavoured to extract the one, deep-lying cause. Yet another dire effect of this mechanist, materialist philosophy is that in several Western countries medicine has already been socialized, hence institutionally depersonalized; and in the Soviet east most of our colleagues are little better than workers in the state health service – and badly paid at that; poor patients! All in all, modern medical progress is a benefit for which we have had to pay dearly indeed. A satisfactory reform of our profession now requires the advent of a revived Humanism: the world does not begin with the Enlightenment and its favoured masters! But, for the present, East and West alike are equally blind in this respect. Are we entering, then, irrevocably, a new nadir in history?

The Happy State of Medicine at the Time of Alfred Nobel: a Brief, Preliminary Outline

A representative view of the state of medicine and physiology at the time of Alfred Nobel could not have been gained without having made a preliminary survey of the modern and contemporary medical movement (which I have also called 'post-modern'[15]) from its origins at the start of the last century up towards the year 2000. It was necessary to recognize where one had come from at the dawn of the 'Belle Époque' and to observe the effects of this movement right up to our own times: in short, it was necessary to draw up the general balance-sheet of modern medical naturalism, with its profits and its losses. But if now we are to consider the decades of medicine and physiology that preceded the institution of the Nobel prize in 1901, we are led to an observation which requires me to anticipate somewhat in order to direct and locate our ideas: The brief period between 1850 and 1900 – which may fairly be extended to 1914 and even a little beyond that date – marked one of the rare moments of happiness, balance and harmony in the entire history of medical science and profession. We must not, of course, go too far in this: by comparison with our own, the therapeutic means of the time were poor, but nobody then could have imagined – not even in his most extravagant dreams – the magnitude of the discoveries to come. Above all, however, an exceptionally propitious conjunction of circumstances led doctors to a point which it is difficult – for us this time – to conceive; and this it is which forms the backdrop, so to speak, to the first Nobel prizes. Let us briefly review the major elements of this propitious conjunction.

The medical faculties, which during the Napoleonic era had been refashioned in several countries, were now intact. The university was celebrating the second high-point in its history, six or seven centuries long. The first peak had been reached with the work of the Church and the mediaeval clerks: let us not forget the role of arbiter conferred upon the University of Paris during the Great Schism of the West (1378–1429)! The new zenith of the university coincided with the rise of the *bourgeoisie* in the nineteenth century and with the hundred years of peace – from 1813 to 1914 – assured for Europe by the great politicians of the Vienna Congress, Metternich and Talleyrand. The Congress of Berlin in 1878, with Bismarck presiding, was almost a replica of the Vienna Congress in its pacificatory intents. Europe at that time seemed to be dominating the world, forever and in all aspects. After 1914, however, we entered into a new Hundred Years War in which the millions who died are no longer counted – the most criminal period in all human history. Today, it is fashionable to decry this nineteenth-century *bourgeoisie*. But it should nonetheless be recalled that, at the time, this *bourgeoisie* was the active wing of society in the most diverse spheres: in science (modern medical naturalism with its optimistic and pragmatic philosophy is a bourgeois product!), in economics and in culture. In the cities, this was a new golden age for the family doctor: lavishing care and advice and tending without charge to the needy, the 'doctor' was regarded with an esteem that equalled or even surpassed that of the university professors, the priests and clerics. Medical science, moreover, was purified. The system-bound spirit and the romantic chimeras of the beginning of the century were no more than a bad memory: they had been replaced by a scientific empiricism that was both positivist and realist. This was also the age of the masters of what was later to be called 'the Classical clinic': Trousseau, Dieulafoy (evoked by Marcel Proust), Widal, Schoenlein, Kussmaul, Naunyn, Garrod, Osler – all masters of medical science and of medical 'art', perceptive and cultivated diagnosticians such as we no longer are. Medicine still enjoyed – though not for much longer – all the advantages of a free and truly independent profession: but for a few exceptions, medical science was cultivated by none but the doctors themselves; the pharmaceutical industry was still in its very beginnings, and it was still with the pharmacist that the doctor collaborated; compulsory health insurance with all its encumbering paperwork appeared only at the end of the century; there were few trials, or none at all, no unions or politicians or malicious journalists. . . .

Yet another favourable element to be considered is that this same medical science, just at the time of Alfred Nobel, was making astounding scientific and practical advances! Once again, we need to make an immense effort of the imagination to reconstitute in our minds the medical reality of those times. For thousands of years, infectious diseases and epidemics had dominated the 'morbid panorama', on account of their extreme frequency and their seriousness, without it ever having been possible for their aetiology to be explained. Theories, each less sound than the last, were bandied about; the contagionists fulminated against the anti-contagionists, who were no less vituperative in turn. Proper microscopes were lacking, so too was the brilliance of Davaine, Pasteur, Koch and their pupils. The end of the century, however, was to see their work, which was one of the milestones of world history (one day, perhaps, as much will be said again when more is known about cancer). Medicine emerged shaken and transformed: pathology, nosology, therapeutics, hygiene, prophylaxis. Pasteur confirmed Semmelweis' empirical and statistical observations, and produced the antirabies vaccine; in 1901, von Behring was to receive the first Nobel Prize in Medicine or Physiology for his antidiphtheria serum. And, last of the advantageous factors: science and technology were appearing only in their best light; 'Hiroshima' was not yet a word to be heard in Europe, and the pleasure boats on the Rhine were still serving salmon that came straight from the river itself. And much more: scholars and scientists were exulting – sometimes to the point of becoming overweening – yielding to the illusion of the definitive and believing themselves close to omniscience. Emile Zola was inventing the 'experimental novel'; and for Virchow, the new medicine was to open the way towards a new politics. . . .

In short, this was a blessed time for medicine and for doctors. Do I exaggerate? Not at all! Listen to the contemporaries speaking. Here is what the German clinician Fritz Munk said in his recollections of the Berlin medical world around 1900:

> '. . . in the second half of the nineteenth century, at the apogee of positivism, the harmony existing in the relations between medical instruction, the practitioner's activity, and the confidence of the people in his work was without doubt unique.'[16]

And Adolf Kussmaul, his elder, was even more enthusiastic when he wrote in his *Memories of Youth* (1899):

> 'I consider myself privileged to be a child of this century [the nineteenth], for none, or almost none, of the innumerable centuries which have been engulfed by time has so much deserved the gratitude of mankind. None has shown so much courage and talent, delved so deeply into the mysteries of nature, or promoted the public welfare with so much imagination and success, with such enhancement and enrichment of life, and none has so resolutely and effectively struck off the fetters of enslavement in all parts of the world.'[17]

These *Memories* must be read to be believed, yet they went through a dozen successive re-editions, which means that they captured the tone of the times. One could go on with the recital of such eulogies, and it would indeed be revealing. Men, of course, remained men: Pasteur, Koch, Ehrlich, all had to fight throughout their lives. Nor was there any lack of 'disturbing' minds: Kierkegaard, Jacob Burckhardt, Dostoevsky and Nietzsche, who all in their own way foretold what later was to be shouted with 'a raucous, sinister voice' by Jaco the parrot in the stories of the French historian Jacques Bainville: 'Ça finira mal!' ('It'll end badly').[18] But the voices of these Cassandras were lost in the desert. When, on 15 October 1913 in Berlin, Max Planck in his Rectoral speech[19] recalled the importance of the modest everyday struggle to his colleagues and students assembled after joyous commemorations of the first centenary of the victory at Leipzig over Napoleonic despotism, there seemed to be no cloud upon the horizon. Less than a year later came the assassination at Sarajevo with its fatal consequences: the world had tilted back into a true 'Middle Ages', an intermediary age, an 'inter-regnum' from which we are not close to emerging.

One may understand, then, that I should have spoken of a *medicina triumphans*[20] to depict the state of our science and our profession around the year 1900 – a medicine which Alfred Nobel was to know towards the end of his days. This judgement may well be a surprise to more than one of you, since our own achievements have in many respects been yet more astounding. But our triumphs, alas, have also been countered by resounding failures. Medicine in the era of the 'Belle Epoque', however, knew nothing of such discords; it was – if such a thing exists in human affairs – an apotheosis, which coincided with what is called the 'bacteriological era'. For us, this premature apogee is one of the great surprises of the history of modern medicine in the proper sense. We shall see, shortly, that there were also others.

The Historical Position of this Medicine. The Anatomo-clinical Tradition (1800–1939) and the Progressive Incursion of Experimental Medicine – the 'Laboratory'

Let us now endeavour to define more closely the historical situation of medicine at the time of Alfred Nobel, for, actually, neither the experimental revolution of François Magendie at the beginning of the nineteenth century nor the zealous work of the bacteriologists during its last decades can provide an adequate picture of the situation. 'Quite the opposite!', one is tempted to say, so different were the principles by which medical practice in the hospitals and private consultation rooms was imbued. In the medicine of that time, experimental reasoning and laboratory research played an almost negligible role. Does not this seem paradoxical? What had happened that might clarify this apparent contradiction for us? This is what we must now look at; and, as I have already said, other surprises await us here.

Magendie had seen correctly, let us make no mistake. His experimental medicine, conceived after the model and upon the basis of the new natural and experimental sciences, physics and chemistry, was not only a daring but also an unusually fertile innovation! The promulgators of our molecular biology and pathology are all – whether they know it or not – continuing the work of the great French physiologist. It is he who made the Revolution in medicine, and not Paracelsus, Vesalius, or Harvey, as has been repeated so often hitherto. Modern medicine in the proper sense was thus born in opposition to the medicine dominated by Hippocrates, Galen and the medical systems of the sixteenth, seventeenth and eighteenth centuries – which is a fresh surprise for us. Far from being swayed by the idea of continuity, Magendie and his first pupils in Paris, Berlin and Leipzig were dominated rather by the ideas of *tabula rasa* and antithesis; this resulted in a counter-reaction to these upheavals, and partly serves to explain why, 150 years later, modern medical science and practice should be of such a clashing and highly contrasted character.

But Magendie was right too early. Physics and chemistry were not yet ready in his time to perform for physiology and pathology the pilot role that had been assigned to them. Listen to the words of our Jacobin: 'It is in the study of vital physics [!] that the future of medicine lies.' Vital physics is nothing but 'the (experimental) physiology of the sick man'.[21] As we well know, biochemistry and biophysics did not begin to move towards their present heights until the middle of our own century! Medicine, however, is a practical science and cannot wait. There were patients to be attended to, and students to be taught. Hence the third and last surprise: it was not the experimental medicine of Magendie, Claude Bernard and the pupils of Johannes Müller which made the direct transfer from Hippocratism-Galenism and the medical systems of the eighteenth century, but another form of medicine which, as subsequent history was to show, assumed the function of an intermediate medical tradition between the two periods.[22] One might describe it as hospital medicine, but I prefer the more explicit expression of the anatomo-clinical tradition. Actually, at the very moment when Magendie in Paris was laying the foundations of his experimental medicine of such rich future prospects, a medical tradition was being set up – again in Paris (for the City of Light was to be the Mecca of medicine between 1800 and 1840) – a tradition which the older among us would still have known at the time of their studies, one based essentially on pathological anatomy and the clinical examination of patients: on the one hand, organic lesions, on the other hand, physical signs (Bayle), deriving from observation, palpation, percussion (Auenbrugger, Corvisart) and auscultation (Laennec). *De sedibus et causis morborum per signa diagnostica investigatis, et per anatomen confirmatis*: thus it was that around the year 1800 Corvisart brought up to the present the older views of Morgagni.[23] Between 1838 and 1839, the discovery of the cells; and in 1858, Rudolf Virchow's celebrated *Cellular Pathology*: macroscopic pathological anatomy had been joined by pathological histology and cytology. It was above all a descriptive medicine and hence 'sterile' – to use Magendie's critical expression –, since it could not comprehend 'the mechanism of morbid alterations', aetiology and pathogeny.[24] But it had an inestimable advantage over its rival, experimental medicine, an advantage it retained for a long time: it was open without delay to great advances. Pathological anatomy and histology and clinical semiology were immense territories to be prospected. There was no need to wait here. Cellular pathology represented the second general and positive theory of medicine, between that of Hippocrates and Galen and the molecular pathology of today.

Having thus marked out the way, we can from now on begin to form a proper idea of the development of medicine from around the year 1800 until the Second World War. There are two simultaneous evolutionary movements that must be distinguished, even though they ultimately interlock. On the one hand, anatomo-clinical medicine progressed by its own strength, acquiring increasing magnitude and dominating the entire nineteenth century and even the beginning of the twentieth. It was the age of Laennec and Virchow. German as a language was gradually gaining the upper hand in medical literature and did not lose its sway until after 1939–1945 in the face of the English of the Anglo-Saxons and the Scandinavians (as is revealed by our Symposium). On the other hand, this hospital medicine was to be progressively encroached upon, until it was ultimately overthrown by the 'laboratory', by 'Magendie' and his experimental medicine. This incursion took place in several waves, four in all, each successively continuing to expand once it had gathered momentum. These four waves may be described as: clinical chemistry, from 1840; bacteriology, from 1860; pathological physiology, which became established as an autonomous discipline only with difficulty, towards the end of the century; and finally, after 1900, a wave of partial doctrinal syntheses – neuro-vegetative dystonia and physiological surgery, for example – emerging within a medicine in which the functional tendency was becoming, not without difficulty, hegemonic.[25]

We realize now that to understand medical science and medical practice at the time of Alfred Nobel from the outlook of the general history of medicine means determining the place of this science and practice within the context of this double evolution. For us today, this evolution has long since been finished. Magendie has won the battle. The laboratory – 'physiologism' – has become so thoroughly imposed upon classical hospital medicine as to have rendered it unrecognizable. The organic lesions of Corvisart, Bayle, Laennec – lobar pneumonia, cirrhosis of the liver, etc. – are no longer the focal-point of our preoccupations; we consider them rather as the effects and sequels of disorders – far more important to science – all of which, invisible to the microscope, began solely at the functional level.

Galileo, Descartes, Magendie and Claude Bernard were the great victors, and at the beginning of this paper I brought out the advantages and disadvantages of their victory. Between 1860 and 1900, by contrast, the situation was quite different. Numerous advances were made in pathological anatomy and clinical semiology. The laboratory, however, remained discreetly withdrawn. Clinical chemistry had only just emerged; and even in 1861, Armand Trousseau in his celebrated *Medical Clinic of the Hôtel-Dieu of Paris* warned his students against the 'exaggerated intermingling of the physico-chemical sciences in our art', which 'has done so much harm and could so regrettably mislead young students of medicine that, in spite of myself, I find myself exaggerating the danger and distancing you from studies from which you have nonetheless derived much useful information'.[26] In Vienna, the great Skoda shared this opinion. Had they foreseen our present deadlock? And towards 1900, bacteriology led medicine to a triumph.

Did all this come from the side of the experimenters? Far from it! The end of the nineteenth century was also one of the great ages of physiology, in which Carl Ludwig, Marey and their emulators multiplied graphic registering methods in order to raise the discipline to the status of an exact quantitative science on the lines of physics and chemistry. Physiology then became the top science of the new experimental trend. More than 200 pupils had passed through Ludwig's laboratory in Leipzig and gone on to spread their knowledge throughout the world[27] – as a result of which there are Nobel prizes in medicine 'or physiology'! Progress was also made in experimental pharmacology, of which Magendie was likewise one of the pioneers. And in hygiene, which in turn became an experimental science, with Max von Pettenkofer (1818–1901) in Munich. But the two lines of development did not yet interfere, or very little. In scientific and hospital practice, it was still the anatomo-clinical tradition that prevailed. Worse still, the clinicians regarded laboratory work with mistrust – Pasteur and how many others were to suffer this depressing experience; but this was understandable, for the defenders of hospital medicine had always been terrified of the medical systems and romantic chimeras of the past: they wanted facts, and more facts, and the least possible 'reflexion'! And, it must be admitted, the experimenters' conclusions were not always accurate: *errare humanum est*. We should not therefore be surprised that pathological physiology and experimental pathology should have stagnated, producing nothing but sporadic research, and particularly that they should have remained under the charge of pathological anatomy, preferring to devote themselves to the analysis of functional disorders resulting from irreversible organic lesions such as valvular defects, renal sclerosis, gastric ulcer and Basedow's disease. On 17 September 1900, the German clinician Bernhard Naunyn delivered one of the many contemporary 'centenary speeches' (1800–1900), in which he enumerated all the medical sciences that were flourishing at the end of the century: anatomy, physiological chemistry, pathological anatomy and the various different clinical disciplines – but he 'forgot' pathological physiology![28] And when, between 1936 and 1939, I was doing my clinical studies in Lausanne, most of my teachers had remained at this point. Instead of endeavouring to assimilate the new experimental discoveries, the pontiffs of the Faculty contented themselves with establishing artificial antitheses between clinic and laboratory. As I said in my *History*, false Coans were engaging in polemic against false Cnidians.[29] It took decades for pathological physiology and functional pathology to liberate themselves from pathological anatomy and become regular research and teaching disciplines; in surgery, the same liberation movement was linked to the name of René Leriche. And since the war, new partial doctrinal syntheses – psychosomatic medicine, stress – have also been developed on a functional base.

The Characteristic Traits of Medicine Shortly before the 'Belle Époque'

Let us move closer in on our subject: What are the characteristic features of this anatomo-clinical medicine which dominated the scene at the end of the last century? It is instructive to compare them in our minds with the traits of our own century. Medicine 'under Alfred Nobel' was morphological, as were almost all the sciences of life in the nineteenth century: anatomy, comparative anatomy, systematic botany and zoology, embryology, ontogenesis and phylogenesis. One concentrated on the forms of beings and of their parts, not only out of necessity but also from taste and predilection (and today we are timidly returning to this in our 'functional cellular morphology').[30] In medicine, one was observing organic lesions 'chez Morgagni' and the typical clinical pictures of morbid entities in the sick wards.

This medicine was also nosological; we, however, have been attempting for four or five decades to break down the clinical manifestations of diseases into 'syndromes' (one of the most poorly defined terms in the medical vocabulary). Trousseau devoted an entire chapter of his *Medical Clinic* to the specificity of morbid entities, shortly to be substantiated by the bacteriologists' observations: acute lobar pneumonia is produced by a different pathogenic agent from pulmonary tuberculosis! And finally, in this predominantly nosological medicine, diagnosis was preferred to the Hippocratic prognosis of individual morbid cases. As each nosological species had its own prognosis, knowing the diagnosis meant *ipso facto* becoming acquainted with the customary pattern of evolution. The great clinician of this age was a skilful diagnostician. He investigated organic lesions and their causes, which might be microbial, physical, chemico-toxic or other causes; heredity and individual constitution did not become of current concern until after 1900. Diagnosis was based upon anamnesis, the *status praesens* and the laboratory (clinical chemistry, microscopic and serological examination). This trio, which is familiar to us, gradually began to form as from 1840. Nosological diagnosis also dominated over therapeutics, as had Hippocratic prognosis, its predecessor – and for the same reason: whether they were dealing with ancient 'fevers' or with new organic lesions, the doctors remained very much at a loss in confronting the illness. They had reduced to *tabula rasa* the pharmacopoeia of the *Ancien Régime* as well as the old medical doctrines (but the young Claude Bernard still went on lambasting theriacs!). The number of truly active remedies, both traditional and recent, could have been counted on the fingers of one hand. Even at the beginning of our century, two French doctors published a *Therapeutics with Twenty Medicines*![31] The fashion of the time was towards scepticism and even towards therapeutic nihilism. Internal medicine was taken to be the science of incurable diseases. The 'grand patron' pronounced the diagnosis, then unconcernedly entrusted the treatment to the family doctor; the older among us will not have forgotten this.

An exception to this were the surgeons. Owing to the advances in pathological anatomy from which they benefited, they were able at last to operate on the internal organs as well, thanks to the inventions of narcosis and asepsis, and their prowess made them seem like demigods. And to think that at the beginning of the nineteenth century they were still excluded from the faculties!

But for medicine as a whole, the decisive turning-point in the modern history of the treatment of illnesses came with the introduction of the new bacteriostatics, Prontosil, Cibazol, penicillin and streptomycin. It was under the impression of their prodigious – at the time almost miraculous – effects that in my first Hippocratic study, back in 1948, I exclaimed to myself: 'A hundred years of pathology is being followed by an age of therapeutics',[32] and today there can no longer be any doubt of this fact.

Complementary Remarks

Let us now consider certain additional features of this age of pathology between 1800 and 1939. Anatomo-clinical medicine observed organic lesions and clinical signs: that was its vocation. Claude Bernard was to criticize this medicine for its 'passive' observation, so less fruitful than the 'provoking' observation of the experimenters.[33] This was a science at a lower level, a purely descriptive science, which made up in its rigorous approach to minutiae for what it lost in general insights. It was also a medicine of disease rather than of health. In fact, the first age of pathology saw the successive emergence of: pathological anatomy, cellular pathology, bacteriology, serology, genetics (hereditary diseases) and psychoanalysis! Public health was, undoubtedly, improving; but there was still little question of personal hygiene and daily nutrition patterns, i.e. of dietetics in the broad sense. It was, moreover, a medicine in which the scientific and social advances (which, being slower than our own advances, were therefore more readily assimilated) still today remain unchallenged. It is worth mentioning at least a few of the social improvements: the institution of the general hospital was expanded; polyclinics, often private at first, were created for the different medical specialities, and lay nursing schools were set up; in 1883 and 1884 Bismarck enforced epoch-making laws concerning insurance against ill health and accident. In the domain of the university, the already medieval tradition of the medical *grandes écoles* had been continued through Humanism, Absolutism and the Enlightenment until the previous century (Paris, Berlin, Vienna, Zurich), and even until the Second World War, before ultimately ending up in our present medical cosmopolitanism with its dual Anglo-Saxon and Scandinavian centre of gravity. (The Hellenistic and Arabic medicines were likewise cosmopolitan.) It was, then, a humanist medicine in the relations between doctor and patient, in which the practitioner's experience, classical education and common sense played cardinal roles, and indeed a very artisanate role in the different surgical disciplines – if need be, the operator would repair his instruments in mid-operation.

Three general remarks on the medical period between 1800 and our days

Let us conclude with three remarks.

(1) As Erwin Ackerknecht has rightly observed, hospital medicine in the nineteenth century – that intermediate medical tradition between the old Hippocratism-Galenism and modern medical experimentalism – is the most 'concrete' medicine there ever was.[34] With regard to catarrhal infections, Hippocratic morbid matter did indeed command attention, but with other fevers it tended rather to remain concealed. And medicine today, entangled as it is in figures and graphs, is becoming abstruse even to the doctors themselves. By contrast, nothing could be more palpable than good organic lesions and their physical signs! Let it be added that they were both integral parts of a fundamentally qualitative medicine.

(2) In the conflict we have described between 'clinic' and 'laboratory', it is noteworthy that the protagonists of the two scientific positions were defending two different visions of the world. And, strange to relate, they sprang up almost at the same time – in the latter days of the Enlightenment – and are thus both modern in spirit. Without being fully aware of it, the purist clinicians of the nineteenth century were sensualists in the wake of Locke, Sydenham, the Encyclopedists and the group of the Ideologists. They aimed, after their own fashion, at being as rigorously scientific as the physicists and chemists of their times, but they were firmly attached to strict observation at the patient's bedside and in the autopsy room. This whole sensualist, sceptic and empiricist current was conceived as a reaction against the system-bound spirit of the preceding centuries. On the other hand, as we saw at the beginning, the laboratory doctors were the successors of Galileo, Descartes and Francis Bacon. They too used observation – for one must indeed begin there – but they wished to go beyond the wall of appearances, of *phainomena*, to discover the laws and mechanisms of nature in order subsequently to modify it and gain mastery over it. Thus it is, by reaching down to the metaphysical and epistemological roots of this antagonism between clinic and laboratory, that we discover its real underlying causes. The medical nineteenth century was not as exclusively scientific as it believed itself to be in its phobia towards the divagations of old. Just as Hippocratism-Galenism had been subjected to the influences of the pre-Socratics, Plato, Aristotle and the Stoics, the positivist doctors of the last century were in some instances sensualists, in others Cartesians and Baconians. And we today must be aware of this in order to be able finally to shake their prejudices.[35]

(3) A third phenomenon is likewise worth commenting upon. Between roughly 1800 and 1840, medicine was dominated by the Paris School, with its macroscopic pathological anatomy and its physical examination (observation, percussion, auscultation); between 1840 and 1939, it was the German

School, with its cellular pathology, its full clinical examination of patients, and its clinical chemistry and microscopy; since 1945 up till the present day, the Anglo-Saxon and Scandinavian countries have been imposing upon the whole world a distinctly technical and experimental medicine based upon the new basic sciences of biochemistry and biophysics. These conclusions should, of course, be tempered by considerations of detail, but they nevertheless hold true if one considers the matter as a whole. One is led, then, to wonder why medicine chose to develop along these lines. Why were these two leaps made, from one country or geographical area to another, and from one form of medicine to another? For each time, both in the Paris and in the German School, there were far-seeing minds capable of ensuring the transition towards the subsequent medical stage in their very own country! Why then did they remain largely unrecognized, and why did they not succeed in winning over the conviction of the majority of their compatriots? It all occurred as though the epoch-making advances achieved in a specific line of research ultimately caused the doctors to become blind witnesses to advances of comparable importance, yet more original, destined to supplant the progress already made, without, however, making it anachronistic; i.e. it was as though past progress represented a mortgage on future science, forcing it to be transformed in other places and under the pressure of foreign innovators. And since the most recent form of medicine ('post-modern', since 1945) has become 'globalized' – with all its lighter and darker aspects – one is faced with the question: from where will the initiative for a new medical movement begin? From isolated individual efforts, scattered almost throughout the world, and based on the model of historical Humanism?[36] But I am not a prophet.

Epilogue: the Exemplary Character of Medicine at the Time of Alfred Nobel – an 'Optimal' Medicine

In this medical movement, lasting almost 200 years, from 1800 until today – which as a historian I have tried to outline briefly for you in as many of its aspects as possible – the medicine of Alfred Nobel's time stands out as a truly high peak, a moment of serenity. It was a medicine with the maximum of advantages and the minimum of disadvantages, and one which in essential respects enjoyed great unanimity. It was a medicine free of imbalance; all in all, an optimal medicine given the times and the human condition of the age. In saying this, I am not attempting to idealize or to play the apologist. Without our ultra-specialized and technical medicine, I should not be here speaking to you, for I was saved by 'post-modern' medicine from a grave illness. We now know too much and are able to do too much, and we cannot sacrifice ourselves to nostalgia and wish reasonably for a return to the past. No, no matter how favoured this medicine of the late nineteenth century may be in so many other respects, it is from a different point of view that it is of importance to us today. There is no need to return now to the modern medical crisis, but we are all aware that it must be overcome in one way or another. We do – globally speaking – need another medicine: the uncouth engineer-doctor's smattering of psychology and sociology is a heresy; medical practice must be changed so that we have a better balance between hospital clinicians, specialists and general practitioners. It is chiefly the latter who must conquer the wide field that university medicine has left to the 'heretics'; but they must be trained to do it. As for our patients – many of whom destroy themselves through a perverse health regime, succumbing to the materialist illusions of the 'right to health' and of an omnipotent medical science –, they too will have to impose on themselves a *metanoia.* But in the midst of all this confusion, the medicine of the years 1860–1900 suddenly emerges as exemplary for us: in the balance it struck between medical science and medical 'art', in the harmony that then existed between the 'doctor's' means and the patient's aspirations, both equally modest. Indeed, but for the most benign cases, diseases strike us in a way that cannot be expressed in mechanistic terms, and the individual human fate resists being reduced to equations. How can doctors imbue their patients with the principles of their medicine if they themselves no longer understand it?! Here a vulgar and simplistic Cartesianist finds himself caught in a dead-end. We may measure the difficulties ahead of us, but we will not escape from the present medical crisis unless we grapple with these difficulties. Otherwise our official medicine will become a sort of scientific sect. I am more and more of the impression that the responsible authorities do not even meet the problems that they are supposed to solve.

Alfred Nobel did not himself live in a state of such alarm. He was encouraged by the state of the natural sciences and the medicine of his time. He followed their progress with enthusiasm, and it was in order to make his own contribution to this progress that he established the prize which is the reason for our being gathered here together.

Notes

1. Lichtenthaeler, C. (1977) *Geschichte der Medizin*, 2nd ed., Vol. 2, Cologne, Deutscher Ärzte, pp. 481 ff.
2. Lichtenthaeler (note 1), Vol. 1, p. 38.
3. Bernard, C. (1865) *Introduction à l'Etude de la Médecine expérimentale*, Paris, J. B. Baillière, p. 6.
4. Bernard, C. (1937) *Pensées, Notes détachées*, Paris, J. B. Baillière, p. 378.
5. Bernard, C. (1880) *Leçons de Pathologie expérimentale*, 2nd ed., Paris, J. B. Baillière, p. 438.
6. See Taillens, J.-P. (1976) Leçon d'adieu. *Rev. méd. Suisse romande*, **96** (6), 425.
7. Lichtenthaeler (note 1), Vol. 2, pp. 503 ff.
8. Or, more exactly: *nam et ipsa scientia potestas est.* Bacon, F. (1597) *Essayes. Religious meditations. Plaies of perswasion and disswasion.*
9. Bernard, C. (1878) *La Science expérimentale*, Paris, J. B. Baillière, p. 48.
10. Lichtenthaeler, C. (1948) *La Médecine hippocratique. I. Méthode expérimentale et*

Méthode hippocratique: Etude comparée préliminaire, Lausanne, Gonin, pp. 39 ff., 55 ff.
11. Lichtenthaeler, C. (1952) The dates of the medical Renaissance. The end of the Hippocratic and Galenic tradition. *Gesnerus*, 9, 16. The first 'medical scholastic' (in the pejorative sense) was, it should be recalled, Aristotelian and Galenic.
12. Lichtenthaeler (note 10), p. 21; and note 1, Vol. 1, pp. 40, 265; Vol. 2, pp. 655, 685 ff.
13. Lichtenthaeler (note 1), Vol. 1, p. 42.
14. Ortega y Gasset, J. (1952) *Schuld und Schuldigkeit der Universität* [1930], Munich, R. Oldenbourg, p. 23.
15. Lichtenthaeler (note 1), Vol. 2, p. 609.
16. Munk, F. (1956) *Das Medizinische Berlin um die Jahrhundertwende*, Munich-Berlin, Urban & Schwarzenberg, p. 36.
17. Kussmaul, A. (1902) *Jugenderinnerungen eines alten Arztes*, 5th ed., Stuttgart, A. Bonz, p. 3 (Einleitung).
18. Bainville, J. (1927) *Jaco and Lori*, Paris, B. Grasset, p. 58.
19. Planck, M. (1949) Neue Bahnen der physikalischen Erkenntnis. In: *Vorträge und Erinnerungen* [1913], 5th ed., Stuttgart, S. Hirzel, p. 69.
20. Lichtenthaeler (note 1). Vol. 2, pp. 561 ff.
21. Magendie, F. (1831) *Leçons sur les Phénomènes physiques de la Vie*, Vol. 1, Paris, J. Angé, p. 310; Magendie, F. (1825) *Précis élémentaire de Physiologie*, 2nd ed., Vol. 1, Paris, Méquignon-Marvis, p. xii; Lichtenthaeler (note 11), p. 24, notes 74 and 75.
22. Lichtenthaeler (note 1), Vol. 2, pp. 516 ff.
23. Corvisart, J. N. (1811) *Essai sur les Maladies et les Lésions organiques du Coeur et des gros Vaisseaux*, 2nd ed., Vol. 1, Paris, H. Nicolle, p. xvi.
24. Magendie (note 21), Vol. 4, p. 6; and Lichtenthaeler (note 11), p. 24 and note 76.
25. Lichtenthaeler (note 1), Vol. 2, pp. 529 ff., 534 ff.
26. Trousseau, A. (1861) *Clinique médicale de l'Hôtel-Dieu de Paris*, Vol. 1, Paris, J.-B. Baillière, p. xv.
27. For the list, see Schröer, H. (1967) *Carl Ludwig*, Stuttgart, Wissenschaftliche Verlagsgesellschaft, pp. 287 ff.
28. Naunyn, B. (1900) *Die Entwickelung der Inneren Medizin mit Hygiene und Bakteriologie im 19 Jahrhundert. Centennialvortrag*, Jena, G. Fischer, p. 17.
29. Lichtenthaeler (note 1), Vol. 2, p. 569.
30. Schröter, W. (1972) Intracellulärer Transport und Membran des endoplastischen Reticulums der Leberzelle. *Mschr. Kinderheilk.*, **120**, 119 ff.
31. Huchard, H. & Fiessinger, C. A. (1910) *La Thérapeutique en vingt médicaments*, Paris, A. Maloine.
32. Lichtenthaeler (note 10), pp. 21, 27 ff.
33. Bernard (note 3). Ch. 1.
34. Ackerknecht, E. (1967) *Medicine at the Paris Hospital*, Baltimore, The Johns Hopkins Press, p. 201; and Lichtenthaeler (note 1), Vol. 2, p. 608.
35. Lichtenthaeler (note 1), Vol. 2, pp. 503–509, 586.
36. Lichtenthaeler (note 1), Vol. 2, Vorlesung 14, pp. 359 ff.

IV

The interchange of technology and society: 1860–1914

THE INDUSTRIALIZATION OF WESTERN SOCIETY, 1860-1914

MELVIN KRANZBERG

School of Social Sciences, Georgia Institute of Technology, Atlanta, Georgia, USA

What is Meant by 'Industrialization'

The period from 1860 to 1914 was one of tremendous vitality in Western society. Traditional patterns of work and thought underwent a great transformation as the modern world came into being. The engine that powered this revolution was what we call 'industrialization'.

What is meant by 'industrialization'? To most people the term conjures up images of change from small-scale, handcraft production to large-scale, machine production: that is certainly a prime requisite; but other major changes are also involved. For industrialization is much more than a change in manufacturing methods and economic organization. It consists of two chief elements: (1) a series of fundamental technological changes in the production and distribution of goods; accompanied by – sometimes caused by, sometimes reflecting, but in any event, interconnected with – (2) a series of economic, political, social and cultural changes of the first magnitude.

Both the technical and non-technical elements must be present. A series of technological changes can occur without producing industrialization – as, for example, the developments in sources of power and in agricultural technology during the Middle Ages. Conversely, there have been many sociocultural changes in history that were not accompanied by concomitant technological development. Only when the two are linked does industrialization occur. By approaching the question through correlation hypotheses rather than through the generic problem of causality, with my definition I attempt to distinguish industrialization from other kinds of sociocultural change or from undue concentration on monistic technological factors.

Industrialization thus embodies a congeries of developments: technological as well as sociocultural, political as well as economic, institutional as well as individual. For most of human history, the hearth and home were the centres of production and social life, and agriculture was the chief occupation of the bulk of mankind. In the industrialized world, the factory becomes the production centre; manufacturing and services replace agriculture to provide livelihood; and the home no longer remains the central institution for the inculcation of educational and social values. These revolutionary transformations in the ways in which people work, live, think, play and pray – and how and where they do it – constitute the phenomenon of industrialization. We are here concerned with this interweaving of many different developments in the half-century before the outbreak of the First World War.

Transfer of the Industrial Revolution from Britain

Traditionally, historians have characterized the period from 1860–1914 as the era during which the Industrial Revolution was transferred from its British birthplace to the rest of Europe. In *The Unbound Prometheus,* David S. Landes wonders why it took so long for the Continental nations to follow Britain's industrial lead.[1] However, when one considers the many non-technical elements involved in industrialization, Landes's question – 'Why the delay?' – might well be turned around to read, 'How were they able to do it so rapidly?' How is it that Germany, still a disunited mixture of monarchies, dukedoms and principalities in 1860, was able to forge ahead to compete with Britain by 1914? How did France, made into a nation of peasant proprietors by the French Revolution, begin industrializing under Louis Napoleon and continue to do so even after defeat in the Franco-Prussian War? How did Belgium convert so rapidly to the new industrial society? How did the Scandinavian nations become viable industrial economies?

The answer to those questions lies in the concatenation of political, economic, cultural and social elements with the all-important technological changes which together produce industrialization. Merely enumerating them in summary fashion, as I intend to do in this paper, gives some idea of the synergistic quality of the technological advances of the times, the complex and varied interdependencies among social and technological factors, and the depth and breadth of the transformation in social institutions and human life involved in the industrialization process.

We start by looking at the technological changes, separating them for analytical purposes into functional areas. But, in actual practice, the technologies were interdependent – developments in one field evoking advances in other and related technical fields, resulting in a synergistic effect. Then we look at the accompanying sociocultural developments which interacted with the technological ones, leading industrialization along certain paths and to further interactions, in an almost dialectical process.

Technological Elements

Materials technology

One major effect of technology is to increase man's control over his physical environment, manifested by his changing ability to use different materials provided by Nature; such changes can involve changes in tools, organization of labour and other factors. For example, the transition from stone to metal by prehistoric man demanded specialized industries, organized trade and a surplus of foodstuffs to support those engaged in metallurgical work rather than in producing food.

Not every change in materials has such far-reaching effects. Although S. Lilley attributed democratic social change to the introduction of iron in early times,[2] the fact is that, for most everyday purposes, men continued to use wood and stone for tools and implements. Metals were still so expensive that their chief use was in weapons: for governments, then as now, spared little expense in armaments.

However, a true revolution in materials occurred after 1860. Although wood continued to serve as a basic material, iron and steel became available cheaply in large quantities through eighteenth-century improvements in smelting and rolling iron and development of the Bessemer (1865), Siemens-Martin (1866) and the Gilchrist-Thomas (1878) processes for making steel. The nineteenth century became 'the Age of Iron and Steel' because technological advances so cheapened production of those materials that they could be widely utilized for civilian industrial purposes as well as for specialized military uses.

At the same time, the application of electricity provided still another method of exploiting mineral resources. Electrometallurgy, applied first to plating metals, made other metals (especially aluminium) and alloys cheaper and more readily available.

The late nineteenth century also witnessed the creation of a whole new body of man-made materials – synthetic materials. It was through the discovery of coal-tar dyes in 1859 that the laboratory's contribution to industry became the foundation of the dye industry, and later of the pharmaceutical industry, and, very near the beginning of our own century, of the production of plastic materials. This marriage of laboratory science to technology proved exceptionally fruitful in bringing forth new materials, first to replace older materials, or to make them cheaper, and finally to produce wholly new, artificial materials which surpassed natural products in meeting specialized needs.

In parallel to the development of new materials was an enhanced utilization of natural products, symbolized by the vulcanization process for rubber. Here can be seen how technologies advanced one another: development of the bicycle increased the market for rubber, and the later development of the automobile made the rubber industry into a major adjunct of transportation technology. At the same time, the utilization of electricity created another market for rubber (and, later, plastics) as insulation material.

Energy technology: fuels and prime movers

The synergistic character of technological developments is best demonstrated, however, by developments in the field of energy, involving both fuels and prime movers, which were closely related to metallurgical developments. Already in the eighteenth century, the shortage of wood had led to Abraham Darby's discovery that coke could replace charcoal for smelting purposes, and had forced the substitution of coal for wood for heating and industrial purposes. The need to pump water from the deepening coal mines had led to Watt's improvements to the Newcomen engine, which led, in turn, to a demand for more coal to fuel the new engines and to produce the metal to make them. The application of steam engines to locomotive transportation also increased the demand for iron and coal, which, in turn, led to the requirement for still more transportation facilities, with consequent requirements for still more metal for prime movers, and then more fuel.

The growing reliance on coal also had demographic consequences, indicating another type of interaction between society and technology. Because smelting requires several tons of coal for each ton of iron ore, ferro-metallurgical industries concentrated near the coal sources. Thus, as a result of technological demands, great industrial centres flourished in the areas of large coal deposits: in Britain's Black Country, in Germany's Ruhr, in Lille in France, and in Pittsburgh and the Great Lakes region in the United States. Political power also accrued to those regions where energy in the form of coal was dominant; just as in today's world, petroleum resources give the OPEC nations great economic and political power.

The use of heat to perform mechanical work, in the form of the steam engine, forced engineers and then scientists to consider the relations of heat and power to work performed. The result was the creation of the science of thermodynamics.

Electricity

If, as has been said, the steam engine did more for science than science did for the steam engine, science's debt to technology was amply repaid when electricity was introduced into the life and work of mankind. This was a case in which scientific investigations led to great technological developments and the introduction of an entirely new energy resource and industry.

The first major application of the new electrical science was for telegraphic communication. This was useful for the fast-developing railway systems, as well as for the rapid communication of news. In 1876, Alexander Graham Bell patented his telephone, enabling human speech to be transmitted over wires; and in slightly more than a decade the range of electrical communications was extended still further by Marconi's wireless. Shortly thereafter, the development of the audion tube made radio into a practical means of widescale communication. And when, in 1890, the US Census Bureau applied Herman Hollerith's innovation of punched cards with electric circuitry to store and manipulate information, the basis for today's computerized information systems was laid.

Electricity's uses were not confined to communications. In 1879, Thomas Alva Edison successfully demonstrated the incandescent bulb. Within a short time, electricity had doomed gas-lighting (itself an industry which had developed during the course of the nineteenth century) and the kerosene lamp, another nineteenth-century development, into obsolescence.

At the same time, electric motors were being developed to convert electrical energy into a prime mover for countless applications. Means for the large-scale generation and distribution of electrical power were successfully worked out in conjunction with an extension of the illuminating system and the development of street railways. During the 1890s, factories began to be electrified; when electric induction motors replaced cumbersome central steam engines, it became possible for factories to be located away from water supplies (for steam boilers or for water turbines) and allowed for the simplification of industrial plant design. By 1914, nearly all new factories were electrically powered. Its unmatched versatility, accurate control, and availability to perform large and small tasks made the electric motor into the 'muscle' of modern industrial society.

The internal-combustion engine and petroleum

The electric motor was not the only prime mover born during this exciting period. Originally developed as a stationary motor, the internal-combustion engine was translated into a power source for self-propelled vehicles; these involved auxiliary technical developments in ignition and cooling systems, in braking and gearing devices. The pioneer work was done in Europe; but development of the motor car as a vehicle for everyday use by millions was due to the mass-production genius of an American, Henry Ford. His moving assembly line of 1914 brought us into today's automotive society.

Just as the electric motor faced competition from steam and water power when first introduced, so the internal-combustion engine at first contended with steam and electric-powered cars. One reason that the internal-combustion engine won out was that it relied on an entirely new and cheap fuel, which also emerged in the period after 1860: petroleum.

The petroleum industry dates back only to 1859, with the drilling of large-scale oil deposits in Pennsylvania. Its major product during most of the later nineteenth century was kerosene, widely used in lamps until displaced by the electric light. Even before automobiles provided a vast market for petroleum products, the industry was making technical advances in methods of refining and in pipelines for transporting oil. Furthermore, as we will see, the petroleum industry lent itself to consolidation and to the formation of trusts, which were to characterize large-scale industrial operations in the latter part of the nineteenth century, and in which Alfred Nobel himself played a part. Within a short time, the petroleum industry had moved into the forefront of the new industrial age.

Let us just step back and look at the nineteenth century's record in revolutionizing energy. For millennia, man's chief power source for performing work was muscle – animal and human; then, the development of water and wind power during the Middle Ages signified a great advance. But during the eighteenth century, the steam engine showed that heat could perform work, and new forms of energy and fuel – electricity and petroleum – came into use following 1860. This energy could be multiplied, divided, subdivided at will – truly lifting the burden off men's backs.

The conquest of scarcity: machines and mass production

There was a concomitant development of machines to utilize the new energy sources and prime movers effectively, thus increasing production with but a small expenditure of human energy. Beginning with the landmark inventions in textile machinery in the eighteenth century, virtually every industry was mechanized during the years before 1914. Here the credit must be ascribed to the machine-tool industry, which ultimately made possible precision tooling, interchangeable parts and mass production.

Although small-arms producers were the first to feel the impact of this new system of manufacturing, it spread during the last half of the century to new products developed during that time – the sewing machine, the typewriter, and the bicycle – as well as to older products, such as shoes.

Because the skill was built into the machine, less-skilled workers could be employed. Furthermore, mass production reduced manufacturing costs and made possible mass consumption of what had hitherto been luxury items.

However, it was not only the development of powered machinery that made mass production possible: the reorganization of work, begun earlier in the textile factories, was carried further during the late nineteenth century. Frederick W. Taylor, an American engineer, attempted to apply 'scientific principles' to the man-machine relationship as well as to the machine itself, in order to 'rationalize' production. His 'Scientific Management' movement broke work down into individual operations, designated the most efficient way to perform each operation, and developed the sequence for doing the work most rapidly and economically. And by separating the planning and design functions from the actual work operations, Scientific Management marked a first step toward a society in which knowledge, rather than manual skill, became the major source of labour productivity.[3]

The great changes wrought in materials, energy sources, prime movers, production machines and the organization of work during the nineteenth century were affected by and reflected in transformations in the functional areas of technology. In manufacturing, as I have pointed out, precision machine tools and a more efficient organization of work allowed the development of mass-production methods and the enormous cheapening of products which had formerly been available only to a select few. Technology thus seemed on its way to conquering the age-old problems of poverty and scarcity. At the same time, advances in transportation and communication provided spectacular demonstrations of technology's ability to conquer space and time.

The conquest of space and time: transportation and communications

In the mid-century, wooden sailing vessels – a very ancient form of transportation – still ruled the waves. But by 1900 these had been almost entirely eliminated from ocean shipping by iron ships propelled by steam. The steam engines themselves underwent vast improvements, allowing for increases in size and speed. In 1900, the Hamburg-American Line's *Deutschland,* weighing 16 000 gross tons, capable of a speed of twenty-three knots and with quadruple-expansion reciprocating engines, held the blue ribbon of the Atlantic. In 1907, the Cunard Line launched the *Mauritania* and the ill-fated *Lusitania*; utilizing steam turbines, these ships weighed 32 000 tons and travelled at twenty-four knots. And just before the onset of the First World War, conversion from coal to oil began, reducing the space required for coal bunkers and allowing greater space for cargo.

Similar improvements occurred in rail transportation: locomotives were made more powerful by the use of superheated, articulated engines and more efficient by auxiliary improvements in braking and signalling systems, roadbed and bridge construction, and the like, enabling heavier loads to be carried with greater frequency over longer distances. Not only did the railroad and steamship provide transportation of raw materials and manufactured goods, but they stimulated further industrialization by their nearly insatiable demands for the iron and coal which formed the material and energy foundations of nineteenth-century industrialization.

In the decades immediately preceding 1914, technological advances in the internal combustion engine, as we have seen, began creating new modes and networks for twentieth-century transportation. However, even more dramatic in its impact upon men's minds, although it had not yet assumed an important place in the world of transportation, was the aeroplane. Although for a time it appeared that lighter-than-air ships would become the means of airborne transportation – Count Ferdinand von Zeppelin flew his first dirigible in 1900 – the future of air transport was to belong to heavier-than-air ships. Although men had long dreamed of controlled flight, it was not until the development of a powerful and light propulsive unit – the internal-combustion engine – and the development of aerodynamic principles that such flight was possible. In 1903, the Wright brothers made their first flights, the longest one lasting only fifty-nine seconds and covering a mere 852 feet (less than 300 meters). Although aviation would not become a significant factor in transportation for many decades to come, the mere demonstration of its possibilities had a tremendous psychological impact. It seemed that technology could accomplish anything – even fulfilment of man's ancient and mythic dream of flying through the air.

Developments in communications were equally spectacular. At the beginning of the nineteenth century, a man's voice could be heard only as far as it could carry; written or printed communications could travel only as fast as could a man on horseback or on a sailing vessel – and those speeds were not significantly greater than they had been for centuries; a man's image and the sound of his voice died with him. By the end of that same century a series of remarkable technological developments had changed all that.

The telegraph was a working reality by the 1840s, but it could transmit only coded messages, not voice sounds. In 1876 came the telephone; and as the century drew to a close, wireless transmission paved the way for the development of radio and the advent of broadcasting to a large audience. Furthermore, a man's auditory memory was extended as well as the range of his voice. In the same year that Bell patented his telephone, Edison invented a phonograph using a cylinder; in the next decade, Emile Berliner improved the device by utilizing a flat disc to impress sounds on his gramophone. Music and speech could now be preserved and heard by millions.

Man's visual memory was also extended when, in the 1830s and 1840s, practical forms of photography were developed in France and England. In the 1880s, technological progress – a change from wet-plate to dry, then the introduction of celluloid film – further revolutionized photography, changing it from a professional art to one pursued by millions of amateurs. Celluloid film also made possible motion pictures. The development of motion pictures provides a splendid example not only of the social nature of modern invention but also of the supportive nature of technological innovations: no single inventor can be credited with the entire system, but a number of innovators contributed to the emergence of satisfactory film, cameras and projectors.

In the midst of these revolutionary developments, older forms of communication were not neglected. Newspapers were made more widely available and cheaper by the development of continuous-roll presses, by the introduction of the Linotype and of paper made from wood pulp. The telegraph and cable enabled the rapid gathering of news from far-off places; new printing methods made it possible to print large editions rapidly; and the growth of a literate public increased the demand. At the same time, there were improvements in illustrations, and wood-block line drawings were replaced by zinc cuts and half-tone pictures; colour printing became commercially practicable by the 1890s.

Written communication also underwent technological change. During the last half of the nineteenth century, J. von Faber mechanized the production of 'lead' pencils; the manufacture of steel-nibbed pens, which had replaced quill pens in the first half of the century, was also mechanized; and by 1914 fountain pens had become a practical reality. But handwriting itself was being replaced mechanically by the development of the typewriter. A side effect of the typewriter was to bring women out of the home and into the office, for their fingers were as agile at the typewriter keyboard as with needle and thread.

Engineering Education: Professionalization and Specialization

As industrial machinery become more complex, there was a need for greater technical knowledge and competence. Few of the well-known inventors of the Industrial Revolution were what we today call 'engineers', with specialized technical knowledge and professional training. However, the demands of a growing industry gave rise to a professionalization and specialization in engineering.

At first there were only civil engineers; but then mechanical engineers, mining engineers, and other varieties began to appear as part of the increasing specialization of function demanded by the more intricate needs of an industrial age. The development of wholly new branches of technology, such as electricity, led to the emergence of electrical engineers, while advances in the chemical industry produced chemical engineers. This process of differentiation and specialization has proceeded apace in our own times, reflecting the emergence of new technologies, as, for example, nuclear engineering.

The old system of apprentice training no longer sufficed, in view of the growing complexity of technological applications. Professional training – engineering education – became necessary. Surprisingly enough, Britain, the greatest industrial power throughout the century, lagged in this respect. Although the Scottish universities emphasized science and engineering, Oxford and Cambridge maintained their classical emphasis; and a whole new category of institutions, known as the 'red brick' universities, were belatedly founded to meet the new demand. On the Continent, on the other hand, state-supported institutions for technological training had grown up, including the *Ecole Polytechnique* in Paris and the *technische hochschulen* in Germany. Not until the Technical Instruction Act of 1889 and the Education Act of 1902 did Britain develop a broad scheme for technological education.

During the first half of the nineteenth century, most American engineers were self-trained or learned from on-the-job experience, supervised by engineers imported from Europe or by graduates of the United States Military Academy. The major stimulus to engineering education in the United States was the Morrill Land-Grant College Act of 1862, for the creation and support of state 'colleges of agriculture and the mechanical arts'. A number of private engineering institutions were also founded as an inevitable accompaniment to the rapid industrialization process and the concomitant need for trained professionals.

Science and Technology

Technical education assumed added importance in the late nineteenth century with the increasing interdependence of technological advance and scientific achievements. Although there has been much scholarly dispute regarding the contributions of science to the origins of the Industrial Revolution, the application of science to industry still remained somewhat fortuitous. However, during the nineteenth century the situation changed. For one thing, certain industrial developments, primarily those in electricity and chemicals, rested upon new scientific findings. Furthermore, the growing complexity of materials and machines meant that technologists had to acquire at least a rudimentary knowledge of scientific principles, to apply mathematics to their studies of stresses and strains and heat-energy-work relationships, and to augment traditional empirical methods with scientific formulations and approaches. As Edwin T. Layton has shown, engineers did not simply 'apply' scientific findings and discoveries, but had to develop their own and parallel scientific approaches.[4]

In the growing development of 'industrial science', Germany took the lead.[5] The benefits deriving from a systematic coordination of science, technology and industry were shown in Germany's rise to industrial prominence in the manufacture of chemical products, optical instruments and electrical equipment. It should be noted that Alfred Nobel was one of the chief practitioners – and beneficiaries – of this multi-pronged and sophisticated approach to technological advance.

Agriculture and Food Production

Among the most important of the functional areas subjected to industrialization during the nineteenth century were agriculture and food production. Technological developments in manufacturing – improvements in machinery, specialization of function, application of new sources of power, and the like – had their counterparts in agriculture.

At the beginning of the nineteenth century, agriculture was still carried on much as it had been done since the Middle Ages. The farmer tilled a small parcel of ground, growing crops for his own consumption and for the local market; the tillage was largely accomplished through the sweat of human beings, some of whom still remained in serfdom or slavery; and 'scientific' methods of cultivation were confined to but a small group of enlightened gentlemen farmers. By the close of the century all this was changing.

For one thing, the mechanization which transformed industrial production from handcraft to factory work also affected agriculture. The steel blade ploughshare was invented in 1830 and the reaper in 1834, by Cyrus McCormick; then followed a series of other mechanical devices to assist in seeding, tillage and harvesting. Furthermore, the burden was lifted off the backs of men by the application of animal power – horses and mules – to agricultural production; and, during the last quarter of the century, mechanical power, through steam-powered implements, became increasingly employed.

At the same time, political events changed the status of many agriculturalists in Europe and the United States; the 1848 revolutions ended serfdom in Central Europe, the Russian serfs were freed in 1861, and the Civil War in the United States abolished slavery.

The scale of agricultural production was magnified by the opening of vast new areas for cultivation, in the United States, Canada, Argentina, Australia and Russia. Large-scale farming stimulated mechanization. By allowing the cultivation of more land with less manpower, mechanization increased cereal crop production in the United States some ten to fourteen times; in harvesting alone, man hours of labour were reduced to one-sixth the time previously required. Furthermore, drainage and irrigation in the American high plains, coupled with the technical innovation of barbed wire, allowed for a shift from the open range to crop production.

Commercial fertilizers were increasingly applied. Improved transportation allowed the importation of Chilean nitrate, Peruvian guano and superphosphates. Then the discovery in Germany (1903) of means of 'fixing' atmospheric nitrogen enabled even greater production and utilization of fertilizer. Commercial pesticides were also introduced; especially important was the Bordeaux mixture (1885), a key weapon in the fight against fungal diseases of plants. Scientific plant- and animal-breeding further contributed to the development of varieties capable of faster growth, higher volumes of production and adaptation to unfavourable environments.

Mechanization, increased use of fertilizers and pesticides, and large-scale farming replaced subsistence farming by the production of cash crops for world markets. A host of secondary technological changes, primarily in the transportation and processing of foods, allowed for further specialization of agricultural production. Crops could be selected that were particularly suited to specific climatic and soil conditions, and advances in transportation allowed the crops to be widely distributed.

Railroads could haul produce faster and over greater distances than canal barges or wagons, so that commercial dairying, market gardening and horticulture could spread out from urban centres. The development of refrigeration allowed for the preservation and transport of perishable food over long distances, thereby breaking up the monotony of diets and providing more nutritious diets to all.

Technical advances in food processing also helped to 'industrialize' agriculture. Canning, introduced at the beginning of the nineteenth century, underwent various improvements, becoming a common method for processing, preserving and transporting foods. Pasteurization, derived from the work of Louis Pasteur, helped for preserving milk and other products. In certain fields, such as meat-packing, industrial methods converted a home industry into a factory. Indeed, the 'dis-assembly line' of the Cincinnati meat packers was said to have inspired the later moving assembly line for manufactured products.

Technological advance in nineteenth-century agriculture was thus of basic significance for industrialization. In 1798, the Reverend Thomas Robert Malthus issued his gloomy prediction of mass starvation as a result of the multiplication of population. Although population growth did indeed occur – Europe's population in the nineteenth century grew by no less than 200 million people – the predicted starvation did not follow. The opening of new agricultural lands – over 400 million acres were brought under cultivation in the United States alone in the period 1860–1900 – plus progress in agricultural and food technology not only kept Western man from starving, but also made him a better-fed and longer-lived species with a wider variety of foods than ever before in history. Alas, the populations in those parts of the world that did not undergo agricultural industrialization continued to grow without concomitant increases in food production, a problem still felt in acute form today.

Non-technological Elements

Economic and political elements

Contemporaneous with the revolutionary technological developments in the latter part of the nineteenth century were a series of transformations in non-technical spheres: economic, political, social, cultural and psychological. It would be wrong to seek direct, linear, causal connections between these elements. Instead, we are dealing with a series of interactive changes, a developing system known as industrialization, in which both the technological and non-technological aspects were essential.

Sidney Pollard calls the diffusion of industrialization from its home in Britain to the rest of Western society a 'peaceful conquest'.[6] In the first stage, before 1880, British industrial technology spread across Europe by the introduction of new forms of labour and the factory system and the creation of new international markets; after 1880, governments played an ever-larger role in industrialization. Even if one does not accept Pollard's periodization, his approach reinforces the importance of exogenous, non-technical factors in industrial development.

Capital investment

Industrialization requires capital investment. The crucial importance of this factor in technical advance has led some observers almost to equate capitalist growth with industrial growth. But capitalism and industrialism are not inevitably or necessarily connected. There had been forms of capitalism at earlier dates in the absence of much industrial change; and industrialization is occurring in many avowedly anti-capitalist countries today. Nevertheless, the advance of industrial technology was historically linked with the creation of industrial capitalism.

The accumulation of capital, formerly derived from commerce and banking, now resulted from industrial production. In turn, the availability of capital assisted in further industrialization, for the amount of capital required to install manufacturing facilities grew as the technology became more complex and as large-scale factory production became the order of the day.

Early factories tended to be small, and their owners were in close and direct contact with their businesses. As larger enterprises were undertaken, requiring more capital than a single individual possessed, ownership passed from individuals and partnerships to joint-stock companies. Such corporative growth was stimulated by the gradual enactment of the legal principle of limited liability. English statutes of 1855–1862 and French laws passed between 1863 and 1867 extended limited liability to industrial corporations, making it safe for an investor to risk capital in one or more ventures without endangering his entire fortune.

The growth of corporate industry made inevitable a rearrangement of property interests. Previously, wealth was represented primarily by land; industrial investment meant that property interests were more commonly represented by securities than by land ownership.

Corporate management

Relationships between owners and workers were also changed. Early industrial capitalists, with their small factories, were likely to know virtually all their employees personally. As enterprises grew in size, that became less possible; and when firms became corporations, with many absentee owners, there was no longer direct contact between owners and employees. The owners were usually represented by a managerial group, who might or might not own some share of the business. Hence, the managerial function assumed greater importance as industrialization enlarged the scope of industrial enterprise.

Commercial growth

Advancing industrialization required a concomitant growth in commerce. To keep factories running smoothly and steadily there had to be a regular flow of raw materials, as well as permanent channels for the sale of mass-produced merchandise to the consumer. Furthermore, consumers required more goods, due to the specialization of production involved by the development of the factory system: Previously, the toiler, in rural isolation, produced all or at least many of the things that he consumed; now, however, he worked in a factory where he produced just one product and was forced to buy everything else he needed.

Commerce was also stimulated by the fact that there were simply more consumers. In 1800, Europe's population was about 187 million, but by 1860 it had increased to 282 million; from 1821 to 1911 Britain's population increased by one hundred percent. The mere physical task of supplying a larger population with the elementary necessities – food, clothing and shelter – forced commercial enterprise to a frenzied pace of activity, and, of course, also provided a stimulus to further industrialization. Not only were there more consumers, but the demands of European peoples altered, in some cases due to improved standards of living, and this too acted as a spur to industry.

Banking and industrial consolidation

The expanded needs of trade and industry led to a growth in financial dealings, for the banks performed the necessary functions of accumulating capital and savings for industrial investment, transferring funds, and providing manifold exchange activities for carrying on widespread commerce. Bankers became a powerful voice in the control of industrial enterprise, leading to Lenin's indictment of 'finance capitalism' as a tool for exploiting the workers.

In those industries in which large-scale investments were necessary for production and distribution, the period after 1860 was one of industrial consolidation, manifested by the growth of cartels, trusts and monopolies. In the United States, bankers played a major role in consolidating the railroads and the steel industry, both of which required large amounts of capital. For example, J. P. Morgan, a banker, collaborated with Andrew Carnegie, a steelmaker, to form the giant United States Steel Corporation. Not all the 'robber barons' were bankers, however. John D. Rockefeller established the Standard Oil Trust, driving his competitors out of business and effectively controlling the production and distribution of petroleum products, until the Progressive Movement in the early twentieth century split the Standard Oil Trust into separate companies.

Cartelization was international in scope, as illustrated by the German dye and chemical industries and especially by the Nobel Dynamite Trust. In revolutionizing the explosives industry and controlling the patents, Nobel opened the way for consolidation of that industry on an international basis. After a short period of free competition, dynamite firms throughout Europe were consolidated in 1886 into the Nobel Dynamite Trust. This trust, run by a carefully constructed Anglo-German board, linked export markets with the home markets of Britain and Germany and controlled virtually the entire dynamite industry of the world.[7]

Power of industrial wealth

The role of governments in promoting – or at the very least, not interfering with – industrial consolidation is indicative of the changed relationships between economic and political power wrought by industrialization. If we look at the connection between economic and political power through the cynical eyes of the historian, we find that, with but few exceptions, political power throughout history has usually been in the hands of those who controlled the wealth – or in terms of the modern formulation of the Golden Rule, 'He who has the gold, rules'.

Thus, during antiquity and the Middle Ages, when the economy rested largely on agriculture, society was organized so that the noble landlords possessed political power. In the early modern period, when commerce and banking became sources of wealth, bankers amassed landed properties and the titles that went with them, and thus joined the landholders in exercising political control.

The growth of industrial wealth brought a new factor into the political equation. In Britain, a compromise in the 1830s allowed the new industrial class to share power with the landed proprietors; and in Germany the junker landowners arrived at an agreement with the great industrialists. Wherever industrialization spread, the political power of the capitalist owners of industry grew to match their new-found economic power.

Not surprisingly, governments began to pursue policies that reflected industrial interests.

The challenge from below

But the power of big business did not go unchallenged from below. Industrial workers, with a growing recognition of their economic power derived from the interdependency of the industrial system and fortified in their beliefs by the development of socialist ideologies and movements, began to institutionalize their demands for political power through labour unions and working-class political parties. To meet the grievances of the workers and to blunt their thrust for political control through socialist movements, the governments of various European states began to intervene in business matters, to curb unrestricted exploitation of workers and to institute social security systems, as Bismarck did in Germany.

Political unification and imperial expansion

Industrialization affected politics in still another way – by playing a major role in the unification of nations. German unification had been furthered by the development of a customs union earlier in the nineteenth century; this helped provide a common market for the German industry, still in its adolescence, as did a semi-unified and coordinated railway network. In the United States, completion of the transcontinental railroad helped to unify the country after the divisiveness of the Civil War; and the Civil War was itself decided by the superior industrial might of the North, which enabled it to triumph over the superior leadership of the Confederate Army.

The interaction between political elements and industrialization was also reflected in the imperialistic competition that dominated international relations during the last third of the nineteenth century. European expansion to other continents was by no means new in history, as the Hellenistic and Roman empires and the mercantilist duel for empire in the early modern period can attest. During the latter part of the nineteenth century, imperialistic rivalry again flared. What role did industrialization play in this revival of colonial ambitions?

In his classic study of imperialism, Nikolai Lenin analysed it as the final stage of a doomed capitalistic system. However, most of the economic arguments to explain imperialism – Marxist or otherwise – such as the need to invest surplus capital abroad, the demand for raw materials or for new markets and the like, have since been disproved by historical economic data, which show that colonies did not really 'pay for themselves'. Nevertheless, at the time, such arguments appeared convincing to special interest groups and politically persuasive to voters and governments. Along with the economic arguments were many other justifications, usually of a sociocultural nature, such as social Darwinism, which justified the domination of the strong over the weak, and ideas of 'manifest destiny' and the 'white man's burden', based upon the moral superiority of the West and the inferiority of 'backward peoples'. Imperialism, therefore, is a psychological and cultural phenomenon, and does not derive simply from political rivalry or industrial competition.

The latest work on the subject considers that the imperialistic thrust of the latter part of the nineteenth century was stimulated by technological advances, and was made possible by those advances. The development of quinine prophylaxis, for example, allowed Europeans to survive in tropical climates; and many diverse products and processes, from pith helmets to battleships, gave concrete quality to the cultural and social rationalization for imperialism.[8] Furthermore, the limitations of nineteenth-century technology account for certain aspects of the colonial enterprise; for example, because steamships and naval vessels required frequent refuelling, nations were impelled to acquire coaling stations to protect their routes to far-flung colonies.

Superior military technology made possible the easy conquest of colonial territories, while developments in transportation and communication allowed the colonial powers to maintain imperial control: Inasmuch as military developments have been intertwined with technology throughout history, that is not surprising.

Military Technology

Although Napoleonic tactics and strategy still dominated military thought during the nineteenth century, technical advances were making them obsolete. Improvements in small arms – the musket gave way to breech-loading rifles and to the percussion cap, allowing for greater speed, accuracy and range – and the development of machine-guns doomed the old-fashioned cavalry charge. Artillery, too, improved in accuracy, range and power of destruction; metallurgical developments allowed the use of steel for cannon, precision tools made possible the rifling of cannon barrels and breech-loading cannon, and advances in chemistry provided Nobel's smokeless powder and nitroglycerine.

Naval warfare also was transformed by improvements in guns and torpedoes and by a transition from wooden sailing vessels to armoured dreadnoughts and destroyers, as well as by the development of submarines. Every aspect of naval propulsion, weaponry, tactics and training was affected by these technical developments.

Advances in transportation and communication affected military concepts of time and space. Railroads enabled the rapid mobilization and supply of huge armies; the electric telegraph provided greater control by generals; and the field telephone allowed for quick tactical response. Instead of spending a great deal of time foraging for food in the countryside, armies could be assured, through canned foods and better transport, of adequate provisions for long campaigns in distant places.

Furthermore, the rapid communication of news from the battlefront helped to raise national consciousness and to create 'total war' – the nation in arms. This awareness was fortified by the need for production on the homefront of munitions and the other accoutrements of warfare – uniforms, food supplies and the like – in great abundance.

The range, power and destructiveness of warfare thus increased immeasurably during the nineteenth century. To some, as we shall see, this meant that special efforts must be made to guarantee peace; to others, it was a challenge to utilize the new military technology to realize political ambitions and national goals.

Sociocultural Changes

The economic and political elements of industrialization described above do not tell the whole story. They, along with the technological elements, were accompanied by revolutionary developments in society and culture.

Delineation of the sociocultural aspects of industrialization would require a listing of virtually every characteristic of modern Western civilization. The change from traditional society, with agriculture as the main source of livelihood and with goods produced by hand-craftmanship, to a society based upon the machine production of goods in factories thoroughly transformed life in Western industrial society. All the paraphernalia of living – material and intellectual – underwent change; here we can mention but a few of the changes and attempt briefly to illustrate their relationship to the accelerated pace of technological advance.

From farm to city

One of the most obvious changes was in the landscape – from the countryside to the urban scene. The factory town replaced the hearth and farm as the environment for family, work and social activities. Improvements in agricultural production made it unnecessary for so many people to devote themselves to growing food; there was no longer sufficient work on the farms, so workers were forced to leave for urban areas where factories might provide employment. Life in the city was far different from what it had been in the country. In farming villages, horizons were limited: People were born, lived and died in the same, unchanging rural environment. In the cities, even in the slums, life was more varied and exciting, opening men's eyes to new and different vistas.

While technological advances in both agriculture and industry helped to create the modern city, the growth of the city in turn required changes in technology. As the urban population grew, primitive arrangements for water supply and sewage disposal no longer sufficed, and new means had to be devised. Transportation systems also had to be developed, as the city grew so large that people could no longer live next to their work; but the development of urban transportation was a cause as well as an effect, since it enabled workers to travel greater distances to work.

First in the United States, and later elsewhere, cities began growing upward as well as outward. Development of the skyscraper entailed development of new methods and materials for construction, as well as auxiliary technologies: elevators, lighting, heating and communications systems.

Factory work broke up the constant 'togetherness' which had characterized rural life throughout the centuries. In the traditional agrarian economy, children worked alongside their parents in the field or in performing household chores. With the advent of the factory system, the world of home and the world of work were separated; the father left home in the morning in order to be at the factory when the whistle blew and returned only at night, while the mother remained in charge of the household and of the rearing of the family. The growth of public primary and secondary schooling during the nineteenth century meant even that the education of children increasingly took place outside the home.

Educational opportunities and social mobility

Public education was an indirect yet necessary consequence of technological growth. With children no longer working on the farm and with social legislation often preventing their other employment, they had more time for education. In a sense, an advancing technology provided a sufficiency of material goods so that children could be withdrawn from the work force to go to school. At the same time, the growing complexity of technology demanded a work force that was literate. By 1914, one had to be able to read and write and do elementary arithmetic to be employed in a factory or office. Education was no longer an ornament or a luxury; it was increasingly a central economic resource of an industrialized society.

Better educational facilities put pressure on the social structure. The opportunities were no longer confined to those of money and rank; they came increasingly to those of education, and the educated man resented class and income barriers which prevented him from rising in society. Hence, the growth of education enhanced the possibility of social mobility.

The darker side

It would be a mistake to view the social impact of industrialization solely in such glowing terms of raising expectations, improving education, social mobility, and the like. In the early factory system, at least, industrialization seemed to deepen human poverty and misery. On the farm, the family could at least grow its own food; but in the factory the worker became a 'wage slave', entirely dependent upon the functioning of the machine for his subsistence.

The litany of abuses perpetrated in the early factory system is a long one. Workers laboured long hours for miserable wages and lived in ugly, insanitary tenements – the slums. Above all, the workers were insecure. The existence of a large pool of labourers, due partly to the employment of women and children and partly to the influx of workers from rural areas, meant that industrialized society was faced with a new social phenomenon: chronic mass unemployment.

In addition to those unable to find work, others were thrown out of work by temporary shutdowns, depressions or business failures; because of old age or impaired efficiency; or because of technological improvements which enabled one machine to do the work of many men. Despite the theoretical possibility of economic advancement in a society in which technological development was accelerating, the average working man had little chance to escape a lifetime of poverty and despair. Even when humanitarian social legislation removed some of the worse abuses, especially in regard to child and female labour, the conditions of life remained harsh and dismal.

Much of the writing on the consequences of industrialization has been affected by this picture of the appalling condition of the factory workers. It is not surprising, therefore, that very early in the industrialization process, socialist theories were formulated to explain the harsh exploitation of the workers, and working-class movements emerged in order to alter these conditions.

The proletariat, the 'social question' and other complaints

Industrialization had thus given rise to a new social class: the proletariat, the urban industrial workers. Improvement of their lot became the focus of economic thought, social philosophy and political action. Something new had appeared on the scene: the 'social question'.

The poverty and misery of the workers was contrasted with the growing wealth and material comfort of the middle and upper classes. With factories pouring forth a flood of goods, social thinkers asked: Why is it that the workers are unable to benefit from the wealth they are creating?

This 'bleak' interpretation of industrialization still prevails among many scholars; but within recent decades some historians have quarrelled with the notion of the misery and poverty of the workers, finding that in certain regions and in certain fields of production the living standards of the workers rose materially. Others, while admitting the misery of the workers during industrialization's early stages, compare this with what they regard as the even more horrible conditions of the common labourer in the traditional agrarian society, claiming that it was truly a step upward. Still others differentiate between the baneful short-term effects and the long-range effects of industrialization in raising living standards for everyone.

This debate will certainly continue, especially as new data are uncovered to provide us with more accurate econometric information. After all, the transition from countryside and cottage to city and factory was bound to create stress and strains. Yet, pragmatically speaking, industrialization offered the potentiality, if not always the actuality, of raising the standards of living of the working masses in the industrialized nations; and the historical statistics currently available indicate that by 1914 it was doing so.

Industrialization has also been called into question on other grounds. Lewis Mumford claimed that it produced a new 'barbarism', in which society shifted from a respect for human values to measuring life in material and pecuniary terms. Others, pointing to the mechanics of factory operations, claimed that industrialization reduces man to a machine and makes the machine the master rather than the tool of man.

Some twentieth-century evaluations of industrialization echo nineteenth-century complaints. Almost from the start of industrialization there was a fundamental tension between the Arcadian attitude toward Nature – expressed by Henry Thoreau in the United States and by European romantics – and the notion of man's dominion over Nature.[9] The former view, resurrected in today's concern over the environment, nevertheless lost out to the notion that industrialization represents progress.

Technology and progress

In the late nineteenth century, the great material prowess of advancing industrialization stimulated thinkers such as Herbert Spencer to great optimism regarding man's future on earth. Applying Darwin's theory of organic evolution to society, Spencer saw man's past as a long struggle upward from the primal ooze, through the Stone Age and progressive stages, to the comforts of the Victorian era. Through industrial advance, Spencer saw for future generations the prospect of a world from which poverty would have disappeared, in which democratic society would spread the blessings of liberty to all mankind, in which machines would be performing all wearisome toil so that men lived in leisure, in which universal education would have blotted out ignorance and superstition, and in which international peace and brotherhood would reign.[10] Industrialization would make all this possible, and it was believed that art, science and morals would advance along with technology.[11]

Writing in 1910, Leo H. Baekeland, the inventor of Bakelite, an early plastic, showed his enthusiasm for what nineteenth-century industrialization had accomplished:

> 'To put it tersely, I dare say that the last hundred years under the influence of the modern engineer and the scientist has done more for the betterment of the race than all the art, all the civilizing efforts, all the so-called classical literature of past ages. . . . The modern engineer, in intellectual partnership with the scientist, is asserting the possibilities of our race to a degree never dreamt of before. . . .'

Baekeland's optimism at the close of the nineteenth century was shared by Alfred Nobel. But, as we know from Kranzberg's First Law – 'Technology is neither good nor bad, nor is it neutral' – technology interacts with the social ecology in ways that are not necessarily inherent in the technology itself and that can lead to many different – and sometimes contradictory – social results.

Technology and the human dilemma

The non-neutrality of technology can perhaps be demonstrated by the thoughts and legacy of Alfred Nobel. In his brief study on *Alfred Nobel and the Nobel Prizes,* Nils K. Ståhle speaks of 'Nobel's lifelong search for philosophical, ethical and spiritual values for the good of mankind', and of his belief that an advancing science and technology would put an end to war, 'the horror of horrors and the greatest of all crimes'.

Nobel saw no incongruity between his contributions to making war more deadly and his desire for peace: 'My factories may well put an end to wars sooner than your congresses', he once wrote to the Baroness Bertha von Suttner.

I am reminded of the cartoon showing a caveman emerging from his cave with a bow and arrow. To his companion he says, 'This new little invention of mine will make war so horrible that man will never make war anymore.' How often have we heard that when new instruments of destruction make their appearance!

My point, however, is that while industrial advance at the end of the nineteenth century seemed to point the way to a new and better world, it still was not able to prevent world conflict. Technical improvements in communications and transportation transcended national boundaries, and the resulting growing interdependence among nations pointed unmistakably toward the 'global village' of later idealistic thinkers. By 1914, nationalism was technologically obsolete.

But, alas, nationalism was not ideologically obsolete; it grasped the minds and wills of both peoples and rulers alike. Despite the upsurge of technological prowess, world politics remained dominated by the struggle of independent nation states for power, prestige and wealth in a condition of global anarchy.[12] The result was two world wars and continuing global upheaval during the twentieth century.

Advancing science and technology since his time would seem to make Nobel's arguments for peace even more imperative. Yet today, some eighty years after the awarding of the first Nobel prizes, we remain in the same dilemma that faced Alfred Nobel. The Nobel prizes reflect human achievements in science and the continuing endeavour to improve the lot of mankind. To the recipients, they are well-deserved awards for great accomplishments; to the rest of us, they must remain a symbol of hope for a better and peaceful world.

Notes

1. Landes, D. (1969) *The Unbound Prometheus: Technological Change and Industrial Development in Western Europe from 1750 to the Present,* Cambridge, Cambridge University Press, p. 126.
2. Lilley, S. (1966) *Men, Machines and History: the Story of Machines and Tools in Relation to Social Progress,* revised ed., New York, International Publishers, pp. 20–26.
3. Kranzberg, M. & Gies, J. (1975) *By the Sweat of Thy Brow: Work in the Western World,* Ch. 16, New York, G. P. Putnam's Sons; Merkle, J. A. (1980) *Management and Ideology: The Legacy of the International Scientific Management Movement,* Ch. 1, Berkeley, University of California Press.
4. Layton, E. T. (1971) Mirror-image twins: the communities of science and technology in nineteenth-century America. *Technol. Cult.*, 12, 562–580.
5. Beer, J. J. (1959) *The Emergence of the German Dye Industry,* Urbana, University of Illinois Press; Haber, L. F. (1958) *The Chemical Industry during the Nineteenth Century,* London, Oxford University Press.
6. Pollard, S. (1981) *Peaceful Conquest: The Industrialization of Europe, 1760–1970,* New York, Oxford University Press.
7. Reader, W. J. (1970) *Imperial Chemical Industry: A History,* Vol. 1, London, Oxford University Press, pp. 18, 96, 126 ff.
8. Headrick, D. R. (1981) *The Tools of Empire: Technology and European Imperialism in the Nineteenth Century,* New York, Oxford University Press.
9. Worster, D. (1977) *Nature's Economy: The Roots of Ecology,* San Francisco, Sierra Club Books.
10. Kranzberg, M. (1975) *Confrontation: technology and the social environment.* In: Zandi, I., ed., *The Technological Catch and Society,* Philadelphia, University of Pennsylvania Department of Civil and Urban Engineering, pp. 59–83.
11. Nisbet, R. (1981) *History of the Idea of Progress,* New York, Basic Books.
12. Gilpin, R. (1981) *War and Change in World Politics,* Cambridge, Cambridge University Press.

THE GROWING ROLE OF SCIENCE IN THE INNOVATION PROCESS

NATHAN ROSENBERG

Department of Economics, Stanford University, Stanford, California, USA

I

My concern is with the growing role of science in the innovation process during the period 1860–1914. At first glance, the major contours of that relationship would appear to be reassuringly familiar. After all, that period represents a high-water mark of industrialization in several ways: During that period, the basic innovations of the earlier British revolution, centring upon the expanding applications of coal, steam and iron, came to full maturity. Their application to transportation in the form of the railroad and the iron-hulled steamship, and with the indispensable additional assistance provided by refrigeration, opened up continental hinterlands and linked the entire world, for many purposes, into a single market nexus. It was during that period that Britain began to be fed and clothed by products grown in the antipodes. It was during that period too that British industrial technologies were introduced into the productive process on a world-wide scale – not only in the overseas offshoots of western Europe, such as North and South America, Australia and New Zealand, but also in Russia and Japan.

In addition to the worldwide diffusion of the technology of the industrial revolution, the period 1860–1914 also witnessed the introduction of entirely new technologies. Some of these new technologies were in older, well-established industries, such as chemicals. Others, however, were so novel that there simply were no earlier industrial counterparts. The most conspicuous example of the latter, of course, is the emergence of the electricity-based industries. It is not uncommon to refer to these new clusters of innovation, together with a major new power source, the internal-combustion engine, and new manufacturing techniques that rely upon precision methods and the use of the assembly line, as constituting a Second Industrial Revolution.[1]

Thus, it seems to be no exaggeration to state that, by 1914, the essential technological fixtures of modern industrial society were already visibly, if not always firmly, in place. The question that needs to be addressed is: What was the role of science in achieving this transformation? More specifically, what role did science play in those technological innovations that created modern industrial societies?

It may be useful, in pursuing these questions, to think of science itself as constituting the central innovation of modern societies. As Simon Kuznets has stated: 'The epochal innovation that distinguishes the modern economic epoch is the extended application of science to problems of economic production.'[2] It is important to appreciate, however, that such a formulation constitutes only the beginning of an intellectual inquiry and not its terminus. What the statement leaves unanswered is how and when and under what circumstances this new role of science developed, and how and to what extent its applications were felt at different times and in different industrial sectors.

Thus, although there is general agreement with the view that the dependence of technological progress upon science has increased substantially in the course of industrialization, there is considerable disagreement, or at least considerable difference in emphasis, upon the extent of that dependence. On the one hand, A. E. Musson and E. Robinson have argued forcefully that technological progress was already heavily dependent upon science in the early stages of the British industrial revolution.[3] Their research has provided a wealth of evidence showing, in the British case, the intimate and multitudinous networks that linked the business community to scientists. Robert Schofield has advanced the claim, based upon a careful study of the Lunar Society in Birmingham, that that society was really an eighteenth-century 'industrial research group'.[4]

On the other hand, there has been a long and influential tradition of historians of technology, such as Gilfillan,[5] who have stressed the crude, trial and error, hit-or-miss nature of technological progress. For the pre-industrial period, it has often been suggested that science and technology were social processes which were almost hermetically sealed off from one another. As A. R. Hall has stated, with respect to pre-industrial technology:

> 'We have not much reason to believe that, in the early stages, at any rate, learning or literacy had anything to do with it; on the contrary, it seems likely that virtually all the techniques of civilization up to a couple of hundred years ago were the work of men as uneducated as they were anonymous.'[6]

Clearly a central issue in the formulation of such divergent views is how one goes about defining science. If by science one means, on the one hand, rigorously systematized knowledge within a consistently-formulated theoretical framework, the role of such knowledge is likely to have been small before the late nineteenth and twentieth centuries. On the other hand, if one defines science more loosely in terms of procedures and attitudes, including the reliance upon experimental methods and an abiding respect for observed facts, it is likely to appear universal in industrializing societies.

In addition, the question of the role of science in the innovative process inevitably raises the fundamental question: Where does this science come from? Can the economic or social historian be content with treating science as some exogenous force, some *scientia ex machina*, which is of interest only for its industrial applications? I would like to suggest that such an approach would fail to capture a large part of what scientific progress is all about. Science, after all, when considered as a social process in a social context, is an activity that is powerfully shaped by day-to-day human concerns. One does not need to be an economic determinist to believe that these concerns are, to a considerable extent, shaped and even often defined by the needs of the economic sphere and by the kinds of technologies that prevail in any given social context. It would be easy to demonstrate that the growth of scientific knowledge has itself been decisively shaped by technological concerns.[7]

Thus, Torricelli's demonstration of the weight of air in the atmosphere, a scientific breakthrough of fundamental importance, was an outgrowth of his attempt to design an improved pump.[8] Sadi Carnot's remarkable accomplishment in opening up the science of thermodynamics was an outgrowth of the attempt, half a century or so after Watt's great innovation, to understand the determinants of the efficiency of steam engines.[9] Approaching our own period, Joule's discovery of the law of the conservation of energy grew out of an interest in alternative sources of power generation at his father's brewery.[10] Pasteur's development of the science of bacteriology emerged from his attempt to deal with problems of fermentation and putrefaction in the French wine industry. In all these cases, scientific knowledge of a wide generality grew out of a particular problem in a narrow context. Such a recitation, however, gives only a very limited sense of the exact nature and extent of the interplay between science and technology. Indeed, that sense is totally suppressed in the prevailing (and, in my view, wrong-headed) formulation of our own time in which it is common to look at the causal relationships as if they ran exclusively from science to technology – and in which it is common to think of technology as if it were reducible to the application of prior scientific knowledge.

On one matter, at least, there will be general agreement. The period beginning in 1860 was one of extraordinary scientific progress. Indeed, if one had to choose any fifteen-year period in history on the basis of the density of scientific breakthroughs that took place, it would be difficult to find one which exceeded 1859–1874. I choose 1859, of course, because that was the great landmark year in modern biology, the year of publication of Darwin's *Origin of Species*. The impact of that book has been so pervasive that it would be an act of supererogation even to attempt to spell it out.[11] In that same year, Mendel was already at work in his monastery garden, performing his experiments with peas, which were to lead, in a few years time, to a mathematical formulation of the laws of genetics. Mendel's findings, which were long neglected, were announced in an article published in 1866, 'Experiments with Plant Hybrids'. During the 1860s, Pasteur was formulating the germ theory of disease, which was quickly followed up by Koch's remarkable work in identifying the microbial basis of some of the most dreadful diseases that afflict the human race.

During the 1860s, Kekulé discovered and described the structure of the benzene molecule, a theoretical breakthrough that was to provide guidance for wide-ranging experimental researches in organic chemistry. In 1871, Mendeleev imposed a wonderful unity upon the entire field of chemistry with his presentation of the periodic table of the elements. So great was Mendeleev's confidence in his assertion of the periodic recurrence of certain chemical properties, that he left holes in his table and predicted that elements possessing specific characteristics would be discovered which fitted into those locations. In 1873, Maxwell published his *Electricity and Magnetism*, the culmination of a line of research going back to Faraday's empirical discovery of the relationship between electricity and magnetism, an announcement which he had described in a paper to the Royal Society in 1831. Maxwell's mathematical formulations and predictions, in turn, pointed the way to further research (such as Hertz' discovery of radio waves), which was to provide the intellectual basis for all the electricity-based industries. Also in 1873, Josiah Willard Gibbs published the first two of a series of papers in which the laws of thermodynamics were applied to chemistry. Truly a remarkable fourteen years![12]

II

Thus, the post-1860 period was one of numerous and remarkable breakthroughs in fundamental science. It was also a period of widespread innovation and productivity growth, sufficient to warrant the expression 'Second Industrial Revolution'. The central question that needs to be addressed is: What was the connection between these two sets of events?

The short, and inevitable answer is, of course, that those connections were complex. By that I mean specifically to assert that no single characterization can encompass the relations between science and technological change during that period. This is not only because the relationships themselves changed over time, and at any moment in time varied considerably from sector to sector – although these statements are both true and important. It is also because – and this is why the question cannot be pursued very far from the side of history of science alone – the relationships depended at least as much upon developments in the realm of technology as in the realm of science. Thus, I will argue that the common observation that technological innovation became closely linked to the major breakthroughs in basic science during the period constitutes a premature and even unwarranted elision. While there were indeed major scientific breakthroughs, and an extensive application of science to industry during this period, the two classes of events were nevertheless much more loosely coupled to one another than is commonly thought.

The construction of a bridge between recent scientific discoveries and technological innovation is typically a very protracted affair, although that is not always the case. In the realm of medical science, specific breakthroughs could sometimes be exploited very rapidly. Röntgen's announcement of the discovery of X-ray phenomena in 1896 led to immediate applications, because, in that particular case, the technology that was required for exploitation of the new scientific knowledge already existed, and its usefulness was immediately apparent. In other fields of medical science, such as the researches of Pasteur and Lister, the necessary technologies were either easily achievable or, often, unnecessary. Knowledge of the ways in which infectious diseases were transmitted would often yield benefits by inciting simple changes in medical procedures, alterations in certain hospital practices, or public sanitation measures. Quite spectacular reductions in the incidence of puerperal fever could be achieved simply by persuading attending doctors or medical students to wash their hands before visiting each of their patients in hospital maternity wards. Nevertheless, the sad experience of Ignaz Semmelweis, who was hounded out of the medical profession of Vienna for suggesting this practice, is powerful testimony to the difficulty of gaining acceptance for new knowledge that runs counter to long-standing practice.[13] Semmelweis' assertion of the connection between unwashed medical hands and the frequency of puerperal fever is also extremely interesting for present purposes, because he never actually explained the vector of transmission. Rather, he simply inferred some connection from extensive observation. One does not necessarily *need* a good scientific theory to adopt some appropriate modification of human behaviour. Lister, after all, was the founder of antiseptic surgery, even though he believed that there was only one kind of germ producing many diseases.[14]

But while some scientific breakthroughs experienced rapid application, in other cases the applications were very slow. Sometimes the reasons were transparently simple. In Mendel's case, his writings were simply ignored for three decades, until his findings were independently rediscovered. An understanding of the laws of heredity was eventually to have profound effects in the twentieth century, especially in agriculture, with the development of new, superior hybrid varieties.[15] But, unfortunately, unread scientific papers exercise no technological consequences.[16]

In the case of Faraday and electromagnetic induction, there were no major technological innovations for several decades after his announcement of the discovery of the phenomenon in 1831, with the exception of the telegraph. There were numerous reasons for this, both scientific and technological. Faraday's discovery did not constitute a very profound level of understanding of a complex phenomenon, and it did not unambiguously point to specific directions for exploitation. Maxwell's formulation, more than forty years later, brought electrical phenomena to a level of understanding that was systematic and mathematical, and that pointed in specific, useful directions. Thus, Maxwell's prediction of the existence of radio waves was experimentally confirmed by Hertz in 1887, and this confirmation led directly to the use of radio waves for the transmission of sound.

In addition to the need for more sophisticated scientific understanding, the technological exploitation of electricity required the development of a complex system of innovations – including the dynamo for the generation of electricity, techniques for the efficient transmission of electricity over long distances, a small and efficient electric motor for converting the electricity into useful work in industry and home, and a range of alloys possessing specialized performance characteristics which were substantially different from anything required by other industries.

The technological exploitation of new scientific understanding has often been a very protracted matter, because a great deal of additional applied research has been necessary before the economically useful knowledge can be extracted from a new but highly abstract formulation. According to Cyril Stanley Smith:

> 'Perhaps the greatest achievement of nineteenth century science was the thermodynamics of Gibbs. This was exploited by metallurgists in the form of the phase rule – but not, it must be noted, until after they had catalogued and puzzled over innumerable structures and changes of structure observed empirically under the microscope in alloys of various compositions and treatment. The simple phase rule not only enabled metallurgists quickly to understand their binary and ternary systems and to limit the range of the possible, but it also provided the background against which various transformations could be studied.'[17]

In other important cases, advance in scientific understanding required, for its exploitation, the development not only of new technology but even an entirely new discipline, such as chemical engineering. Perkin's accidental synthesis of mauveine, the first of the synthetic aniline dyes, in 1856, rapidly gave rise to a new synthetic dyestuffs industry and exercised a powerful impact upon research in organic chemistry. At the same time, the exciting breakthroughs in the science of chemistry were often remote from the capacity to produce something on a commercial basis. Discovering or synthesizing a new material under laboratory conditions was usually only the very first step toward the possession of a marketable final product. The discipline of chemical engineering arose during the period as a systematic way of designing and manufacturing industrial process technologies, without which many chemical discoveries would have remained no more than laboratory curiosities.

Even when an important class of technological breakthroughs was based on recent scientific development, all subsequent improvement in that field was not necessarily directly science-based. Even though chemical science was indeed vitally important to industrial developments during the period, much of the actual timing of innovation turned upon developments in chemical engineering. Speaking of the problems of the chemical industries during the First World War, L. H. Baekeland, who had patented the first synthetic resin, Bakelite, stated

> 'When the present crisis came, we had no trouble finding chemists and chemical engineers in the field of acids and heavy chemicals. Our lack of experience was mostly evident in the newer problems of coal-tar dyes and other organic industries. Here, like in other fields of chemical industry, it is not enough to know a chemical reaction on a laboratory scale. It makes an enormous difference whether you are manufacturing by the ounce or by the ton. In a laboratory, operations can be performed in little glass vessels, or in porcelain or expensive platinum. On a manufacturing scale, all this becomes totally different and the difficulty is no longer the chemical reaction itself, but the vessels and the methods of carrying it out. Acids and other substances which attack iron and other metals have to be handled in machinery which can withstand their action and insure not only the highest yields but great purity, and exclude the possibility of accidents. An entirely new industry had to be created for this purpose – the industry of chemical machinery and chemical equipment.'[18]

Finally, it needs to be emphasized that the coupling between science and technological innovation remained very loose during this period because, over a wide range of industrial activity, innovations did not require scientific knowledge. This was true of the broad range of metal-using industries in the second half of the nineteenth century, in which America took a position of distinct technological leadership. Indeed, following the American display at the Crystal Palace Exhibition in 1851, the British came to speak routinely of 'The American System of Manufactures'. The distinctive aspect of this American System was the production of large quantities of metal components to a high degree of precision by the use of a sequence of specialized machines. The high degree of precision so attained made it possible to avoid the costly process of fitting, which consumed so much time and effort in craft-dominated industries. Components produced under the American System were sufficiently precise and standardized that they could be easily assembled rather than laboriously fitted.[19] In the second half of the nineteenth century, America provided the leadership in developing a new production technology for manufacturing such products as reapers, threshers, cultivators, repeating rifles, hardware, watches, sewing machines, typewriters, bicycles, etc. These developments also necessitated a great deal of innovation at the level of machine tools – turret lathes, universal milling machines, precision grinders, etc. In some cases, such as the well-known Lincoln miller, the machines themselves were produced by the same mass production techniques to which their own development had given rise.[20]

The development of this new machine technology rested upon mechanical skills of a high order, as well as considerable ingenuity in conception and design. It did not, typically, rely upon a recourse to the scientific knowledge of the time. The notion of 'Yankee ingenuity' has been much overworked by generations of authors who have approached the writing of American history in a celebrationist mood. The negative aspect of the notion, at least, is worth recalling in this context: Whatever the extent or the source of that ingenuity, it did not draw in a substantial way upon a scientific education.

III

The burden of the previous section has been to suggest reasons that account for the fact that, although great progress was being made at the scientific frontier after 1860, the research at that frontier was still only rather loosely coupled to industrial innovation. This is not to say that science was only loosely coupled to innovation but, rather, that *recent* scientific research was loosely coupled to innovation. To appreciate the importance of this distinction, we need to approach the interaction from the side of industry and the needs of the innovative process in major sectors of the economy. When we do so, it appears that, over a wide range of the economy, the knowledge required for moving out the *technological* frontier was rather elementary scientific knowledge of a kind that had been available for a long time. This is true in those central sectors of the First Industrial Revolution: metallurgy and the metal-using and metal-shaping sectors. In those sectors, the main needs of the innovation process were such that there was a very high payoff to elementary science and its applications. It is under those conditions that science first entered much of the industrial establishment in an organized way, and it is out of those simple and prosaic beginnings that the more sophisticated scientific research of the twentieth century eventually evolved.

The transformation of the iron and steel industry – and in fact its transformation from an iron industry to a steel industry – began with Bessemer's striking announcement, to the 1856 meeting of the British Association for the Advancement of Science, of a new technique for refining iron. Bessemer's process offered opportunities for great cost reductions, in part by taking advantage of the impurities present in pig iron and using them as a fuel. Indeed, Bessemer's paper bore the rather compelling title, 'Manufacture of Malleable Iron and Steel without Fuel'.[21]

Bessemer had developed his process without the benefit of any training in the chemistry of his day. Neither he nor his contemporaries had a very precise idea of the chemical transformations that were going on inside the converter. In fact, the Bessemer process failed to produce a satisfactory product when it was first adopted in Britain – to the considerable indignation of ironmakers who had paid handsome royalties for access to Bessemer's innovation. The blowing of air through the molten iron removed some of the impurities – the carbon and silicon. Unfortunately, as was eventually established by chemical analysis, it did not remove the phosphorus. As a result, the Bessemer process was confined for some years to the use of non-phosphoric ores. The fact that Bessemer's process could refine only materials that fell within certain narrow limits of chemical analysis had, of course, important economic consequences, imparting a strong comparative advantage to those regions possessing the non-phosphoric ores. Britain's use of the (acid) Bessemer process grew rapidly with the exploitation of her large deposits of non-phosphoric haematite ores located in the Cumberland-Furness region, supplemented by imports of non-phosphoric ores from the Bilbao region of Spain. Germany and France had only very limited deposits appropriate for the Bessemer process, and Belgium had none. The Bessemer technique was useless for the exploitation of the massive deposits of high-phosphorus ore in Lorraine and Sweden.

As was eventually established, Bessemer had conducted his original experiments using a highly pure form of pig iron, Swedish charcoal iron. It was the determination to establish the cause of the failure of the technique with certain British ores which eventually led to a prolonged, systematic analysis of the chemical processes involved in iron and steel production.

Other, closely related forces at work after Bessemer's innovation also dramatized the importance of chemical analysis: Users of Bessemer steel eventually encountered additional problems of reliability. In particular, Bessemer steel tended to become brittle as it aged; and this deterioration in quality had much to do with the declining popularity of Bessemer steel, especially as other methods of steel production became available and as new products were introduced which required steel of high quality and reliability. Although nineteenth-century metallurgists never succeeded in establishing the cause, it turned out that the blowing of air through the molten iron not only removed impurities, but added one of its own: The brittleness was caused by the presence of minute quantities of nitrogen from the air, which dissolved into the iron during the blowing.

Thus, the peculiarities of the post-1860 steel industry placed a great premium on quite basic chemical knowledge. This knowledge was of great economic value, not only in the assurance of quality control in manufacturing, but in the selection of the raw materials – the coal and the iron ore – to be employed in the productive process. Again, one must distinguish between what is 'interesting' to the historian of science and what has a high economic payoff. Manufacturers who made use of this elementary chemical knowledge could gain significant economic advantages over their competitors through a more informed selection of raw material sources. Andrew Carnegie could scarcely conceal his delight at the benefits that accrued from the chemical assays performed by his first trained chemist:

> 'We found . . . a learned German, Dr Fricke, and great secrets did the doctor open up to us. [Ore] from mines that had a high reputation was now found to contain ten, fifteen, and even twenty per cent less iron than it had been credited with. Mines that hitherto had a poor reputation we found to be now yielding superior ore. The good was bad and the bad was good, and everything was topsy-turvy. Nine-tenths of all the uncertainties

of pig iron making were dispelled under the burning sun of chemical knowledge.

'What fools we had been! But then there was this consolation: we were not as great fools as our competitors. . . . Years after we had taken chemistry to guide us [they] said they could not afford to employ a chemist. Had they known the truth then, they would have known they could not afford to be without one.'[22]

Carnegie, with his customary business acumen, had quickly perceived the vast possibilities of chemical analysis in evaluating ores and in attaining better control over metallurgical processes.

The increasing introduction of steel into new uses, such as structural engineering, created a compelling need for precise information concerning the performance of structural steel members under a variety of highly specific circumstances. The existing theory of the behaviour of steel did not provide sufficiently detailed guidance for design and construction purposes. It was the need for such detailed guidance, in numerous specific industrial contexts, which provided a powerful impetus for the establishment of new testing facilities.[23]

It seems fair to say that, although the seminal research of Henry Clifton Sorby had begun at the very beginning of our period, and was eventually to do so much to place twentieth-century metallurgy on a sophisticated scientific basis, the great metallurgical innovations of the period drew primarily upon elementary science, when they drew upon science at all.[24] None of the three great technological innovations in ferrous metallurgy in the second half of the nineteenth-century — Bessemer's converter, Siemens' open-hearth method, or the Gilchrist-Thomas basic lining, which made it possible to make steel with high-phosphorus ores — drew upon anything but elementary chemical knowledge which had already been available for a long time. Indeed, only Siemens had had the benefit of a university education.[25] Sidney Gilchrist-Thomas, for example, was a clerk in a London police court, with no more than the smattering of knowledge of chemistry that he had gained by attending evening classes. To be sure, Thomas eventually called in his cousin, Percy Gilchrist, who was employed as a chemist in South Wales. Gilchrist came to assist Thomas in some experimental trials in 1877, but Thomas had concluded, at least as early as 1875, that '. . . a lining conducive to the formation of a basic slag which would combine with and hold phosphoric acid was the chemical basis for the use of phosphoric ores in steelmaking; and he . . . had concluded that the best material for the purpose was lime or a substance with similar chemical affinities. . . .'[26] Thus, the exceedingly simple chemical insight, that the addition of lime would result in the removal of phosphorus from molten iron, resulted in a complete redrawing of the industrial map of Europe by the end of the nineteenth century, by making feasible the exploitation of the huge high-phosphorus ore deposits of the Continent.

Although the introduction of basic steelmaking involved the beginning of the use of scientific knowledge in metallurgical innovation, it is also true that the development of the science of metallurgy in the second half of the nineteenth century owed much to *prior* metallurgical innovations. That is, it arose because of the great commercial importance attached to accounting for the behaviour of metals that were already being produced by the Bessemer and post-Bessemer technologies. A particularly fruitful source of scientific research lay in trying to account for specific properties of steel produced by certain technologies or in exploiting certain specific resource inputs. Such phenomena as the deterioration of certain metals with age or the brittleness of metals made with a particular fuel were, simultaneously, particularly intriguing to scientifically trained people and objects of great concern to manufacturers and users of the metals.

At the same time, the transition to a science of metallurgy – or to 'materials science' as it has recently come to be called – was a slow process and was by no means completed by the outbreak of the First World War. New alloys continued to be developed by essentially unaided trial-and-error methods, or by sheer accident, well into the twentieth century. A new, superior aluminium alloy like Duralumin was more or less accidentally developed and used for years before anyone really understood the phenomenon of 'age-hardening'. That understanding came later, and only with the introduction of improved instrumentation, including X-ray diffraction techniques and the electron microscope.

In 1906, Wilm discovered the phenomenon of age-hardening (a finding of major significance in metallurgy) as a result of apparent inconsistencies of hardness measurements in some aluminium alloy specimens. At the time, it was impossible to connect age-hardening with structural changes which could be observed under a microscope, and no satisfactory explanation of the phenomenon was forthcoming.[27] Nevertheless, Duralumin underwent considerable exploitation in the aircraft industry (including the construction of Zeppelins) during the First World War. Age-hardened and precipitation-hardened[28] alloys were utilized in a widening circle of commercial applications during the interwar years, and the obvious commercial value of new and superior alloys gave a powerful impulse to fundamental research which would link performance characteristics of the alloys with underlying structure at the levels of crystallography and atomic structure. Nevertheless, it was the practical metallurgists who made available to the engineers an array of new materials of much greater strength, superior strength-weight ratios, strength-conductivity combinations and magnetic properties many years before performance could be explained at a deeper level. Indeed, the determination to account for the performance characteristics discovered by the metallurgist and already incorporated into numerous industrial practices was a major incentive to fundamental research. 'From these studies have arisen a better insight into deformation and strengthening mechanisms, additional support for dislocation and magnetic theories and verification of the existence and importance of lattice vacancies.'[29]

I suggest that, even well into the twentieth century, metallurgy can be characterized as a sector in which the technologist typically 'got there first', i.e. developed powerful new technologies in advance of systemized guidance by science. The technologist demanded a scientific explanation from the scientist of certain properties or performance characteristics. Such major technological breakthroughs in the metal-shaping sectors as Taylor and White's development of high-speed steel (1898), and the subsequent development, in the 1920s, of sintered tungsten carbide, are classic instances of technological improvements that preceded and gave rise to scientific research. Indeed, Frederick Taylor had been concerned with questions of shop management and organization, and was not even familiar with the rudimentary metallurgical knowledge available to the technologists of his own time.[30] Nevertheless, the discovery of the heat treatment that was necessary to impart 'red hardness' to cutting tools resulted in remarkable improvements in productivity throughout the machine-tool industry.[31]

This sequence, of technological knowledge preceding scientific knowledge, has by no means been eliminated in the twentieth century. Much of the work of the scientist today involves systematizing and restructuring in an internally consistent way the knowledge and the workable, practical solutions and methods which had previously been accumulated by the technologist. Technology has shaped science in important ways because it has acquired some bodies of knowledge first and, as a result, provided data which, in turn, became the 'explicanda' of scientists, who attempted to account for or to codify these observations at a deeper level.

Thus, if we are to deepen our understanding of how science has influenced and shaped technological innovation, we must give attention also to the increasingly interactive nature of the relationship between those two sets of forces, and we must recognize as well that the technological needs of a rapidly-expanding industrial society were often catered to by very simple science, as opposed to knowledge at the forefront of scientific research. This requires an approach to the history of science very different from that which dominates the specialist historian of science, for the direction of scientific research in the late nineteenth century, and the uses of scientific knowledge, were increasingly being dictated by the specific needs of a growing industrial establishment. In industry, the average scale of output was growing;[32] marketing and distribution requirements as well as the requirements of the productive process itself were leading to greater uniformity and standardization; and a new group of industries was emerging with very precise, specialized requirements for the high-performance characteristics of their components, including electrical industries,[33] chemical processing plants such as petroleum refineries, automobile and aircraft manufacturers, and a widening range of capital goods manufacture. Science could often make great commercial contributions by such academically 'low-brow' activities as quality control, inspection practices, elementary chemical assay and analysis, materials testing, the establishment of exact materials specifications, determining the causes of rust, corrosion, fractures, metal fatigue, etc.[34] Such activities constituted the day-to-day work at the industrial laboratories which appeared, in the United States, in large numbers in the opening decades of the twentieth century, but which had already begun to make isolated appearances in the late nineteenth century.

It is easy to be disdainful of the kinds of research that were performed in those early industrial laboratories. They were not engaged in research at the frontiers of science. Their findings hold little interest for the historian of science who is concerned with important enlargements of the stock of scientific knowledge. But such disdain is highly inappropriate for the economic or social historian, for the historian of technology – and perhaps even for the historian of science. For not only did these primitive laboratories engage in activities that were of great economic value; many of them eventually evolved into highly sophisticated research organizations which later made substantial contributions at the scientific frontier.

Notes

1. Landes, D. (1969) *The Unbound Prometheus*, Cambridge, Cambridge University Press, pp. 4, 235.
2. Kuznets, S. (1966) *Modern Economic Growth*, New Haven, CT, Yale University Press, p. 9.
3. Musson, A. E. & Robinson, E. (1969) *Science and Technology in the Industrial Revolution*, Manchester, University of Manchester Press. See also, A. E. Musson's (1972) extended introduction to *Science, Technology and Economic Growth in the 18th Century*, London, Methuen & Co.
4. Schofield, R. (1963) *The Lunar Society*, Oxford, Oxford University Press, p. 437.
5. Gilfillan, S. C. (1970) *The Sociology of Invention* [1935], Cambridge, MA, Massachusetts Institute of Technology and *Inventing the Ship* [1935], Chicago, IL, Follett Publishing Co.
6. Hall, A. R. ( ) *The Historical Relations of Science to Technology*, Inaugural Lecture,
7. Rosenberg, N., unpublished material.
8. Cohen, I. B. (1948) *Science: Servant of Man*, Boston, MA, Little, Brown & Co., pp. 68–71. The scientific finding, in turn, led immediately to a new scientific instrument, the barometer.
9. Cardwell, D. S. L. (1971) *From Watt to Clausius*, Ithaca, NY, Cornell University Press.
10. Crowther, J. G. (1936) *Men of Science*, Ch. 3, New York, W. W. Norton & Co.
11. The rush to establish priority in scientific discovery, so vividly described in James Watson's *The Double Helix*, would certainly have been very alien to Darwin's temperament, although one commentator also suggests that 'It was characteristic of the age that although Darwin first conceived his theory in 1838 he made no attempt to publish it, but devoted the next twenty years of his life to the intensive collection of evidence to support or refute it. Even when, in 1858, he received from Wallace a communication containing the gist of his own theory, he was reluctant to claim priority. Eventually joint publication was arranged through the Linnean Society. Darwin then marshalled his vast collection of evidence in his *Origin of Species*, unquestionably one of the greatest contributions to human thought ever written.' Williams, T. (1970) *The New*

Cambridge Modern History, Vol. 2, Ch. 3, *Science and technology*, Cambridge, Cambridge University Press, p. 81.

12. Bode, H. W. (1965) *Reflections on the relation between science and technology*. In: *Basic Research and National Goals, A Report to the Committee on Science and Astronautics, US House of Representatives*, Washington DC, National Academy of Science; and Sharlin, H. I. (1966) *The Convergent Century*, Ch. 8, New York, Abelard-Schuman, Ltd.
13. Semmelweis was forced to resign his professorship and moved to Budapest, where he published his book, *The Cause and Prevention of Puerperal Fever*, in 1861. Oliver Wendell Holmes had been teaching essentially similar ideas in Boston. He too encountered scepticism and hostility, but it was less intense than that which drove Semmelweis to an early and tragic death.
14. Speaking of intestinal-tract diseases, Burnet and White point out: 'The method of preventing such diseases is obvious to everyone. Indeed, the one indubitable blessing of modern civilized life is the development of the technical methods and mental attitude required to eliminate them. . . . Decent sewage disposal, pure water supply, pure food laws, control of milk supply and pasteurization, plus the cult of personal cleanliness have rendered most of these diseases rare in any civilized community. An outbreak of typhoid fever or a high infantile mortality from diarrhoeal disease is rightly regarded as a civic disgrace.

    'It is interesting to look back on the development of such a civic conscience and to realize that its all-important beginnings were based on a completely wrong idea of how infections like typhoid fever arose. In the early nineteenth century it was recognized that typhoid fever and filthy drains went together, but the stress was laid more on the smell than on possible infective principles (bacteria were, or course, unknown) in the objectionable drainage. By some transference of ideas, bad-smelling drains were also blamed for the incidence of diphtheria, and the incentive to remove these two diseases was largely responsible for the development of modern sanitary engineering. The process was under way by the time the infectious nature of typhoid fever was clearly recognized, and it was a good deal later that the typhoid bacillus was found. Both these discoveries, of course, accentuated the necessity for keeping drinking water and sewage, to put it bluntly, out of each other's way.' Burnet, M. & White, D. O. (1972) *Natural History of Infectious Disease*, Cambridge, Cambridge University Press, p. 106.
15. Here, too, it is worth recalling that selective breeding was a very old practice, carried on with considerable success long before a scientific analysis of the laws of heredity was available.
16. Gibbs' astonishingly complete formulation of the theory of 'chemical thermodynamics' in the 1870s attracted no scientific interest whatever for some time. This may have been due partly to its obscure place of publication, the *Transactions of the Connecticut Academy of Arts and Sciences*. It was only when Ostwald translated the work into German, in 1892, that it began to command the attention it so richly deserved.
17. Smith, C. S. (1961) The interaction of science and practice in the history of metallurgy. *Technol. Cult.*, Fall, p. 363.
18. Baekeland, L. H. (1917) *Metall. chem. Eng.*, 17 (7), 394.
19. See Introduction to Rosenberg, N., ed. (1969) *The American System of Manufactures*, Edinburgh, Edinburgh University Press.
20. Referring to the Lincoln miller, Fitch states that, between 1855 and 1880 '. . . nearly 100 000 of these machines or practical copies of them have been built for gun, sewing-machine and similar work.' Fitch, C. (1880) *Report on the manufactures of interchangeable mechanism*. In: *Tenth Census of the United States*, II, p. 26.
21. Writing on the use of the Bessemer process in the American rail industry, Temin has pointed out: 'It took 7 tons of coal to make 1 ton of cast (crucible) steel from pig iron by the old method: 2½ tons of coal to make blister steel and 2½ tons of *coke* to make cast steel. The same process with a Bessemer converter required 2/10 of a ton of coal to heat the converter plus about 6/10 of a ton to melt the iron. One part fuel

in the Bessemer process equaled 6 or 7 parts in the old method of steelmaking, and a comparison of labor and machinery requirements would yield similar results.' Temin, P. (1964) *Iron and Steel in 19th Century America*, Cambridge, MA, Massachusetts Institute of Technology Press, p. 131 (emphasis in original).

22. Quoted in Livesay, H. (1974) *Andrew Carnegie*, Boston, MA, Little, Brown & Co., p. 114.
23. As Timoshenko has pointed out: 'The introduction of steel into structural engineering produced problems of elastic stability which became of vital importance. . . . The simplest problems of this kind, dealing with compressed columns, had already been investigated theoretically in sufficient detail. But the limitations under which the theoretical results could be used with confidence were not yet completely clear. In experimental work with columns, inadequate attention had been paid to the end conditions, to the accuracy with which the load was applied, and to the elastic properties of material. Hence, the results of tests did not agree with the theory, and engineers preferred to use various empirical formulas in their designs. With the development of mechanical-testing laboratories and with the improvement of measuring instruments, fresh attacks were made on experimental investigations of columns.' Timoshenko, S. (1953) *History of Strength of Materials*, New York, McGraw-Hill, pp. 293–294.
24. By 1863, Sorby had developed a technique for examining metals under a microscope by the use of reflected light. This technique opened the door to an understanding of the microstructure of steel. The vast significance of this technique has been described by Cyril Stanley Smith: 'From steel, the technique spread to show the behavior of microcrystals of the nonferrous metals during casting, working and annealing. By 1900 it had been proved that most of the age-old facts of metal behavior (which had first been simply attributed to the nature of the metals and had later been partially explained in terms of composition) could best be related to the shape, size, relative distribution and inter-relationships of distinguishable micro-constituents.' Smith, C. S. (1965) Materials and the development of civilization and science. *Science*, May 14, p. 915.
25. As Bernal has pointed out: 'None of the scientific ideas used by Bessemer, Siemens and Thomas were more recent than 1790. . . .' Bernal, J. D. (1953) *Science and Industry in the 19th Century*, London, Routledge & Kegan Paul, p. 110. What was true in an old industry like iron and steel was not necessarily true elsewhere. The Hall-Heroult method of aluminium production, developed simultaneously in America and France in the 1880s, made use of an electrolytic technique which, in turn, had been made possible by more recent developments in science. The same was true of the Mond nickel process.
26. Carr, J. C. & Taplin, W. (1962) *A History of the British Steel Industry*, Cambridge, MA, Harvard University Press, p. 99.
27. 'Although he must be considered a successful applied researcher and careful experimentalist even by modern standards, Wilm did not express any deep curiosity concerning the reasons for the hardening and preferred to leave to others even the speculation concerning its nature.' Hunsicker, H. Y. & Stumpf, H. C. (1965) *History of precipitation hardening*. In: Smith, C. S., ed., *The Sorby Centennial Symposium on the History of Metallurgy*, New York, Gordon & Breach Science Publishers, p. 279.
28. An alloy that ages at room temperature is considered an age-hardening alloy, while one which requires precipitation at a higher temperature is classed as a precipitation-hardening alloy. Alexander, W. & Street, A. (1968) *Metals in the Service of Man*, Harmondsworth, Middlesex, Penguin Books, p. 176.
29. Smith, C. S., ed., *The Sorby Centennial*, (note 27) p. 309. See also O'Neill, H. (1965) *The development and use of hardness tests in metallographic research*. In: Smith, C. S., ed., *The Sorby Centennial* (note 27).
30. Frederick Taylor is, of course, better known as the father of scientific management, and his discovery of high-speed steel was the outcome of some protracted time-and-motion studies. His extraordinary report on his twenty-six years of experimentation, in his presidential address to the American Society of Mechanical Engineers in 1906, is entitled, significantly, 'On the Art of Cutting Metals'. On the opening page, Taylor states: 'There

are three questions which must be answered each day in every machine shop by every machinist who is running a metal-cutting machine, such as a lathe, planer, drill press, milling machine, etc., namely:

'a. What tool shall I use?

'b. What cutting speed shall I use?

'c. What feed shall I use?

'Our investigations, which were started 26 years ago with the definite purpose of finding the true answer to these questions under all the varying conditions of machine shop practice have been carried on up to the present time with this as the main object still in view.' Taylor, F. W. (1907) *On the Art of Cutting Metals*, New York, American Society of Mechanical Engineers, p. 3.

31. The operation of high-speed steel was demonstrated at the Paris Exhibition of 1900. 'At that exhibition they exhibited tools made of this steel, in use in a heavy and powerful lathe, taking heavy cuts at unheard of speeds – 80, 90 or 100 feet per minute, instead of the 18 to 22 feet per minute that previously had been the maximum for heavy cuts in hard material.' (1905) *Metal-working machinery*. In: *Special Reports of the Census Office*, Part IV, p. 232.
32. The increase in scale of operations also created certain opportunities that did not exist on a smaller scale – such as the utilization of waste materials. As one sharp-eyed contemporary pointed out: 'In the chemical industry . . . excretions of production are such by-products as are wasted in production on a smaller scale. . . . The general requirements for the re-employment of these excretions are: large quantities of such waste, such as are available only in large-scale production; improved machinery whereby materials, formerly useless in their prevailing form, are put into a state fit for new production; scientific progress, particularly of chemistry, which reveals the useful properties of such waste. . . .' Marx, K. (1959) *Capital* [18?], Vol. 3, Moscow, Foreign Languages Publishing House, p. 100.
33. In the electrical industries, for example, the development of new alloys became increasingly 'fine-tuned' to an array of highly specialized functions of individual components. The filament of the incandescent lamp eventually came to be made of tungsten carbide; high-silicon steel substantially improved the efficiency of much electrical equipment (such as motors, generators and transformers); steel blades of high chromium content were introduced into steam turbines; magnetic apparatus would be made of steel with additions of tungsten, cobalt or chromium; resistance elements for electrical heating apparatus would employ alloys of chromium or nickel, etc.
34. From the vantage point of 1940, it could be pointed out that 'Research in corrosion and in protective coatings, and the development of alloy steels, have more than doubled the average life expectancy of all iron and steel in the last 50 years. In 1890, the average life was 15 years, in 1910 it was 22 years, and in 1935 it was 35. A considerable part of this increase is due to higher and more uniform quality, with fewer early failures.' (1941) *Research – A National Resource.* II. *Industrial Research (Report of the National Research Council to the National Resources Planning Board)*, Washington DC, p. 169.
35. For a detailed documentation of such activities in the railroad sector, see *The Life and Life-Work of Charles B. Dudley, 1842–1909*, Philadelphia, American Society for Testing Materials. Dudley went to work in 1875 at one of the first chemical laboratories in American industry, established in 1874 by the Pennsylvania Railroad in Altoona, PA. The volume is highly instructive in identifying the kinds of research that were of great value to the railroad system. Around the end of the First World War, the Altoona laboratory had a staff of more than 600 men, '. . . half of whom are engaged in inspection work'. Greene, A. M. (1919) 'Conditions of research in US', *Mech. Eng.*, July, p. 588.

DISCUSSION

TORSTEN HÄGERSTRAND

Department of Social and Economic Geography, University of Lund, Lund, Sweden

As I read Professor Kranzberg's presentation on 'The Industrialization of Western Society, 1860–1914', a treasure on the top shelf of my grandfather's bookcase reappeared in my mind's eye: it was *The Book of Inventions*, printed in 1889–1894. One of the first things I used to do, when I came for a visit, was to ask that the six gold-tooled volumes be lifted down to me. I think I spent weeks of my childhood absorbed in the marvels described and depicted in that work. The things I read about were still in progress in my own world. The richly ornamented kerosene lamp above my head had recently been converted to a new wonder, a couple of electric lights (which one had to turn off when leaving the room). Emigrants returning to the neighbourhood had just brought home the first gramophone from America. Thomas Alva Edison was a living hero. It is not difficult for people of my generation to enter into the spirit of enthusiasm with which the actors of the end of the nineteenth century viewed their achievements. Let me share with you some of what I read.

'And what an impetus has been given to the times by the steam engine, invented by James Watt and developed by innumerable successors! With every cartload of coal conveyed, so many looms are set running, and so many meters of cloth are produced, with no further action of the human hand. In this way, coal has become a far more important standard of value than gold and silver; coal is the worker, and iron his tool. If the world were satisfied with the same quantity of labour as there was one hundred years ago, no human being would now have to strain a muscle in the mechanical manifestation of power. The steam engines now at work would be more than adequate to produce the power needed. But we are no longer that contented. We want to use more labour than we are able to yield, and this the steam engine makes possible. It breaks rocks, unites hill with hill, mountain with mountain, builds bridges over rivers and towns, builds houses, spins wool, cuts trees, melts iron, moves goods, takes us to foreign countries and lets us admire their beauties; it takes the shuttle out of the hand of the weaver, the pen from the scribe; it drains fenlands, transforming them into fertile fields, sows them, cuts the crop and makes it into bread. If a city lacks fresh water, the steam engine procures it from deeper layers of the earth, which could not be reached before; it strikes mint; it makes everything, everything; and before it is worn out, like a rational being, it makes provision for the creation of new machines which work beside and replace it. Not many books would be printed if the printing-press were not driven by a steam engine! Would newspapers have reached their present position without it? We could continue endlessly with such questions, touching equally important circumstances of our lives, but the answer would always be: "No, none of all this without the steam engine!" '

The generation born after the Second World War will probably find this kind of almost religious excitement rather strange. Few are now amazed by any of the discoveries of science and technology, even those that are vastly more sophisticated and powerful than the steam engine or the electric light bulb: Weber-Fechner's psychophysical principle appears to be creeping to the surface even in this realm. In addition, we have now become more and more aware of the costs of technological change, be they environmental, social or disastrous, as military applications could become. Perhaps all of us, old and young alike, are becoming more sober than our fathers and grandfathers about the whole matter of progress.

One sign of this is a growing interest in the history of science and technology, visible not only among professional historians but also among practising scientists and engineers. It is as if we are now trying to come to grips with the present and the future by asking questions about times past.

The problem is, what kind of historical methodology will lead to deeper insights? The voice of the authors of *The Book of Inventions* did not falter on this issue: 'Progress is an organic growth, independent of chance; the course of inventions is determined in advance, the results are necessary because the problems to be solved are put consciously to Nature and according to a rational plan.' This is sheer determinism: Things have become as they are because they had to. The role of the historian is simply to identify and laud those people who became the instruments of destiny.

In reality, such a mechanistic world picture has not been applied seriously to human affairs even by those natural scientists and engineers whose interest turned towards the history of their own fields. The history of science and technology written from the 'inside' has tended instead to go to the other extreme: to view development as the achievement of independent great individuals, standing on each other's shoulders. This perspective is perhaps the ultimate root of the prevailing myth – which Professor Rosenberg has so effectively questioned – namely, that scientific discovery necessarily precedes technological application. The facts tell a different story, at least as far as the period before the First World War is concerned: Isolated chains of consecutive causes and effects do not explain what happened during that time. These no doubt represent an overly narrow means for understanding historical processes during any period.

Professors Kranzberg and Rosenberg agree in emphasizing the interlocking nature of events in history. The *ceteris paribus* principle does not apply here; every event must be seen in its proper context – and the context is more complicated than simply that which is happening simultaneously. Time-lags impose a long perspective on how one particular thread of development is woven into the whole piece of material. In order to bring home their approach, those authors use such terms as the 'interactive nature' of forces or their 'synergistic quality' – to mention only a few.

I have no diverging opinion on this matter. It is essential to stress the need for such a broad view, not the least in connection with the history of science and technology, an area in which it is perhaps particularly tempting for the scientist- or engineer-turned-historian to isolate his subject matter from its surrounding world. But only the full contextual approach makes up the kind of history that can help us to evaluate the present and the future in any meaningful way. Having said this, I should like to remark that the methods for investigating history in contextual terms and for rendering the findings understandable to a reader are still relatively underdeveloped. This is too long a story to enter into here, so I shall indicate by a couple of points how the matter could be approached.

History, as it is usually written, is about phenomena that have taken observable and comparatively durable shape, primarily because that is the material that is readily available. There is also a quite understandable preference for winners in the battle of life. It must also be remembered – and I think everyone knows it by experience – that many intentions lead to nothing, and many thoughts and things that come into being remain disregarded or are soon discarded. For example, referring to Gilfillan, Professor Rosenberg mentioned the hit-or-miss nature of technological progress.

This is a standard feature, I am sure, of many realms other than technology. The hits, then, whatever they are, emerge as a selection from a population of several, perhaps many, efforts made earlier or simultaneously. The reason that something becomes a success is not necessarily due entirely to its own characteristics. Both hitting and missing are part of the total historical process. I believe that we relinquish important knowledge by not paying as much attention to what fails as to what succeeds. Admittedly, failures are less likely to leave clear traces in the historical sources than are successes. A full picture is clearly beyond reach; that is in the nature of things. But all material may not be lost forever: Think, for example, of the thousands of once unsaleable products that are depicted in old catalogues or the many scientific reports that have never been quoted. These and similar items should be given a proper place in empirical as well as in theoretical work.

Coming finally to the communication of contextual understanding, I feel that ordinary language is an inadequate means for this purpose. Language is made for telling stories about consecutive events; it cannot easily clarify the intricate pattern of simultaneous and time-lagged interrelationships. It is characteristic that Professor Kranzberg – despite his very strong emphasis on interactive processes – had to describe technological changes first by separating them into functional areas and then by dealing with the accompanying socio-cultural developments.

This is not meant as a criticism of the presentations, but only as a memento. Ordinary language and its taken-for-granted expressions condition our thinking and shape many of our concepts in the form of simple cause-effect reasoning and not as contextual understanding. It takes a great effort to convey ideas with such content; but it is in the latter form that deeper historical understanding is to be found, as was so admirably shown in the two presentations under discussion. I believe we must first become fully aware of this difficulty and then try to find ways of overcoming it.

V

Specialized aspects of scientific and technological change

ENGINEERS IN INDUSTRY, 1850-1910: PROFESSIONAL MEN AND NEW BUREAUCRATS. A COMPARATIVE APPROACH

ROLF TORSTENDAHL

Department of History, Uppsala University, Uppsala, Sweden

I

How did engineers enter industry? There were engineers before industry, before mechanized industry in any case. The professional men who built roads, bridges and fortresses, and those in civil architecture and mine engineering in the eighteenth century were closely connected with state bureaucracy in the countries in which they lived. In the early nineteenth century, some states – Austria-Hungary, several states in Germany (e.g. Prussia and Baden), France, Sweden and others – established engineering training in order to stimulate private industry and keep up quality.[1] This training did not go hand in hand with industrialization: on the contrary, there was a discrepancy between one educational policy that went ahead of and another that lagged behind social development. There is thus good reason to ask what really happened when engineers got involved in industry. A lot of interest has been evinced in those who became entrepreneurs; but how did the others, the first generation of salaried employees in an industrialized economy, adjust to their position? Information is available from different countries; what follows is a comparison of the histories of industrial engineers in Britain, the United States, France, Germany and Sweden before 1914.

Two specific questions are asked in relation to the historical development. First, what relation did engineers have to their profession, and how did they become professional? Much literature has been devoted to this question in its widest sense; here the scope will be limited primarily to industrial engineers. Second, what relation did they, as professionals, develop to the industrial firms that employed them, with their hierarchical administrative systems? This question can be differentiated into at least three separate problems: identification with an organization or profession; professionalism as a replacement for a career in the hierarchical decision-making system (with the introduction of staffs into the bureaucratic structure of industrial firms); and work alienation among professional engineers in industrial bureaucracies.

The conflict between bureaucracy and professionalism is not a new theme in social science. It was discussed with a great deal of enthusiasm and intensity

STSAN - R

by sociologists in the 1960s, and a first part of this paper is devoted to some of the main points of that discussion and to the refinements that the sociologists brought to it. A second part deals with the questions presented above, on the basis of various historical studies. The available data cannot be compared in all respects, and these data were obviously not generated to fit the hypotheses that have been developed by sociologists in attitudinal surveys. Thus, expressions of behaviour and attitudes are not always comparable, and actual behaviour has to replace attitudes in discussions of the historical social realities.

The intention of this paper is to draw attention partly to the common traits of the societies that were industrialized before 1914, and partly to the history of the bureaucratization of the salaried industrial engineer in the late nineteenth and early twentieth centuries.

II

'It is commonly assumed that bureaucracy, or more accurately, complex organization, clashes with professionalism.'[2] Harold Wilensky thus summed up the earlier literature in an article in which he attempted to restrict the concept of professionalization to phenomena of a certain kind. He expressed no preconceived certainty that the bureaucratic principle was in conflict with professionalism. He emphasized that professional autonomy and a service ideal were central to the concept of professionalism, but he was not sure that professional autonomy and hierarchical structure were mutually exclusive: 'In brief, perhaps bureaucracy enfeebles the service ideal more than it threatens professional autonomy.'[3] Wilensky's standpoint depends heavily on his wish to deal with professions in general. His example of professional autonomy within hierarchical structures are gleaned from medical service, from which he also raised the fear of threats to the service ideal. He and many other sociologists covered a wide range of professions, which are lumped together because they are denoted by the same word in English;[4] this linguistic prejudice may be explained by the dominance of American sociologists. Professionalization, which, according to Wilensky, had four aspects – full-time occupation, training schools, professional association, a code of ethics – was combined by Richard H. Hall with another set of criteria for defining a professional attitude. His set of five criteria included Wilensky's two: autonomy and a service ideal.[5] Hall's conclusion was that both professionalization and bureaucratization were graded; bureaucratization was even graded into different dimensions, which reduced all conflicts between the two into questions of relative strength and eliminated all dichotomous conflict.

The general ideas of professionalization and bureaucratization led to more precise ideas about salaried industrial employees and their relation to their employing organizations and professions. Robert Merton first suggested that the typical attitude of professionals was cosmopolitan, while that of bureaucrats was local.[6] In the 1960s, the unidimensionality of this pair of concepts was rejected both by certain American researchers, who found that their respondents answered in a rather self-centred vein,[7] and by some French sociologists. Michelle Durand analysed the expression of cosmopolitanism as opposed to localism. She found that attitudes that could be characterized as cosmopolitanism or localism arose only if the institutional professional goals were strong and were incompatible with the company goals; otherwise there was either a double identification with both company and profession or a pragmatism similar to the self-interest which the American researchers had found.[8]

The rejection of an unconditional conflict between cosmopolitanism and localism, and thus between attitudes connected with bureaucrats and professionals, is of far-reaching importance. The pragmatist's attitude of non-allegiance in well-conceived self-interest already characterized the 'new entrepreneurs' in C. Wright Mills' idea of the 1950s;[9] but the introduction of double loyalties was something quite new. Durand made an important assumption in this connection when she graded the compatibility of professional institutional goals with different types of industries: the compatibility should be greater with the chemical than with the electrical industry, and greater with the electrical than with the metal industry. The exact grounds for making this assumption are not presented, and it is of course questionable whether such an assumption would hold over a long time. According to her, however, a conflict between cosmopolitanism and localism would tend to appear much more frequently in the metal industry than in the chemical industry.

Durand, as well as Georges Benguigui in an article published some years before hers,[10] made explicit use of various theoretical hypotheses regarding the relation between professionalization and bureaucratization that had been developed for professionals in general, but she applied them to salaried employees and technicians in industrial firms. Used in this way, the hypotheses were not valid to the same degree as for other groups of professionals. This is true, for example, of Alvin Gouldner's hypothesis regarding the direct link between professionalism and cosmopolitanism in college teachers.[11] It also holds for Peter M. Blau's and W. R. Scott's hypothesis, based on an investigation of health and social workers, that professionalism is a means of changing a career, which could not otherwise be done without changing the occupational duties central to the profession.[12] It holds further for F. H. Goldner's and R. R. Ritti's hypothesis – which was, however, generated for industrial firms – that professionalism was a means of manipulating salaried employees such that they would experience a prestigious side of their careers without having to be taken into the company management.[13] Obviously, this is a problem that should be looked at in its historical perspective, especially since the problems encountered in industrial engineering are generally of another kind than in other professions.

Differentiation between groups of professionals may be of great importance, as shown by George A. Miller, who discussed work alienation among industrially active scientists and engineers. Organizational control, freedom of research choice and professional climate were associated with degree of work alienation in the expected manner. Another result was unanticipated: the length and type of professional training was of importance. Among professionals (both scientists and engineers) with long training, alienation from work was more strongly associated with organizational control and company encouragement than with other factors, while among those with a shorter training period the degree of work alienation was primarily connected with freedom of research choice and professional climate.[14]

III

In a discussion of the professionalization of the French engineer, Terry Shinn divided the process into several phases. Full professionalization of the industrial engineer comes only in the last of these phases, from 1880 to 1920.[15] In spite of the very long history of the French engineering profession, engineers did not become a recognized group of salaried employees in industry until about the turn of the century. But there was certainly a prehistory of the fully-fledged industrial salaried engineer. Shinn recognizes two distinct earlier phases, closely connected to educational development. First, in the period of the *Restauration* and the July Monarchy, a new kind of school arose, the *Écoles des arts et métiers*, which were completely different from the prestigious *École polytechnique* in which state bureaucrats were trained. The graduates of the new schools, the *gadzarts*, went into the industrial sector; but since the syllabus of the *Écoles des arts et métiers* was very untheoretical, they had very little that was technically new to offer to the French industry of that period: they were more qualified workers than technical experts. Those who graduated from the *École centrale des arts et manufactures*, which came into existence in 1829 and had a more theoretically oriented programme of studies in the tradition of the *École polytechnique*, were, typically, of high-class bourgeois origin and went into entrepreneurial and managerial posts, which were often inherited. Thus, when the *Grandes Écoles* got their counterpart in the area of engineering education in the late nineteenth century, they could offer something that suited people who wished to go into industry in technical posts – an engineering education that had both a theoretical and a practical content, designed for salaried employees in industry.[16]

Shinn emphasized the typical traits of engineering careers, and his general picture could be qualified. As emphasized by M. Lévy-Leboyer, and, later, G. Ahlström, the contribution of the *École centrale* to industry in the late nineteenth century was a weak one: this is also true for the other traditional institutions of higher technical education in France. The different posts held by former pupils of the *École centrale* in 1863, described originally by Camberousse in his history of the school, show the emphasis on managerial and entrepreneurial positions, although the numbers are presented in categories that prevent any thorough analysis.[17] Ahlström has pointed out, however, that the contribution of traditionally prestigious schools such as the *École polytechnique* and the *École centrale* to French industry was small in numbers but important in creating an infrastructure and in providing examples for others.[18] This conclusion is fully compatible with Shinn's contention that a real stratum of professional engineers as salaried employees in the industry was first created at the *École supérieure de physique et de chimie* and the *École supérieure d'électricité*, that is, late in the nineteenth century.

There is nothing astonishing about this rather late appearance of the professional engineer in industry in France. Salaried engineers may have appeared still later in specific branches of industry, as maintained by Etienne Dejonghe with regard to northern French coal-mining districts. He divided the relation between engineers and society in that area into three phases: before 1914, when engineers were managers and the top layer of the local society; 1914–1950, when engineers became 'officers' and acted as representatives of technocracy and superior interests; and, finally, after 1950, when engineers accepted their position as salaried employees in an industry that had become nationalized.[19]

There is little information from other countries as to the first appearance of the salaried engineer in industry. Germany was a pioneer in education for industrial engineering, with strong movements both to get theoretical men to recognize the value of practical application and to give practical men some theoretical schooling.[20] If anywhere, salaried engineers ought to have appeared earlier there; but, according to one technician who tried his luck, 'knowledge without capital found no use': Peter Lundgreen's analysis of the Prussian labour market for technicians during early industrialization shows that it was not easy for the educated technician to find a position in industry unless he went into an entrepreneurial activity. There were few salaried posts in industry, and technicians who were ambitious to rise above the ordinary workers were virtually non-existent.[21] In the recession of the 1870s and 1880s, educated German technicians had to rely on such occupations as locomotive drivers and stokers.[22]

In Sweden in the middle of the nineteenth century, it was impossible to draw a distinction between men with and without a technical education. This was partly due to the fact that, until 1851, most technical education in Sweden was more that of the *Écoles des arts et métiers* rather than a qualified, theoretically-based scientific training, except in very specialized schools. Although the degree of theoretical sophistication in education grew rapidly after 1851,[23] the technician generally still became either an entrepreneur or was on the level of workers and apprentices; in the Swedish industry of that period there seem to have been very few salaried posts for engineers.[24]

Sweden was then only poorly industrialized; but that this was not the sole reason for the weak position of technicians on the labour market can be inferred from the situation in England, where industrialization had not brought a specialized profession of industrial engineers into existence. 'Engineers' tended to be connected with engine operation, and engineering was transmitted through learning-by-doing on the shop floor rather than through theoretical analysis in schoolrooms; there were only low-level institutions for formal education, intended primarily for workers, to back the standing of engineering in society.[25] Lyon Playfair pleaded in 1851, without success, that industry be raised to the rank of a profession.[26]

Even in the United States there seems to have been no salaried industrial engineer to speak of in the middle of the nineteenth century. Monte Calvert, in his monograph on the American mechanical engineer, when he contends that only some of the mechanical engineering elite were entrepreneurs and that many of them 'were "pure" engineers with only marginal financial interest in the companies for which they worked or acted as consultants', he is referring more to the end than to the middle of the nineteenth century.[27] There is also a question of terminology: as in England, machine operators in the United States were called 'mechanics'. The mechanical engineer was an offspring of this and other categories, as Calvert has shown; but in the 1880s the National Association of Stationary Engineers had great difficulty in getting itself taken seriously by the Society of Mechanical Engineers, since they were regarded as mere operatives rather than as engineers.[28] Yet a professional engineering education was available in the United States from the 1860s.[29]

In sum, in spite of the enormous variation in the degree of industrialization in Britain, the United States, France, Germany and Sweden in the earlier half of the nineteenth century, there seems to have been no great variation in the social development of the salaried industrial engineer. It must be emphasized that the literature on this subject is scattered; but, such as it is, it seems to indicate that the engineer as a salaried employee in industrial firms came into existence not much earlier than 1880. Engineers as something other than deputies of the capitalist-entrepreneur (works-manager, *patron, dirigent*) were a product of late nineteenth-century industrial development, which created a system of management and administrative machinery in industry in which engineers could be used in different subordinate capacities.

IV

When engineers began to enter industry as salaried employees in Europe and the United States in the late nineteenth century, an almost simultaneous development, perhaps brought about by their new position in society, was that they began to be regarded as engineers, not as managers, foremen, mechanics or holders of any other specific occupation. This was a general phenomenon in the countries treated here. Technicians had been employed in such capacities before, but only occasionally and in small numbers. Some had received an education at one or other of the available technical schools: in Germany and the United States, the polytechnical institutes; in France and Sweden, the schools for higher or lower technical education; and in Britain, the few specialized technical schools. However, it is important to note that, even after 1880, many people in technical capacities in industry had not been educated as technicians in schools but had received a practical training. English educational philosophy laid a strong emphasis on such practicality;[30] the *Écoles des arts et métiers* in France reflected the same view;[31] and there were strong trends towards practicality in technical education in the United States[32] as well as in Sweden,[33] even though the effects on the training offered in schools were not the same in these countries. It is doubtful that such people, without a technical schooling but employed in technical occupations in industry, were really regarded as engineers. Some of them 'usurped' the title of 'mechanical engineer' or something in the same vein; but many of these men were thought of by others and by themselves as belonging to one or other of the technical specialities.[34] This was natural, since before 1850 'engineer' by itself was not a title: it had to be qualified with other forms, for instance, by 'civil' or 'mechanical' in the English-speaking countries and in Sweden, and by 'civil' (in another sense) in France.[35]

It is difficult to pinpoint when the industrial engineer established a firm relation with the profession. The authors of books devoted to the professionalization of the engineer have generally reversed the question and have described when the prerequisites of professionalization became available among 'engineers'.[36] In Britain, France, the United States and Sweden (and probably in Germany too, although the existing literature provides no data), the extension of the title of 'engineer' to mechanical engineers who had certain kinds of education was a crucial step in the professionalization process. The link between education and title and between professional organization and title was close,[37] and the typical pattern of professionalization appeared. Only in the United States, however, did the profession of mechanical engineer undergo a real, exclusive professionalization process; in other countries, the mechanical engineer was professionalized through his connection with the wider genus of 'engineer' (of different species). Nowhere does the industrial engineer seem to have developed a professional spirit of his own. Being a salaried engineer in industry was regarded not as a profession but as part of the profession of engineer (including mechanical engineers and, soon, chemical and electrical engineers), and many differences in occupation existed. The relation of the industrial engineers to the profession of engineer therefore was never clearly defined.

V

Before we go into the problem of the relationship between the industrial engineer and his industrial organization, something should be said about the relationship between the industrial engineer and professional education. Two possibilities are available with regard to the non-managerial industrial engineer:

1. The development of industry meant that entrepreneurs did not need a diploma from a technological institute: their success was their diploma, and if they were unsuccessful their inability was obvious. The second generation, who took over the business, required certified technical knowledge, to convince not their parents but their subordinates of their ability. When, finally, the business was transferred to the hands of the employees, a person with a recognized technical education, in addition to social and capital assets, was the preferred choice. Thus, managers came to require a technical education, and their evaluations would make it indispensable for subordinates who wanted to make a career. The demand for a technical education in industry thus increased gradually.

2. Two phases are discernible. During the first phase, it was obviously impossible for all people in industry to receive a formal education, and equally impossible to fit all posts in industry with formally educated, suitable people. Emphasis at this time was thus placed on knowledge rather than on a diploma, since knowledge represented the ability to solve the theoretical and practical problems that arose in industry. During the second phase, when society had become used to industrialization and had adjusted its educational system accordingly, formal education was available for technicians in industry, and it was possible to select educated technicians for posts. In this phase, diplomas became preferred to knowledge (as such).[38]

On close inspection, these two alternatives are not mutually exclusive; however, the emphasis is on gradualism in the first and on phase difference in the second. One important factor to note in the second interpretation, however, is the emphasis on informal education as an alternative to formal education during one period of industrialization.

Data about the careers of engineers are not so readily available that it would be possible to choose between these two interpretative models. The first model fits well into what is known about managers, and this seems to hold true of managers in countries that were latecomers in the industrial revolution, as well as for Britain;[39] to what extent it holds true for subordinate engineers is not known. It was shown earlier that data on engineers in Sweden who had a formal education fit into the second interpretative model; there are various indications that engineers in Swedish society were less dependent on their education in the phase before the 1920s.[40] What Shinn has shown about the careers of the *polytechniciens* indicates that in industry there was no self-evident value in having a diploma from the *École polytechnique* before the First World War, and that industrialists looked for knowledge rather than for a diploma.[41]

The validity of the second model may depend on the rate of industrialization, its period and the commercial relations between one country and another. Sweden was an extreme latecomer in the industrialization process, and industrialization rapidly transformed the country between the 1870s and the 1920s. The other countries considered here had more (Britain a great deal more) experience of the industrializing process during that period. The two models could be merged if the sequence 'knowledge to formal education' could be validated as a general career basis for subordinate engineers.

VI

How and when did engineers develop a relationship with the industrial firms in which they were employed? This important question can be treated historically only if it is recognized that engineers first became involved in industry as the heirs or deputies of entrepreneurs and their families, although in this capacity they were not acting as engineers in a professional sense, but rather as pioneering professional managers. Strictly speaking, technical management preceded the professionalization of engineers, if it is taken to be the result of the combined efforts of organization and education. What, then, about the bureaucratization of engineers – when did it start and what effect did it have on relations between engineers and their industrial organizations?

Numerous monographs on industrial firms have treated the rise of industrial bureaucracy relatively superficially and have considered it a process with no fundamental importance in the production or productivity of the organization. Yet this process, which C. Wright Mills dates to around 1900 in the United States,[42] was of the utmost importance for all white-collar employees. Bureaucratization meant a transfer from the shop floor to the office. In the office, the homogeneous white-collar environment meant constant contact between engineers and other salaried employees, such as cashiers and mercantile agents; and with this contact a white-collar loyalty germinated.

Jürgen Kocka analysed the origin of the white-collar world within a big industrial organization, the Siemens & Halske firm, and the transformation of groups in this office environment. Overall, it may be true that the engineer was bureaucratized from the start and that his involvement in industry came with the emergence of the large-scale corporation (in the United States, from 1870 onwards),[43] but Kocka has shown a more complicated picture. The private entrepreneurship of the Siemens & Halske company was already combined with a bureaucratic tendency in the 1860s; this made it impossible to treat the company as part of a personal and private domain. Inherent factual as well as external societal complications were the grounds for this early bureaucratization. A fundamental change occurred in the 1890s, when functions that had earlier been carried out in a personal manner were institutionalized, when the new scientific nature of technology and the standardization of fabrication made written reports on all matters routine between departments and officials, when work regulations were issued for both workers and salaried employees (which made an end to the personal feeling of responsibility to the company). This is the period when huge white-collar departments rapidly grew up and the office technician appeared.[44]

Kocka pointed out clearly, with concrete detail and with theoretical exactitude, the role of the engineers in this process. In the Siemens & Halske company, technical innovation played a big role from the beginning, but the entrepreneurs themselves were self-made men without theoretical technical schooling. The technical personnel of the company were primarily the works-managers, until, in the 1880s, the design department (*Konstruktionsbüro*) was developed by Werner Siemens, who conducted experiments in his private laboratory.[45] In the decade following 1881 (when the management of the firm was reorganized), technicians for projecting, calculating and installing electrical equipment were engaged in great numbers,[46] and offices were expanded. Workers expressed their astonishment over the big office buildings and wondered what engineers could be doing there all day.[47] The deep economic recession, starting in 1873, affected the labour market for engineers as well as for workers; and when the Siemens company, as well as many others, declined employment applications from persons trained at universities and technological institutes some engineers took operative jobs, such as locomotive drivers or stokers.[48] It is evident that during the 1890s neither years of employment nor education clearly determined the level of salary in the Siemens company.[49] Thus, education did not suffice for the engineer to be recognized in private industry; nor did membership in the occupational organization of engineers. The role of engineers' organizations was partly to differentiate between levels of education and to adjudicate demands for specialized posts from those with higher education.[50]

Allegiance to the company was a cornerstone of the Siemens entrepreneurial strategy. In a first phase, it was promoted by personal liaison among all important salaried employees and the owners. Economic rewards for accomplishments in the service of the company were, however, gradually transformed into standing benefits for certain categories of employees. A forceful execution of lines of command went hand in hand with such material stimuli, and thereby loyalty was directed to the ultimate governors of the company. Personal control was finally abolished in the 1890s, when qualitative was combined with quantitative bureaucratization. The status of a salaried employee remained an important stimulus to loyalty.[51]

Three processes thus took place simultaneously, and gradually transformed the Siemens company. Groups of salaried employees, especially engineers, were professionalized through standardized education and professional organizations; professional groups were differentiated with a separation of education and professional organization; personal reliance and loyalty to the owning family were replaced by a fully developed bureaucratic order of decision-making, with standardized, impersonal stimuli to make the employees feel allegiance to the hierarchical order that was the company. Professionalization and bureaucratization thus came to full development simultaneously, and did not in any obvious way prevent each other's development.

VII

In one important way, the differentiation of professional groups was a way of combining professionalization with bureaucratization and of dissolving their opposing tendencies. As mentioned above, Kocka points out that in the 1890s an organization of engineers with a higher education put forward demands for certain better posts in industry. He also says that nothing came of it, because it would have counteracted the principles of efficiency and profitability in industry. However, in France, Shinn has shown that around the turn of the century the *Société des ingénieurs civils* began to favour the entry into industry of engineers other than the *polytechniciens*. Their struggle, however, was directed against use on a diploma of the name of the school that was issuing it, not against the general idea of having diplomas to certify the degrees and level of education attained.[52]

Effects of a social selectivity of this kind can also be traced in Sweden. Among engineers educated between 1880 and 1910, those with a lower level of formal education compensated for it in their careers with other kinds of experience and knowledge; however, those in later cohorts could not do so to the same extent. Thus, the differentiation between groups with different levels of education had certain consequences for social selectivity in the labour market in Sweden from around 1920.[53] Bureaucratization in Swedish industry may thus have taken at least part of a step that Kocka considers German industry shrank from some decades later.

Kocka's study of the Siemens company shows that the engineer need have no difficulty in adopting a double identification, as analysed by Michelle Durand, with a strong, enforced allegiance to his company and loyalty to his profession. It is more probable that engineers had this double identification rather than a pragmatic attitude to defend and make useful their technical education. The Swedish example shows that professionalization and bureaucratization were not only compatible but could fit closely together in the economy of organized capitalism.

In the 1870s, technical education began to be thought of as a means to balance the needs of society and to supply industry with the different manpower resources it needed. A system of differentiated levels of technical education had existed since the 1850s, but it was revised according to a new philosophy in which each man was kept in his place. Even if this plan was never fully carried out as such, the fundamentally bureaucratic principle that there was a correct schooling for each occupational capacity was inherent in the technical school system.[54] However, in Sweden the gap between social needs for manpower and the output from the educational system grew during the decades after 1870, when rapid industrialization transformed Swedish economy. This gap was wide at the beginning of the twentieth century, but there are indications that it closed gradually from the second decade onwards, indicating that the educational system had adapted gradually to the new needs of society. It also indicates that formal education was again replacing (unformalized) knowledge as the basis for careers in society.[55]

VIII

In the historical development of professionalism and bureaucratization, there was no direct opposition between them; rather, they joined and reinforced each other. However, two questions may be raised: (1) Are there historical differences between a cosmopolitan and a localist attitude, and can the origins of a cosmopolitan attitude be traced? (2) Are there indications that professionalism has functioned as a substitute for careerism or as a way of manipulating professionals who do not fit into a hierarchical career pattern?

The origins of cosmopolitanism are hard to trace. Generally, technical education has played a large role in fostering a feeling for the general rather than the specific content of industrial occupations. Kocka also mentions that Werner Siemens regarded technology as something general, which meant that people with a technical education could be useful in different capacities; but he points out that this generalist attitude was also motivated by the monopolist aspirations of the company, which gave the only specialized training for electrotechnicians in Germany at that time.[56] Even when this goal had been abandoned, engineers could be used in different capacities; if they could be so used, they would have varied ambitions, with implications for their careers. As there are few standards with which to compare these data, one can only state that an investigation of the careers of Swedish engineers who took their diplomas between 1880 and 1910 shows that they were highly mobile both as regards the type of business they went into and as regards the character (or function) of their work. It was quite common to change both job and type of business and character of work, not only in the first years of their occupational career but also (though to a lesser extent) in the later part (fifteen to thirty years after taking their diplomas).[57] These Swedish engineers were almost equally mobile as regards kind of business and character of work, whatever their level of occupation – excepting those who had become managers and who mostly remained in administrative work, although they might change kind of business. Many of them started in the metal industry but went into others later; changes from operative work to design, or the reverse, also took place quite frequently during the first half of a career, and later stabilization did not exclude further changes.[58]

Thus, there are at least some indications that cosmopolitanism was inherent in the technical occupations. Of course, cosmopolitanism refers more directly to a firm of employment and is the opposite of a localism which does not signify a constant character of work but rather an allegiance to the firm. The Swedish data were not analysed for this question, and the degree of localism proper cannot be determined from them; however, changes in kind of business and in character of work often imply a change of employer.

What, then, is professionalism in the career of an engineer? Did it function as a substitute for a 'real' career or as a way of manipulating engineers? The genesis of the industrial laboratory had both advantages and drawbacks, according to Wilhelm Treue, the advantage being, foremost, to provide a possibility for surpassing the old crafts level of industry, the drawbacks being restrictions on individualism and bureaucratization of research, with accompanying secrecy of research results.[59] It is, however, highly questionable that secrecy had anything to do with the form of organization of research. It is more probable that industrial research was not secret by principle but was made secret by the intensified competition of the big firms in organized capitalism.[60]

In order to determine the role of 'professional careers' in industry, a vast amount of empirical material on engineering careers is needed, along with interpretations of the development of industrial organization. These are not immediately obtainable. Different authors have explored various aspects of changes in business history; and the interrelationship between the separation of management from ownership, functional specialization, 'rationalization', separation of research and development and creation of the line-and-staff system is neither obvious nor fully ascertained. Further, the changes of organization differed between the United States and Europe: In the American railways, already by the middle of the nineteenth century, Alfred Chandler argues, there was an organizational structure later known as the line-and-staff system; and by the beginning of the 1920s both General Motors as well as the Du Pont firm were already organized in divisions.[61] The latter organizational pattern did not develop in Germany until the 1960s, according to Hannes Siegrist, and this seems to hold true for Sweden as well.[62] For the time-span considered here the diffusion of functional specialization is more pertinent. Really professional careers were not possible in industry before industrial companies took research and development seriously. As mentioned above, Kocka showed the importance of the design department (*Konstruktionsbüro*) in the Siemens company from the 1870s. Hannes Siegrist has shown that the Georg Fischer Company established a line-and-staff organization and a laboratory at the turn of the century, and generally industrial companies did not create any separate departments for research and development until the First World War.[62] This coincides approximately with the replacement of the pure line system with the line-and-staff system; and Fritz Croner emphasized that the functional separation of this innovation was quite in line with F. W. Taylor's scientific management.[63] Taylor's functional approach,[64] and its development during and after the First World War, fall, however, outside of the present discussion.

General organizational development thus makes it probable that internal professional careers, as distinct from hierarchical careers in decision-making positions, lacked the fundamental prerequisites in most industrial companies before 1914. This was the case in the main Swedish industries;[65] and monographs on international big business suggest that organized research and development departments were sporadic before 1914.[66] In Sweden, research, development and design occupations formed a general starting-point for engineers with a low level of education; fewer of those with a higher education started their careers with occupations of that type, although later in their careers the situation was the opposite. This is due to the fact that most of those with a low level of education, who started their careers in occupations of a research character, had to take routine jobs and were promoted away; while some of those with a high level of education found their way into research-and-development occupations after some years in production-line management. The latter category seems to fill the outward criteria for a 'professional career' in industry, although perhaps manipulated by industrial management. It is thus important to note that occupations of that character generally saw a high turnover of engineers, indicating that experience gained in this kind of occupation was desirable, but only to a certain extent. Generally, the proportion of those who had occupations of this character fell gradually in cohorts of engineers with diplomas; only the 1906–1910 cohort of those with a high level of education is an exception. The actual numbers of engineers in research and development were rising all the time.[67] The evidence that there were professional careers in industry before 1914 is thus very weak: both organizational policies and aggregate career analysis make it improbable that internal industrial professional careers could have played an important role in industrial history before 1914. Such careers were therefore not inherent in industrial development.

Conclusions

Data collated by historians from different countries regarding the history of the industrial engineer have been put together into a comparative perspective, and only the scattered nature of the material has prevented further comparison. Further, sociological data, primarily for the 1960s and regarding the professionalization and bureaucratization of industrial employees, have been submitted to a historical analysis. These two approaches have given the following results.

Engineers have existed as engineers since at least the eighteenth century, but with different relationships to the growing industry. Engineers as a group of salaried employees in industry appeared in greater numbers, independently of the rate of industrialization and the history of engineering, almost simultaneously in Britain, the United States, France, Germany and Sweden in about 1880. The salaried industrial engineer was obviously not a product of industrialization as such but of the economic organization of industry during the phase of industrial capitalism which had appeared to connect the industrial countries from about 1870 (a little later for the latecomers to industrialization).

The relationship between the new, salaried industrial engineer and education is not quite as obvious as some previous literature on professionalization tempts us to assume. While the entrepreneur, either educated or not, could convince the world of his ability only through the success of his firm, the employed manager had to convince the owner(s) of his ability before he could start work; and the general salaried engineer had to convince superiors of different backgrounds and positions of his ability before he had the chance to show it. One way of making others believe in one's capacity was to submit a certificate of education, which then attained the function of a social charter of competence, besides assuring certain levels of knowledge. What has been argued here, primarily on the basis of some German and Swedish evidence, is that the growth of administrative organization in industry favoured the social recognition of education as such and gradually diminished the direct value of knowledge in careers.

The professionalism of engineers contributed to this process. Professionalism thus denotes the attitude of self-consciousness of professional groups. Education and organization were prerequisites of professionalization. Engineers outside of industry, especially civil engineers, were professionalized before mechanical engineers; but these, too, organized between 1870 and 1890 in the countries considered. Professionalization was, however, combined with differentiation, which meant that highly educated groups separated themselves from other engineers and advanced pretensions that were not acceptable to the others.

Professionalization and bureaucratization were simultaneous processes in industry and had a certain interdependence. Gradual concepts of professionalization and bureaucratization (R. Hall) fit the historical reality much better than dichotomous concepts, especially if the latter are considered to be in direct conflict. Kocka's results of a detailed examination of one company fit well into the vague framework established by other monographs. Personal loyalty of a salaried employee to an owner or his family vanished with bureaucratization, which had two dimensions: The hierarchical order was expanded and its levels accentuated, and an impersonal allegiance to the organization was stimulated through material rewards and prestige. There is then good reason to believe that many engineers both in the past and now have a double identification with both profession and organization (M. Durand). It is less easy to find historical evidence of a pragmatic attitude.

Cosmopolitan attitudes – typical of the professional – can be traced to some extent in the career patterns of Swedish engineers on the aggregate level. Comparison with other countries is not possible because of lack of comparable investigations. There are, however, historical reasons to believe that professionalism had not become an impasse for the professional engineer before 1914. The overall organizational pattern of industrial administration is one such reason, and it seems to be valid for all the countries compared. Only in exceptional cases were research staffs and fully developed research and development departments flourishing before the beginning of the twentieth century; and the staff-and-line system became general after the First World War. The career patterns of Swedish engineers do not indicate that specialized professional careers could serve to manipulate professionals in industry.[13]

Notes

1. For a general but not reliably detailed survey, see Emmerson, G. S. (1973) *Engineering Education: A Social History*, Newton Abbot; see also Schnabel, F. (1925) *Die Anfänge des technischen Hochschulwesens*, Karlsruhe; and literature cited below.
2. Wilensky, H. (1964–1965) The professionalization of everyone? *Am. J. Sociol.*, **70**, 137–158, on p. 146. I could not apply Etzioni's definition by goals and compliance of organizations, e.g. professional organizations. See Etzioni, A. (1961) *Complex Organizations. On Power, Involvement and Their Correlates*, Ch. 4, Glencoe.
3. Wilensky (note 2), p. 148.
4. Benguigui, G. (1967) La professionnalisation des cadres dans l'industrie. *Sociol. Trav.*, **9**, 134–143.
5. Hall, R. H. (1968) Professionalism and bureaucratization. *Am. Sociol. Rev.*, **33**, 92–104.
6. Merton, R. (1957) *Social Theory and Social Structure*, 2nd ed., Glencoe, pp. 441–474.
7. Goldberg, L., Baker, F. & Rubinstein, H. (1964–1965) Local-cosmopolitan: unidimensional or multidimensional? *Am. J. Sociol.*, **70**, 704–710.
8. Durand, M. (1972) Professionnalisation et allégeance chez les cadres et les techniciens. *Sociol. Trav.*, **14**, 185–212.
9. Mills, C. W. (1959) *White Collar. The American Middle Class* [1951], New York, pp. 93–100.
10. Benguigui (note 4).
11. Gouldner, A. W. (1957–1958) Cosmopolitans and locals. *Adm. Sci. Q.*, **2**, 281–306, 444–480.
12. Blau, P. M. & Scott, W. R. (1963) *Formal Organizations*, London, pp. 68–70.
13. Goldner, F. H. & Ritti, R. R. (1966–1967) Professionalization as career immobility. *Am. J. Sociol.*, 72, 489–502.
14. Miller, G. A. (1967) Professionals in bureaucracy: alienation among industrial scientists and engineers. *Am. Sociol. Rev.*, 32, 755–768.
15. Shinn, T. (1978) Des corps d'état au secteur industriel: genèse de la profession d'ingénieur, 1750–1920. *Rev. fr. Sociol.*, **19**, 39–71 [also printed as an annex in Shinn, T. (1980) *École polytechnique, 1794–1914*, Paris].
16. Shinn (note 15). See also Shinn, T. (1980) The genesis of French industrial research, 1880–1940. *Soc. Sci. Inf.*, **19**, 607–640, esp. 620–622.
17. Lévy-Leboyer, M. (1974) Le patronat français, a-t-il été Malthusien? *Mouvement soc.*, **88**. Ahlström, G. (1978) Higher technical education and the engineering profession in France and Germany during the 19th century. *Econ. Hist.*, **21**, 51–88, esp. 62–67.
18. Ahlström (note 17), p. 73.
19. Dejonghe, E. Ingénieurs et sociétés dans les houillères du Nord. Pas-de-Calais de la belle époque à nos jours. Communication to a meeting 23–25 October 1980 (unpublished).
20. Treue, W. (1964) *Das Verhältnis der Universitäten und Technischen Hochschulen zueinander und ihre Bedeutung für die Wirtschaft*. In: Lütge, F., ed., *Die Wirtschaftliche Situation in Deutschland und Österreich un die Wende vom 18. zum 19. Jahrhundert*, Stuttgart, pp. 223–237.
21. Lundgreen, P. (1975) *Techniker in Preussen während der frühen Industrialisierung, Ausbildung und Berufsfeld einer entstehenden sozialen Gruppe*, Ch. 6, Berlin, pp. 227–272.
22. Kocka, J. (1969) *Unternehmensverwaltung und Angestelltenschaft am Beispiel Siemens 1847–1914. Zum Verhältnis von Kapitalismus und Bürokratie in der deutschen Industrialisierung*, Stuttgart, pp. 271 ff.
23. Henriques, P. (1917) *Skildringar ur Kungl. Teknisk Högskolans Historia*, Vol. 1, Stockholm, pp. 176–179, 304–322; Torstendahl, R. (1975a) *Teknologins nytta. Motiveringar*

för det svenska tekniska utbildningsväsendets framväxt framförda av riksdagsmän och utbildningsadministratörer 1810–1870, Uppsala, pp. 61–74.

24. Gårdlung, T. (1940) Teknik och tekniker i den tidiga svenska verkstadsindustrin. *Ekon. Tidskr.*, 42, 179–191; Gårdlund, T. (1942) *Industrialismens Samhälle*, Stockholm, pp. 219–223.
25. Emmerson (note 1), pp. 60–63, 91–131; Armytage, W. H. G. (1965) *The Rise of the Technocrats. A Social History*, London, pp. 96–98; Harrison, J. F. C. (1961) *Learning and Living, 1790–1960. A Study in the History of the English Adult Education Movement*, London, pp. 57–89; Kelly, T. (1957) *George Birkbeck, Pioneer of Adult Education*, London.
26. Argles, M. (1964) *South Kensington to Robbins. An Account of English Technical and Scientific Education since 1851*, London, p. 17.
27. Calvert, M. (1967) *The Mechanical Engineer in America, 1830–1910*, Baltimore, p. 71.
28. Calvert (note 27), pp. 25–27, 29–40, 189–191.
29. Bledstein, B. (1976) *The Culture of Professionalism*, New York, esp. pp. 193–196.
30. Argles (note 26), pp. 10 ff.
31. Artz, F. B. (1966) *The Development of Technical Education in France, 1500–1850*, Cambridge, Mass., pp. 205–208.
32. Calvert (note 27), pp. 13–15, 69–85.
33. Torstendahl (note 23), Ch. 2; Gårdlund (note 24).
34. Calvert (note 27), pp. 14 ff.
35. Calvert (note 27), pp. 13–27; Calhoun, D. H. (1960) *The American Civil Engineer. Origins and Conflict*, Cambridge, Mass., esp. pp. 82–90; Armytage (note 25), pp. 99 ff; Shinn (note 15), pp. 40–47; Runeby, N. (1976) *Teknikerna, vetenskapen och kulturen. Ingenjörsundervisning och ingenjörsorganisationer i 1870-talets Sverige*, Uppsala, Chs 3 and 4.
36. For instance, Calvert (note 27), Ch. 6; Runeby (note 35), Ch. 3; Shinn (note 15); and, partly, Kocka (note 22). Note the important contributions of Runeby, who summed up much of the international discussion on professionalization of engineers.
37. In addition to the works mentioned in note 36, see also Layton, E. (1971) *The Revolt of the Engineers. Social Responsibility of the American Engineering Profession*, Cleveland, Ohio; Noble, D. (1977) *America by Design. Technology and the Rise of Corporate Capitalism*, New York; Manegold, K.-H. (1970) *Universität, Technische Hochschule und Industrie. Ein Beitrag zur Emanzipation der Technik im 19. Jahrhundert unter besonderer Berücksichtigung der Bestrebungen Felix Kleins*, Berlin.
38. For further information about the second alternative, see Torstendahl, R. (1975b) *Dispersion of Engineers in a Transitional Society. Swedish Technicians, 1860–1940*, Uppsala, Ch. 7 and 9.7; and Torstendahl, R. (1981) The social relevance of education. Swedish secondary schools during the period of industrialization. *Scand. J. Hist.*, 6, 77–89.
39. Pollard, S. (1965) *Genesis of Modern Management. A Study of the Industrial Revolution in Great Britain*, London, pp. 104–159. Lévy-Leboyer, M., ed. (1979) *Le Patronat de la seconde industrialisation*, Paris, esp. Torstendahl, R., *Les chefs d'entreprise en Suède de 1880 à 1950: sélection et milieu social*, pp. 37–50; Kocka, J., *Les entrepreneurs salariés dans l'industrie allemande à la fin du XIXe et au début du XXe siècle*, pp. 85–100; Lanthier, P., *Les dirigeants des grandes entreprises électriques en France, 1911–1973*, esp. pp. 101–116.
40. Torstendahl (1975b) (note 38), pp. 191, 196, 289.
41. Shinn (1980) (note 16), pp. 158, 169 ff.
42. Mills (note 9), pp. 77–87.
43. Calhoun (note 35), esp. p. 54; Layton, E. (1969) *Science, business and the American engineer*. In: Perrucci, R. & Gerstl, J., eds, *The Engineers and the Social System*, New York, esp. pp. 51–54.
44. Kocka (note 22), pp. 131–145, 254–258, 289–303, 463–490, 547–552. Siegrist, H. (1981) Vom Familienbetrieb zum Managerunternehmen. Angestellte und industrielle

Organisation am Beispiel der Georg Fischer AG in Schaffhausen 1797–1930, Göttingen, Vandenhoeck & Ruprecht, p. 95. Kocka recently published an article in which he generalized his results and emphasized bureaucratization as the starting-point for the professionalization of German engineers: Kocka, J. (1981) Capitalism and bureaucracy in German industrialization before 1914. *Econ. Hist. Rev.*, 38, 453–468, esp. p. 463. D. Rüschemeyer emphasizes in a related vein that the State played a more prominent role in German than in British or American professionalization. Rüschemeyer, D. (1980) Professionalisierung. Theoretische Probleme für die vergleichende Geschichtsforschung. *Gesch. Ges.*, 6, 311–325, esp. p. 324.

45. Kocka (note 22), pp. 239–241.
46. Kocka (note 22), p. 255.
47. Kocka (note 22), p. 299.
48. Kocka (note 22), p. 271.
49. Kocka (note 22), p. 284 ff.
50. Kocka (note 22), pp. 271, 279.
51. Kocka (note 22), pp. 111–116, 238, 246–250, 259–262, 291–297, 465–466, 483, 549–558.
52. Shinn (1980) (note 16), pp. 129 ff.
53. Torstendahl (1975b) (note 38), pp. 180–185, 241–247.
54. Runeby (note 35), esp. Chs 7 and 9.
55. Torstendahl (1981) (note 38).
56. Kocka (note 22), pp. 275–277.
57. Torstendahl (1975b) (note 38), Ch. 3.3. and 4.3.
58. *Ibid.*
59. Treue (note 20), p. 230.
60. Torstendahl, R. (1980) Industrial research and researchers in Sweden, 1880–1940. *Soc. Sci. Inf.*, 19, 641–661, esp. 650.
61. Chandler, A. D., (1977) *The Visible Hand*, Cambridge, Massachusetts and London; Granstrand, O. (1979) *Technology, Management and Markets*, Göteborg.
62. Siegrist (note 44), pp. 99, 103.
63. Croner, F. (1951) *Tjänstemannakåren i det moderna samhället*, Stockholm, pp. 79–100.
64. Of the vast literature on Taylorism, see especially Moutet, A. (1975) Les origines du système de Taylor en France. Le point de vue patronal (1907–1914). *Mouvement soc.*, 93, 15–49; Runeby, N. (1978) Americanism, Taylorism and social integration. Action programmes for Swedish industry at the beginning of the twentieth century. *Scand. J. Hist.*, 3, 21–46; De Geer, H. (1978) *Rationaliseringsrörelse i Sverige*, Stockholm.
65. Granstrand (note 61), esp. Ch. 4, see also Ch. 10; Eriksson, G. (1978) *Kartläggarna. Naturvetenskapens tillväxt och tillämpningar i det industriella genombrottets Sverige, 1870–1914*, Umeå; Torstendahl (1980) (note 60).
66. For example, Chandler, A. D. & Salsbury, S. (1971) *Pierre S. Du Pont and the Making of the Modern Corporation*, New York.
67. Torstendahl (1975b) (note 38), Ch. 4.3–4.4.

CHEMICAL INNOVATION IN PEACE AND IN WAR

L. F. HABER

Department of Economics, University of Surrey, Guildford, Surrey, UK

Introduction

The purpose of this Symposium is to exchange views on the interaction between science, technology and society. No one denies that they are connected, but which of the three claims precedence in the causation of historical change? I do not propose to investigate this particular example of the familiar chicken-and-egg problem because it does not lead to a conclusion. Indeed, if we had a theory of growth that could explain the chemical industry's development, then all would be clear, and my paper would be unnecessary. But, so far as I know, no such theory has yet stood the test of time and of empirical research. My contribution today will be a look at the continuing debate on causes and consequences in terms of Anglo-German rivalry in the early years of the twentieth century. That approach makes it possible to extend the technical and economic analysis into the broader areas of entrepreneurial attitudes, political conflict and the pressures imposed by war.

What is innovation? I have adopted the broad definition used by Christopher Hill,[1] which extends from the creation to the commercial diffusion of a new product or process. Under such a wide umbrella, invention, production and marketing can be safely accommodated; and, as will be seen, all three had their different roles to play.

Characteristics of the Rivalry before 1914

At the opening of the century, the chemical industries of Britain and Germany were not evenly matched. Those of the former seemed to be looking back and stayed with the old-established alkali and bleaching-powder trades — hardly dynamic, rather technologically static. The latter was showing signs of specializing in the newer branches of applied chemistry, specifically in coal-tar dyestuffs and in the organic intermediates from which they are derived. Although the two industries were drawing apart in rate of innovation, they were at the same time showing many structural similarities. Large enterprises were emerging, the barriers to entry were rising and there were many restrictive trade practices.

From this summary description — based on such data as are available and reliable – one would be bound to conclude that the differences between the two countries were not due solely to technical or economic factors: other forces were also at work accelerating or retarding change. The most significant were attitudes to the business of chemical manufacture and, particularly, to risk-taking. In England, the professional chemist usually played a subordinate role and, as far as policy-making was concerned, he was ignored. We would now consider such a state of affairs intolerable, but in late Victorian Britain the pursuit of science had a higher standing than its application. The ownership and management of most chemical works were predominantly in the hands of laymen who failed to appreciate the value of chemical innovation. One result was that potential cost savings from process improvements were neglected, another that monopoly profits through the introduction of new products were rare. Hence, manufacturers were forced to look for cheaper inputs, principally cheaper fuel, raw materials and labour; but there were limits to such savings. Alternatively, they hoped for a change in commercial policy, in the expectation that their inefficiencies would survive behind a tariff barrier. The first solution works successfully as long as price relativities favour the producer, but it would be imprudent to count on it in the long run, and it has proved to be an unfailing recipe for industrial decay. The latter, soon to form part of the 'infant industries' case, was out of the question at a time when Free Trade was still a powerful political force.

Everything pointed the other way in Germany. Contemporaries frequently drew attention to the practice of chemistry (as distinct from its academic study) and to its rapid commercial application in the Wilhelmine Empire. There, Schumpeter's 'creative destruction' and his vivid metaphor that '. . . technological possibilities are an uncharted sea'[2] could be justly applied to innovation in chemicals. We have seen more recently similar forces at work in Japan, and the widening gap between 'leaders' and 'laggards' has been noted elsewhere. What is surprising in this context is the rapidity of the German advance and its wide diffusion.

The Germans did not, at first, rely on home-grown inventiveness, nor on the monopoly power of patents, nor on the economies of scale or of vertical integration. They started small, but grew fast because they tried harder, and because those who directed the enterprises were qualified chemists, receptive to technological change and prepared to take risks. They were also lucky because organic chemistry turned out to be an exceptionally fertile soil when cultivated by energetic innovators. This last aspect needs to be stressed. Much of the data on development in the early dyestuffs industry were secret and inaccessible. The results of basic research were in the public domain; ignorance was no excuse. The difficulties and the financial risks, but also the opportunities lay in the cost of bringing the new products to the market. But the risk-takers also had their reward, for when the first commercial impetus was spent – that is, when the characteristic S-shaped logistics curve approached zero-growth –

another product came along to supply a fresh onward thrust. Hence, diversification occurred within the dyestuffs industry itself and, by 1900, into dyeing auxiliaries, drugs, agriculture and photochemicals.

I have described the details of this process elsewhere.[3] Suffice it here to note that within a matter of ten to fifteen years the gap between the two countries was very wide. The differences were a matter of frequent comment on both sides of the North Sea, and the wonder is only that up to the First World War the British response was so slow. The differences were indeed so great as to invite the question why British chemical manufacturers paid so little attention to the economic indicators that were available to them. Consular reports and foreign trade statistics signalled a continuing deterioration. Both countries were exporting more than they imported, but it was the composition of the trade that mattered: Britain typically exported alkalis, bleaching materials, coal-tar distillation products, some manufactured fertilizers, explosives and patent medicines. The trade increased slowly because erstwhile customers were protecting their own manufactures and so bought less, while new customers, particularly in the Far East, only gradually made up for the lost North American sales. Even more seriously, the unit value of chemical exports was relatively low.

Germany exported a similar range of chemicals, as well as potash salts, which were at the time a natural monopoly; but its exports grew faster, and, above all, there was the dyestuffs business, expanding impressively and characterized by very high unit values. Comparison of the export trade was not flattering to the British. There was another awkward fact: the British textile finishers were then the largest single group of customers for dyestuffs, and, as a result of mergers, a handful of dyers and printers had emerged as valuable customers; these the Germans pampered with free technical advice and quantity rebates. The British producers of colouring materials lacked the means to copy the German sales support methods and had to follow the German price lead. Price-takers have to struggle harder, especially when their research and development effort is too small to improve quality through process improvements or to extend the range through product innovation.

Here was a dilemma, but there was a solution — 'if you can't beat them, join them!' There was nothing novel in the response, but it was atypical for the leading manufacturer, Levinstein Ltd, to go its own way. The incident, which is of some significance in the history of the British chemical industry, serves to underline the role of the individual in decision-taking, even if the decisions taken were the wrong ones. At any rate, Dr H. Levinstein, the founder's son, was convinced that he could beat the Germans at their own game. Unfortunately for him, the business was under-capitalized; and, failing to get support from the banks, he called on the government for help. That move turned dyestuffs into a political issue, for Levinstein, and those who were like-minded, warned of dire consequences to the country if indigenously-made dyes were not protected from unfair German competition. It was not unusual then (and it has become common practice since) to consider the growth of particular industries as being in the national interest and to treat a high rate of innovation as a manifestation of technological patriotism. If the Germans and others did so, why not the British? The problem for Levinstein was how to achieve his ends: a tariff was the obvious solution. There were no import duties at the time, and the rate would therefore have to be fixed at a level high enough to give British manufacturers sufficient protection to justify home production. The post-war experience of other countries suggested that a rate of fifty to one hundred percent on the point entry value would be needed. If that were done, the Germans would transfer manufacture to Britain and take advantage of the higher duty paid on prices. This would have been of no benefit to British firms, and under the circumstances obtaining at the time the proposal was a political non-starter. That left the possibility of patent law reform, specifically a working clause, as used in all other countries. This sensible measure was widely canvassed (not least by Levinstein), and the new Act of 1907 made provision for it. Within two years the effectiveness of the legislation was destroyed by an adverse judgement, which did not even concern a chemical compound.

These details serve to illuminate the legal and economic background, as well as personal judgements, which combined in those prewar years to handicap chemical innovation. Some of the obstacles could have been overcome or their effects mitigated, but there were other more deeply-seated problems, which needed to be tackled first, if the gap between the industries of the two countries was to be narrowed. The first of these was the role of the chemist or chemical engineer in the strategy of the firm; the second, much less easy to remedy, was the vigour of the marketing effort.

One could argue at one extreme and assert that the professional scientist should be represented on the board of the enterprise and, at all levels, should participate in making and implementing policy. That was the position in Germany. At the other extreme was the typical British management structure just before the war, in which two or three chemists were on tap, but not on top. The rewards were commensurate, and there were, not surprisingly, frequent complaints from the professional chemists. The kernel of the argument, however, lay elsewhere: was further and higher education responsive to the needs of the manufacturing sector of the economy? If it was, there would be enough chemists coming forward to seek a career in business; if it was not, then employers were, through no fault of their own, prevented from recruiting the people they needed. I am aware that this approach tends to exaggerate the Anglo-German attitudes to applied chemistry. It was a fact that an Oxbridge education was better suited to the formation of academic chemists, while German universities turned out a man who was more versatile. But that, on its own, is not a sufficient explanation; and the relationship between universities and chemical enterprises was, in Germany, a major factor in promoting innovation.

An arrangement, now in common use, was then practised only in Germany, and consisted in farming out research to universities or polytechnics. Businessmen were thereby enabled to transfer some of their research overheads to the educational system, but it is likely that the other benefits were more important. Research students needed dissertation subjects, young and ill-paid lecturers secured extra income as supervisors, and professors obtained funds to pursue their specialities and gain fame and consultancy fees. The companies, as paymasters, were able to choose among a wide range of investigations, secure in the knowledge that the donkey-work would be done thoroughly under conscientious supervisors. It wasn't very exciting, but much chemical research was purely routine, and the results were likely to gather dust as yet more uninspiring PhD dissertations. In principle, there was little cross-fertilization between university and business, for information flowed one way and financial support the other. But a great deal of scientific hack work got done; there was some talent spotting; and the senior members of the academic staff played a valuable role as go-betweens.

The Great War

Chemical retardation and neglect of research were considerable handicaps to the British upon the outbreak of war. The rivalry with the Germans had reached a point at which grave fears were expressed as to the survival of British industry, while, on the other side, many Germans overrated the significance of their chemical superiority. Events were to confound the forecasts made by each side. The effect of war as an accelerator of technical change was not understood at the time, and the contribution from chemistry was a totally unknown factor, the more so as General Staffs had no chemical advisers, and chemists were unfamiliar with military ways. All therefore depended on how quickly the professionals could get together and harness the industry's products to the requirements of war. It was soon realized that home-produced goods had to be developed quickly to replace imports. The size, versatility and innovative capacity of the industry would therefore determine the speed and success of this substitution.

The Germans were faced, within weeks of the outbreak of war, by unexpected problems in the supply of nitrates. Their own indigenous production of nitrogenous materials was insufficient for the combined demands of farmers and munition factories, and imports soon dried up. The story of how the crisis was overcome by enlarging the factories operating the arc, the cyanamide and, above all, the Haber-Bosch process has often been told and need not be repeated here, except to note its relevance to points already touched upon in this paper. In the first place, there was the academic aspect: the advances in physics and chemistry were a precondition and had reached a stage in the closing years of the nineteenth century when industrial development could be envisaged, and even attracted some risk capital. But it soon became obvious that very large resources would be called for. It is also interesting to observe that the time-consuming investigations into the equilibrium conditions for the formation of ammonia from its elements were contracted out to Walther Nernst by Chemische Fabrik Griesheim-Elektron and to Fritz Haber by Badische Anilin- & Soda-Fabrik (BASF). Without research contracts, the development of nitrogen fixation would have taken even longer. Five years passed between the issue of Haber's patent and the start-up of the BASF commercial-size ammonia plant at Oppau; and another two to three years were necessary to design and build the equipment for the large-scale production of hydrogen and nitrogen. A lot of money and determination were needed to carry the project through, but the rewards during and especially after the war were enormous. At that time, only one firm in the British chemical industry, Brunner, Mond & Co., was of a size to carry out such a project; and we may note here that, unlike BASF, it chose to expand by way of takeovers of similar, but smaller, businesses, instead of diversification through innovation.

Diversification, that is, the development in wartime Germany of substitutes (pejoratively called *Ersatz*), was the hallmark of the industry's growth from 1914 onwards. Some of the synthetics were technically unsatisfactory and had to be discarded; others were high-cost products that could only maintain themselves because there were no imports; yet others were cheaper and better than traditional materials, and generated new business. What matters is to think of these innovations as part of a continuing process, stimulated by war, in which one development might lead to another and to another. As a result, the Germans, although they were weakened economically by the War and although their prewar specialities were subject to greater competition than before 1914, were, nevertheless, in a stronger position as regards the latest technological advances.

The outbreak of war created precisely those conditions favourable to growth that Levinstein and other manufacturers had demanded: imports cut off, 'unfair' competition in third country markets stopped, German patents seized and a patriotic zeal mingled with business expectations to produce in England what had hitherto been 'Made in Germany'. But within weeks difficulties emerged and became a brake on innovation. The obstacles were, in order of appearance, muddle, finance, know-how and incompetence. The responsibility for the chemical industry during the war was divided between several departments of State, of which the Board of Trade and the Ministry of Munitions were the most important. The documents in the Public Record Office show that the bureaucrats were not so much ill-intentioned as unable to perceive simple solutions and carry them out. Since the German war machine also made many mistakes and was similarly bound by red tape, both sides were equally handicapped.

More serious was the lack of finance. Who was going to support the enlargement of dyestuffs works to meet the anticipated demand? Who was going to pay for the equipment needed to convert coal-tar distillation products into intermediate compounds required by dyeworks, pharmaceutical producers, munition factories and others who had hitherto bought these organic chemicals abroad? There were plenty of volunteers, eager to make money out of import substitution. But the quality of their goods and their financial reputation were held in such low esteem that neither the users nor the investing public were prepared to hazard their money without some sort of official, copper-bottomed, guarantee that these British enterprises would be protected against foreign competition after the War. That raised many questions on risk-taking in a rapidly changing technical environment, on the rate of return from innovation, and on the role of government in supporting manufacture through loans, equity finance or promises that might be redeemable at some time in the future. These were novel issues at the time, and they have become more acute since then. Yet it is difficult to see how such 'key' products can be developed on a manufacturing scale and their range widened unless there is some assurance that 'infant industries' will be protected until they can withstand imports from larger and technologically more advanced producers. The free-trade case was well known at the time and commanded support even after 1914, but the nature of the German competition was misunderstood. Their chemical producers maintained their hold over the business not merely by reason of their innovative skill and thrust, but because the cost of entry was very high and the market-sharing arrangements were acceptable to all the members of the oligopolistic group. The combination of these conditions did not exist in prewar Britain, and, as we shall see, their gradual emergence took longer than expected.

The lack of know-how was underrated at the time. It was not sufficiently appreciated that access to patents did not ensure trouble-free production or marketing. The information provided in the patent description concerning the critical stages of the invention was often incomplete (this was sometimes done deliberately) and had to be supplemented by experience derived from previous innovation in the same general area. University chemists and their laboratory assistants were, in these respects, of little use; what mattered was getting the patent to work in a pilot plant or on a commercial scale, and merely increasing the laboratory equipment by an order of magnitude was not enough. 'Learning-by-doing' had taken the Germans many years, and this had been accompanied by the emergence of chemist-engineers and a skilled labour force trained in-house. The Americans found a solution in a growth in the number of professional chemical engineers in the 1910s and 1920s; but the lack of such cross-disciplinary specialists was a major handicap to the British industry's war-time development.

The fact that many of these obstacles were overcome within four or five years is a tribute to the effort put into the industry. Progress would have been more remarkable if the acquisition of know-how had not been held up by managerial incompetence, which appears to have been ineradicable throughout the War. Examples abound, but if we confine ourselves to the very simple case of two chemical warfare agents, phosgene and dichloroethylsulphide (or mustard gas), we can see these inefficiencies in action. Both had been described in the chemical literature on several occasions; the former and the final intermediate of the latter were produced in Germany at the rate of several hundreds of tons a year by processes that were known in England. The War Office first called for phosgene in the summer of 1915 and for mustard gas two years later. But by the time of the Armistice, the quantities delivered were still well below those ordered, and the irregularity of shipments to France was a frequent cause of complaint. How can one explain these failures? The answers, as they emerge from the official histories and the documents on which they are based,[4] lie in poor management generally and, specifically, in the failure to maintain rigorous quality controls. In particular, the carbon monoxide used to make phosgene and the ethylene required for mustard gas were frequently impure, so that the reactions were incomplete and the batches spoilt. It seems extraordinary that weeks and even months passed before the faults were remedied. The United Alkali Company took greater care when it was threatened with cancellation of the phosgene contract; but that threat could not be exercised in the case of mustard gas, for which the Ministry of Munitions had direct operational responsibility. In the end, shift-extra chemists were recruited to monitor the ethylene units twenty-four hours a day until the quality was, at all times, up to specification. A few weeks later, the War ended!

Conclusions drawn from particular instances often lead to faulty generalizations, but in this sector there were many examples of managerial failures. They cannot be attributed to financial misjudgements, but were usually caused by a feeble or even negative response to the difficulties normally encountered in the innovation process. It is easy to be wise after the event, but the roots of the trouble were diagnosed fairly quickly at the time, and could have been removed if there had been the will to do so. I do not underrate the costs involved in transforming a system of management which tolerated inefficiencies and insufficiently rewarded those who overcame the problems and applied the remedies. It may be noted in passing that no one at management level in poison-gas manufacture was sacked because of dilatoriness or on other grounds of incompetence, but some outstanding failures were moved to less-demanding jobs.

That raises the interesting question whether chemical innovation can be accelerated or slowed down by the quality of the development chemists and process engineers, and whether better and quicker results could be achieved if the innovators were kept separate from the management of the enterprise. Such a segregation of chemical policy-makers from those whose particular skills lie in the implementation of innovation strategies appears to have been tried out first in the United States, and in the interwar years in large German and British concerns. It must not be assumed that such an organizational system established itself quickly, and that the flow of information to and from development groups and marketing departments worked smoothly and was correctly interpreted and swiftly executed.

It is worthwhile summarizing and analysing the innovatory practices and observing whether they led the chemical industries of Britain and Germany to converge in their development and characteristics.

Consequences of Wartime Innovations

In 1919, Britain had a dyestuffs industry capable of supplying a broad range of colours, and which exhibited most of the characteristics that the German industry had had twenty years earlier, such as strong academic links, well-staffed research departments, technical service for customers and the production of many intermediate compounds. The quality, however, still left much to be desired. Government support had been essential; could the industry survive without it? The customers of the British Dyestuffs Corporation (BDC) – the leading business – were eager to buy from the Germans once more or to continue trading with the Swiss; HM Treasury wanted to be rid of its investment; the Board of Directors was beset with quarrels; and the Corporation's employees were understandably worried about the future. The substitution of British-made goods for similar, imported colouring materials had been achieved in volume terms, but the doubling of prewar capacity was not matched by a corresponding rise in demand – after all there is a limit to the amount of dyes that can be used, especially if the fashion is for pastels instead of wartime blacks and khakis. Under these circumstances, BDC was unlikely to survive German competition, aggravated by the additional supplies sent as reparations under the Treaty of Versailles. A solution was urgent, both in the national interest and to safeguard the Treasury's holdings, and the Board of Trade devised an unusual system of protection: there was to be no import duty on finished dyes, but an import licence had to be obtained; organic intermediates suffered a thirty-three percent tariff but were free from quantitative restrictions.

As can be imagined, that particular method of protecting 'key' or 'infant' industries was administratively complicated. It was misunderstood at first and caused some difficulties to the textile industry; but after five or six years, the Dyestuffs (Import Regulation) Act, 1921 became accepted, and was shortly to be praised on all sides. Successful government intervention is rare, and in this particular instance it owes as much to the good sense of the civil servants as to the improving relations between makers and users. The Act had provided safeguards of quality and price for the textile finishers, and implementation of the rules necessarily brought both sides into regular contact in the presence of a neutral chairman. At these formal meetings, the users could demand a licence, and the makers could oppose it on the grounds that they could supply a similar dye at a price not exceeding three times (later 1.75 times) the imported colour. A quality test usually settled the matter. The procedure imposed constraints on both sides but provided an incentive to improve quality and customer service.

As to intermediates, the high tariff discouraged imports of the simpler types and so helped to maintain that branch of organic chemical manufacture.

Without such protective legislation, the companies – and in particular BDC – would have reduced their research and development effort, which led, in the early 1920s, to the discovery of a novel range of dyes for acetate rayon and a variety of dispersing agents. But protection did not ensure profitability, and BDC's financial results were unsatisfactory. The government wanted to pull out of its investment and involvement in the commercial strategy of the dye-makers, and, after lengthy negotiations and the writing down of BDC's capital, it was included in the amalgamations of 1926 from which Imperial Chemical Industries (ICI) emerged. As a branch of a very large concern, the future of British dye-making was secure; and under the umbrella of ICI its bargaining power, weak in the 1910s and 1920s, became stronger. But dyes no longer occupied the same relative importance in both countries; innovation had moved on and the pattern was changing. As the Red Queen said to Alice, '*here* it takes all the running *you* do, to keep in the same place, if you want to get somewhere else you must run at least twice as fast as that'.[5] And that indeed was the rub; for the diffusion from dyes and their auxiliaries into a great variety of other products was accelerating, so that while the Germans lost ground with one group of products they gained it with another. But the War had done more than accelerate diffusion, it had dispersed innovation as a whole, created many centres of research outside Germany and reduced German market dominance from a lead of twenty or even thirty years, to ten or at the most fifteen. The pattern of industrial rivalry was thereby decisively altered.

Hitherto I have interpreted rivalry mainly as a British effort to catch up with the research and technical progress of the Germans. Broadly speaking, that aspect became less important after the mid-1920s, whereas the commercial rivalry became, if anything, even sharper than before the War. It seems at first sight surprising that that should be the case when, under the leadership of German and British chemical concerns, markets throughout the world were divided between numerous cartels, and quantity or price controls were carefully monitored. But this state of affairs, though important, was only a reflection of a major shift in the strategy of firms. On the one hand, not only in Germany but even in Britain there was much evidence that chemists had been accepted as professionally qualified members of senior management; on the other hand, little effort appears to have been made to use them as salesmen.

The Germans were quicker off the mark than the British to remedy this weakness, perhaps because the economic chaos of 1919–1923 had blurred occupational class distinctions, though it is more likely that some far-sighted directors anticipated the need for some chemical marketing experts. As long as people thought of chemicals as intermediates, the consumption of which depended on the demand for cotton, wool or washing powder, business was largely a matter of a few gentlemen meeting infrequently to transact it. But as

BASF and, later, ICI were to discover, nitrogenous fertilizers, cellulosic paints and plastics brought a handful of producers into frequent contact with a fragmented market in which sales depended not only on the absolute price but on price relativities, and so on the demonstration effect of the novel products.

Demonstrating the superior qualities of the novelties added greatly to the cost of sales, and, henceforth, marketing problems multiplied: instead of scores of dyers, there were now hundreds of agricultural merchants and cooperatives and thousands of individual farmers. Sales support, ranging from experimental farms to trade shows and advertising, became an unexpectedly large overhead aggravated by the seasonal nature of the trade, which entailed the expensive construction of large storage silos. Paints represented problems of a different kind: it was necessary to work alongside the car manufacturers to help them devise the spraying equipment and to persuade the extremely conservative and fragmented furniture trades to try out the new cellulose finishes.

Chemical retardation had been ascribed to a variety of causes, such as the inadequacies of higher education, the lack of professionally qualified staff, the failure of lawyers or accountants to invest in new chemical ventures and, on the contrary, to retain obsolescent processes. These explanations or excuses no longer sufficed in the 1920s when the rules of the game were changing. Technical weaknesses and scientific handicaps mattered less – the geographical diffusion of applied science made it that much easier to buy the technology – whereas the responsiveness to the changing characteristics of diversifying markets mattered more. Swift commercial judgement on where and how to deploy marketing resources became an important factor in a company's financial success. If the demonstration effect proved that the new was superior to the old, customers were gained and they were found to be very responsive to successive price cuts: supply created, as it were, its own demand. There was, of course, another side to the story – the cost of entry became higher and the risks correspondingly greater. Mergers and take-overs (buying the management along with its innovations) checked competition, and, as if that were not enough, the chemical trade as a whole was riddled with cartels and other restrictive practices.

But not everything was controlled, and the handicaps also provided opportunities. The dyestuffs cartels – among the best-organized in the industry – were designed to promote quality improvements and higher unit values. Market-sharing devices allowed plenty of opportunity to innovate in organic chemicals. So, once again, risk-takers and risk-avoiders influenced the rate of innovation in the newer branches of the industry. The bureaucratic obstacles that accompany Big Business made for delay but could not prevent change. The development of plastics during the interwar years is a case in point. What is significant in the context of this paper is that Americans and Germans were quicker off the mark in perceiving and reacting to the opportunities, whereas the board of ICI was rather slower. The difference was more one of degree (and of the attitudes prevailing in the councils of the large concerns) than of kind. But cumulatively, over the years, the differences began to show. A decade or so after the Great War, the chemical industries of Britain and Germany were still showing remarkable contrasts. The structure was no longer as dissimilar as in the 1900s; the rates of growth also resembled each other, but the striking difference was the intensity of effort devoted to market development — in Britain the challenge of the new evoked a lesser response than it did in Germany. The rate of innovation still divided the rivals.

Notes

1. Hill, C. T. (1979) *Technological innovation: Agent of growth and change*. In: Hill, C. T. & Utterback, J. M., eds, *Technological Innovation for a Dynamic Economy*, Ch. 1, New York & Oxford, Pergamon.
2. Schumpeter, J. A. (1961 ed.) *Capitalism, Socialism and Democracy*, London, Allen & Unwin, pp. 83, 118.
3. Haber, L. F. (1971) *The Chemical Industry 1900–1930*, Oxford, Clarendon Press.
4. *History of the Ministry of Munitions* (undated; 1921?), xi, Part II, *Chemical Warfare Supplies*, London, HMSO, pp. 48–52; Public Record Office, London, *Quinan Papers*, under reference SUP/10/6.
5. Carroll, L. (1939) *Through the Looking Glass* [1872], London, Nonesuch Press, p. 152.

THE CONTRIBUTION OF FRENCH SCIENTISTS AND ENGINEERS TO THE DEVELOPMENT OF MODERN MANAGERIAL STRUCTURES IN THE EARLY PART OF THE TWENTIETH CENTURY

MAURICE LEVY-LEBOYER

Department of History, Université de Paris-X Nanterre, Nanterre, France

Introduction

There is no need to be reminded of the gap that has existed since the 1930s between men of science and men of business. At that time, it was due partly to a contraction in job opportunities, which prevented transfers of individuals from one sector to another; in terms of unemployment rates, engineers were then among the hardest hit by the depression, so that barriers to entry in industry tended to be high. In subsequent years, this situation persisted; and, along with it, a lack of mutual understanding developed which is currently ascribed to differences in training, career patterns, systems of values, social origins and similar factors. One might be tempted to attribute these elements to earlier periods, as if they had always had the same influence, and to make the breach between the two groups a permanent feature of industrial societies. But this would not be justified. In a broad historical perspective, the division was a novelty, and most of the arguments that are used to give it a rational foundation would not have applied in the past.

Firstly, industry was never a closed field. For most of the nineteenth century, it is true, few engineers were able to attain executive positions in manufacturing firms: they were scientists of repute, appointed to the boards of old companies (Gay-Lussac, for instance, acted as chairman of Saint-Gobain in the 1840s); or state engineers, who took part in the building of the railroads or entered the coal and steel industries; or pioneers of new technologies, such as Alfred Nobel, Ernest Solvay and the founders of the German dyestuffs industry, i.e. individuals who succeeded in developing large-scale enterprises on a multinational basis in the 1870s and 1880s[1] (Table 1). But for the generations who aspired to leadership in the closing decades of the century and during the First World War (columns 2 and 3 of Table 1), executive appointments of that kind became the rule. There were only 20 000 to 25 000 engineers working in French and

TABLE 1. *Educational levels of French business leaders in the twentieth century, by year of birth (%)*[a]

Academic level	Pre-1853	1854–1873	1874–1893	1894–1913	Post-1914
Polytechnique 1	6.0	13.5	17.9	12.6	21.1
Polytechnique 2	3.0	15.7	13.8	17.0	7.0
Ecole centrale	13.4	27.0	8.1	6.7	1.8
Other engineering schools	16.4	13.0	13.0	19.3	17.5
Law, trade schools, etc.	31.3	17.8	30.1	34.8	49.1
Secondary or less	29.9	13.0	17.1	9.6	3.5
One school	92.5	77.3	70.7	57.8	35.2
Two schools	6.0	21.1	25.2	33.3	64.8
Three or more	1.5	1.6	4.1	8.9	17.5

[a]From Lévy-Leboyer, M., ed. (1979) *Le Patronat de la seconde industrialisation* (*Le mouvement social No. 4*). The sample was made up of managers of the largest industrial corporations, active between 1912 and 1973. The group 'ingénieurs des corps des Mines et des Ponts' is classified as '*Polytechnique* 1'.

German industries in 1870 (divided equally between the two countries); but there were 80 000 (two-thirds in Germany) at the turn of the century, and up to 100 000 or 120 000 on the eve of the War.[2] Among them, large numbers of managers who had both academic training and some administrative experience were leading the new international firms that were to dominate modern industry: they made up the teams of directors, set up by (among others) E. I. du Pont in 1903 and by Carl Duisberg at I. G. Farben in 1906–1916, the technicians who patented and applied the first electrochemical processes in France and Switzerland, those who built up the mechanical industries, and, in the interwar years – at least on the Continent – most of their successors and colleagues.

Sample studies of university graduates among business leaders born after the 1890s show the extent of this professionalization: graduates account for forty-one percent of the total in Britain, for sixty-two to sixty-seven percent in the United States, for seventy-one percent in Germany (in samples limited to salaried executives) and for ninety percent in France (when the largest corporations are surveyed) (Table 1). It was not unusual to find among leading company chairmen individuals who had had teaching experience at one point in their careers.[3] This move can be accounted for by the use of techniques that were more complex than in the past, and by the development of research laboratories within industrial firms. (No fewer than 5000 chemists were employed in Germany by 1913.) Furthermore, the movement of business concentration and the higher scale of operations made it possible to cover the cost of a professional group whose already high salaries had increased by some sixty percent between 1890 and 1913, i.e. from a ratio of 4:1 to one of 5:1 in relation to average industrial wages.[4] As has been well documented by historians, the great step forward after the 1890s in scientific education and research had its counterpart in industrial life: both sectors involved the same personnel, trained in the same schools and working in close cooperation.

Secondly, insofar as education could be held to be a substitute for older privileges, opportunities and social recruitment, as well as codes of ethics, should have been similar for the two groups. Of course, it is often assumed that an unfavourable family background severely restricts access to the business elite; no such obstacles ever existed in academic circles, making social mobility uneven. However, a distinction should be made between medium and small firms in the traditional sectors, in which no real progress was registered, and large corporations, where expansion called forth new talents. Quantitative surveys have shown that in the latter part of the nineteenth century there was a true increase in upward mobility in the modern sectors of industry for people of the middle classes, i.e. for the sons of civil servants, army officers, professionals and even of white-collar workers, who made up forty-five to fifty percent of the pre-1939 business leaders in Western countries.[5] This was not simply because management then took precedence over capital ownership, but also because the prestige of older industries and family ties proved to be a source of inertia. Members of privileged groups did not respond quickly to opportunities in periods of rapid technological change; and the length and difficulties of scientific curricula were less easily accepted by people of the upper classes, since, by carrying on their studies, they would have to forego the higher incomes that would accrue to them if they entered business, in comparison with less fortunate students, whose careers would otherwise start at a lower income level.

The democratization of the engineering profession – as can be seen by comparing recruitment at the *Écoles centrale* and *polytechnique* (Table 2) – was a function of the scientific content of entrance examinations and courses: in the years 1880–1925, socially mobile students made up only two-fifths of the classes at the first school, which was somehow less demanding in advanced mathematics and deductive physics, and three-fifths of those at the *polytechnique* – probably a higher percentage than that of university professors at the same time.[6] Later, in the interwar period, the trend may have been reversed, as business management required more specialists in law, taxes and finance, matters that were usually taken up by less progressive students. But, even then, mobility was not hampered, because training in those fields involved a varied formation; and fifty-five percent of business leaders in the French sample, who had thus attended more than one graduate school (Table 1, lines 8–9), were of the lower classes. In short, if scientists are looked upon as a meritocracy, the term should also be applied to a large part of the community of corporate managers in the early part of the century.

Thus, external factors cannot be held responsible for the separation of science and industry: Professionals in the two groups initially had the same training, the same social background, very much the same urge to develop applied sciences and industrial concerns, and none would have imagined these could develop

TABLE 2. *Social origins of French engineers, 1895–1973 (%)*[a]

Socio-professional activity of parent	*École centrale* (year of graduation)		*École polytechnique* (year of graduation)		*Polytechniciens*			Business leaders		
								Engineers		Others
	1895	1925	1895	1925	Army officers	Administration	Private sector	*Polytechnique*	*Centrale*	
Industry, banking etc.	24.4	22.7	14.9	16.6	11.9	15.2	31.4	25.0	51.9	47.9
Higher civil service	10.4	13.1	18.2	19.1	17.5	24.2	15.7	11.7	13.2	8.3
Landlords, fund-holders	28.2	17.2	20.4	2.3	16.1	17.7	12.4	–	–	–
Privileged group	60.8	53.6	53.5	38.0	48.4	48.0	57.7	36.7	65.1	56.2
Professions	18.5	23.1	7.5	9.5	8.7	8.6	7.8	28.9	16.3	20.2
Middle management and business	14.3	16.2	24.6	33.1	26.8	25.7	22.1	27.3	10.9	19.3
Retailers, craftsmen, workers	6.4	7.1	14.4	19.4	16.1	17.7	12.4	7.0	7.8	4.3
Mobile population	39.2	46.4	46.5	62.0	51.6	52.0	42.3	63.3	34.9	43.8

[a]From Mercié, C. (1972) *Les Polytechniciens, 1870–1930. Recrutement et activités* (M.A., Nanterre); Lhomer-Deslandes, C. (1973) *Etude sur les ingénieurs de l'Ecole centrale des Arts et Manufactures, 1880–1939* (M.A., Nanterre); Lévy-Leboyer, M. (note 3).

The sample of graduates from the *École polytechnique* (columns 5–7) is taken from the school records of 1886–1929; the students listed under 'private sector' (column 7) represent only those who chose not to enter the public service when they left school; they should not be confused with the whole body of *polytechniciens* employed in the private sector, since a great number had resigned after a period in administration. The years covered by columns 8–10 are the same as in Table 1, i.e. 1912–1973.

without the close cooperation of all. To take one instance, in a series of meetings held in Paris in 1917 by leading officers in government services, the *grandes écoles,* the faculties of science and some of the major firms to study the possible introduction to France of the German *technische hochschule*, not a single voice of dissent was heard with regard to the common goal – which was to hasten modernization of the country.[7] Still, in view of the schism that finally came about, the idea must be accepted, as an alternative hypothesis, that cooperation between scientists and manufacturers did end in failure – a disappointment that explains their later breach. The literature on the subject is unfortunately scanty, and historical works are more concerned with the positive side of the story – the internalization of markets by large corporations, the formation of new social structures, and related matters. This paper is perforce limited in scope: it examines, on the basis of the French case, the contribution of scientists and highly trained engineers to the development of modern management; and, in a broader, comparative view (since many countries went through the same experience), the effects that this involvement had on the engineering profession.

French Managerial Structures

Economic recovery arrived in France in about 1900, after a severe and protracted depression that had held down technical progress for more than twenty years. As a consequence, the flow of engineers into the advanced sectors of industry reached unprecedented proportions. These people were former students of the many schools that had opened in the latter part of the nineteenth century, at the request and expense of local authorities and manufacturers eager to promote on a collective basis the teaching of applied sciences and to raise the level of skills. It has been estimated that the student body attending lower technical schools was 45 000 in 1913 (compared with fewer than 20 000 in 1892) and that some 2400 engineering degrees (of an intermediate or higher level) were awarded each year.[8] The transfer of state engineers from government services also assumed an unusual scale: first in the late 1890s, when many senior officers joined firms in the newly developing industries; again after 1905, as the saturation of military posts compelled young graduates of the *Polytechnique* to resign and seek employment in the private sector; and lastly, after the War, when virtually entire graduating classes – which had been doubled to compensate for war casualties – made the same choice as older Army officers and civil servants, to take the opportunities offered by peace and prosperity. At the end of the 1920s, some 8000 *Centraliens* and 3600 *Polytechniciens* (compared with 1700 in 1905), not to mention graduates from the lesser schools, were working in French industry.[9]

This inflow of highly trained personnel and their unusually young age (forty-one percent were less than forty years old in 1930) brought about a sharp change in the management of industry. Manufacturers became more inclined to use the mass-production techniques that had been developed during the War or were being imported; and over the period 1905–1930 French economic growth, measured in terms of production per caput, was unmatched by its nearest competitors, Britain and Germany. In the modern sectors, in which the younger generation tended to congregate, there was greater readiness to adopt the managerial principles that had been devised by, among others, Frederick Taylor and Henri Fayol, and brought to the French public in 1907 and 1908. Their practical influence had been limited at first, in spite of the keen interest they had raised among metallurgists and pioneers in the automobile industry; only a few repair shops and shipyards and a dozen plants, at best, had experimented with them. But from 1915, under the pressure of demand, scientific management became something of a fashion: scores of manufacturers and engineers crossed the Atlantic to assess the new practices and work technologies; American consultants opened up agencies in Paris; while various schemes were started by industrial firms and engineering schools, including the *École centrale,* to train technicians in shop management. By 1925–1927, half-a-dozen major associations had been set up to stimulate interest, circulate information and create a national consensus.[10]

Given the role assumed for twenty-five years by engineers in the industrialization process, in what way did they shape French managerial structures? Did they try to rationalize production and step up its technical content, or did they have different objectives?

1. Their major concern – and in this respect the *Polytechniciens*, as a group, were a decisive factor – was primarily to develop large-scale production units. They had the ability to do so, not simply because of their scientific status, but rather because of their administrative experience. As opposed to most engineers, who had gone directly from school to industry, members of pre-1890 classes at the *Polytechnique* (those, for instance, who pioneered electricity in northern France and Paris) had spent some ten to twenty years running public services – a canal, a port or a railroad division – for the account of the State, before taking their chances in the private sector. Of the later generations who had tended to by-pass that stage, many were called back during the War to organize and operate armament factories or to assist members of the government, and in particular, Louis Loucheur (himself of an older class), when he was in charge of the Ministry of War Industries and later of the reconstruction.

Thus, in a country where firms were still predominantly small and often backward, the group brought a new spirit, a broader view of the economy and the will to rebuild it on an extended basis. When planning for peace, in 1917–1919, members of the group were ready to use the power of the State to enforce a complete rationalization of output: they wanted, for instance, to reduce machine-tool enterprises to two or three firms per branch; and, in the automobile sector (the scheme was conceived by André Citroën), to set up a pool of interchangeable parts and to limit the number of assembly firms to half-a-dozen (instead of the 150–200 in activity on the eve of the War). In the mid-1920s, they were still urging manufacturers to enter into horizontal agreements and cartels to pool financial resources, to develop larger production facilities and to expand markets, thus attaining economies of scale and reducing costs.[11] According to this school of thought, integration was to displace outmoded forms of competition.

In the 1920s, these views proved determinant. The leading firms registered a major step forward, in terms of size (30 000 workers per firm and 40–50 000 shareholders were no longer uncommon), and in terms of concentration ratios (since three or five companies in the modern sectors were covering up to sixty percent of their market). This progress was due, of course, to successful individual policies, but above all to a series of mergers in which the same group of engineers was instrumental. In 1917–1921, for instance, most steel mills in northern and eastern France were enlarged and their operations extended; the main electrometallurgic plants in the Alps were amalgamated into two major groups (Ugine and Péchiney, the latter under Gabriel Cordier and three other *Polytechniciens*); all the dyestuffs industry, which had been developed by the State, was brought together under Kuhlmann; and northern collieries and large chemical corporations were starting joint ventures to enter new fields. In 1929–1931, as part of a complete reorganization of the electrical industries, departments in the same line were merged branch by branch; and a single company (Alsthom, under Auguste Detoeuf) leased most of the French heavy equipment plants.

2. Although French engineers were not deeply concerned with the shaping of corporate structures, it was a field in which their contribution was not insignificant. Thanks to Fayol's publications, contemporaries were taught the necessity of allocating responsibility along elaborate vertical structures, of separating functions between staff and line, of leaving to central executive officers the final choice on policy decisions (apart from their implementation and control), and – in conformity with Fayol's preference for authoritarian hierarchies, which he forcefully expressed in the closing years of the War – of keeping management stable and independent of political pressures, so as to preserve unity of command and direction.[12] French managers were also kept informed of experiments being conducted abroad – in particular by Alfred Sloan at General Motors – to decentralize routine operations, but to control more firmly financial policies and long-term planning. Sloan's divisional structure of management was temporarily put into operation at Saint-Gobain and, through the mediation of the Lazard Bank, at the Citroën works in the late 1920s. But all this to no avail: managerial structures remained relatively poorly coordinated and unprovided for.

Two circumstances may explain this situation. First, there had been a tradition, justified by the competence of engineers, to decentralize industrial operations within firms, to such an extent that dynasties of directors remained in charge of the same division in the railroads and large corporations during the second half of the nineteenth century. This trend could but develop further with the many opportunities that technical change opened up from the 1890s. Corporations branched out in all directions, using joint-ventures and other financial devices to enter new fields and to experiment with new patents. Of the larger ones, Saint-Gobain moved, in the 1920s, from making glass and fertilizers to the manufacture of all sorts of chemicals, to hydroelectricity, and to oil drilling and refining; Schneider moved from heavy industries to making electrical equipment, steel pipes and automobiles; and both subsequently assumed the status of (unstructured) holding companies. Second, authoritarian hierarchies were not compatible with the rapid growth of new, innovative firms. In the electrical industries, for which detailed samples have been collected and analysed,[13] vertical structures are not found, but instead there were executive teams, with ten to twelve members on an equal footing, which led new corporations, like Thomson-Houston and CGE (General Electric), when they were building their first markets at the turn of the century. In the interwar years, as the production and distribution of electricity required more external financing and a closer cooperation with local authorities, a national network of firms was built; but again by small groups of engineers from the same schools (fifty to sixty percent from the *Polytechnique*, compared with thirty-five percent in 1913), each of whom held an average of eleven to fifteen directorships in subcompanies (many had three to five times that number), and who substituted their communal view, derived from the same training, to bring about a more formalized structuring of the industry.

3. The same impression of incompleteness prevails when shop management practices are examined. Although the Government had requested that use of Taylor's methods be made general in steel mills (by December 1915) and in arms factories (from April 1916), rationalization of industry seems to have been slow to develop after the War, even in the automobile sector where the ground work was left to the care and initiative of production engineers who had a more practical training than the rest of the profession. Of course, it would be unrealistic to look for a single cause of this delay – the more so that the detailed historical studies that have been conducted in various plants do not provide a unified picture. Still, it would appear that unit tasks analyses – the search for 'the one best method' – which was the essence of Taylorism, did not fulfil the purposes of French manufacturers.[14]

Work on the shop floor in factories and assembly plants had been changed radically by integration of tasks: in the past they had been allocated (or subcontracted) to independent teams; now they were part of a continuous production flow. Time studies were needed, therefore, to segment processes, to measure the length of each sequence, and to gauge the maximal load for machinery. At first, however, they were seldom done for individual unit work. Scientific management, according to André Citroën, was to be used 'to apportion tasks in a logical order and give each foreman [not each worker] his share of the work'.[15]

The saving of time was an obvious goal, which could be achieved by a better distribution of manpower. However, since factories in the early 1920s were badly organized, often overmanned, and with equipment dating back to the War, the first objective of first-line engineers, was to eliminate excess capacity, to reallocate tools and machinery, to build reserves of spares, to extend the use of interchangeable parts, to bring quality under control, and, first and foremost, to prevent excess work, stoppages and the involuntary accumulation of materials, which were major sources of financial loss and labour conflicts.

Production factors were different in France from those abroad: in America, skilled hands were in short supply, and shop management was used to help migrants enter industry. (A similar situation had occurred in France when women were called upon to work in factories during the War.) There was no such deficiency in France, since industry there had been based traditionally on the high quality of its labour, and training had resumed in technical schools. Although automobile companies had moved into foundry work, skilled hands still made up seventy-one percent of the work force at the Renault plants in 1913 and sixty-five percent at Peugeot in the 1920s. Thus, some autonomy could be left to labour on the shop floor: vehicles moving along assembly lines (which at first were fixed, wooden rails) were simply pushed by hand, not by a mechanical conveyor; and in 1926, when the lines were brought under the control of a central office at the main Citroën factory in Paris, the work of the entire line was allotted to teams, whose leaders negotiated the total wage for their men, determined the number of hands necessary, and supervised the process and the rhythm.[16] This was very far from Taylor's principles.

In short, industry was being thoroughly modernized under the leadership of engineers, but with this qualification: that with the continuous prosperity and high expectations of the years 1900 to 1930, no rigid managerial principles were applied, and the new corporations still relied on middle management (given the decentralized structures of firms in the modern sectors) and on first-line engineers and skilled labour (in industries using labour-intensive technologies).

Social Response

The whole experience took a new turn, however, with the worsening of economic conditions: exports were slashed (with monetary adjustments in 1926 and 1931), domestic demand contracted (as a consequence of the 'scissors' between agricultural and industrial prices), and investments became negative (due to business uncertainties and severe government deflation). To a certain extent, employment was protected; but existing priorities could not be retained as long as output and labour costs had to be reduced. By an unfortunate coincidence, just at the time when prior investments (in new foundry works, automatic machinery with expanded load, etc.) were reaching maturity, increased use of assembly lines and time controls (the latter were extended to a great many firms and sectors in 1926–1930[17]), increased productivity and more rigorous shop-management practices were replacing increased production as the new objectives. All the fears that had beset labour in other industrialized countries – diluted skills, increased time loads, loss of autonomy at work, etc. – came to the fore; and the ineffectiveness of French managerial structures became all the more apparent.

Partly because the time had been short, and perhaps because many felt that academic training and degrees had given them precedence,[18] no participative structures had been devised that might alleviate the tensions. Almost alone during the last years of the War, members of the Electrical Manufacturers' Association had urged the enactment of mixed committees, in firms of more than one hundred workers, to deal with social and economic problems, and that of the closed shop to bring non-political members into the unions and to clear the way for some form of cooperation. In 1925–1926, the matter was again taken up, at the initiative of William Green, then president of the American Federation of Labour, and of the French representatives to the International Labour Organisation in Geneva,[19] but with no practical results. Industrial relations were thus left to deteriorate.

All periods of full employment had brought times of high labour turnover, which proved most harmful in terms of time spent in training, scrapped parts, etc. This reached crisis proportions in America in 1910–1920 and in Germany in the mid-1920s; and experiments with personnel management were hurriedly begun to teach new loyalty to the firm and to the community at large.[20] French manufacturers experienced the same phenomenon in 1915–1916 (to guarantee that one worker would stay at the Renault plant, four hands were taken on), and in the latter part of the 1920s. But, perhaps because they were looking for different types of labour, or were simply mistaken in their interpretation (they were primarily interested in Lahy's research on fatigue and boredom in repetitive tasks), they did not raise the problem of human relations in work but limited themselves to welfare schemes, for the benefit of women during the War and for the general work force after 1927.[21] The study of industrial psychology was postponed for a generation.

With the worsening of the depression in sectors in which wages had always been high (in percentage of primary cost), methods more rigorous than Taylor's were resorted to, specifically by firms that had experienced drops in productivity during the upswing. The northern collieries, for instance, called in consultants to review the tasks of all workers, skilled and unskilled, and to reduce them to multiples of a single average per unit of time ('le point Bedaux'). This made it possible, from 1933, for management to close pits, redistribute the work force, break up teams and increase productivity, at the cost of embittering industrial relations.[22] There are many examples of the implementation of similar policies and of resulting hard feelings during the 1930s in sectors in which skilled hands were submitted to new constraints on time.

As in other countries – though later than in Germany, where a sharp fall in activity and in morale had occurred at the end of the War – members of the engineering profession were not spared the effects of the changed circumstances. During the years of prosperity, they had asserted their authority on the double basis of their technical achievements, and the charismatic power they had borrowed from civil servants (the *privatbeamte* figure in Germany) or from Army officers, in the tradition exemplified by Lyautey. Their claim to be the intermediary between absentee (or diluted) capital and labour was often well substantiated. But these images had lost much of their relevance: with investment at a standstill, technical change, not to say research, was at a low ebb, and unemployment was spreading (as early as 1927 in Germany) among both the older and the very young engineers; intermediation had given way to subservience to financial capital – what Veblen, in his campaign to enlist support against the machine process, had called 'the predatory instincts of the idle class'.[23] There was a general revulsion against American rational methods, which had deprived staff members of their power of initiative and of their contacts with the workers, and against concentrations of business that had multiplied levels of decision and stultified the willingness of individuals to assume new responsibilities, etc. Unexpectedly, engineers' professional journals and corporate reports in the 1930s testify that there were fewer employment opportunities and at the same time a decline in loyalty and a fall in work standards previously unheard of. Unemployment and demoralization were reinforcing each other.[24]

None of the means French engineers had devised in the past to protect their profession could cope with this new situation: they had lost prestige (as a consequence of the new bureaucratic rules that applied at every level), status (they lost many social benefits and the right to take part in industrial bargaining), and standard of living (even though unemployment was more widespread in the coal and metal industries). Together with white-collar workers, they felt outnumbered and excluded when, in 1936, strikes swept the country. Hence, the spontaneous development of unions – Catholic, Socialist and others – as a means of reasserting their independent position and of supporting their claim to be treated like other salaried personnel. Hence, further, the attempts that were being made in France and Germany to build a new image for engineers, more humanistic and professional, and to lengthen their training in order to redress the balance between future classes and market demand.[25]

The unionization of some 20–25 000 engineers and technicians was short-lived, but the changes in curricula (a fourth year at the *Écoles d'arts et métiers*; research courses at the *École centrale,* etc.) were to last. Nowhere, however, was the change so striking as at the *École de physique et chimie,* which had been fashioned in 1882 on the model of a German *technische hochschule*: under the guidance of Paul Langevin, the director from 1926, it gave precedence to mathematics and specialized scientific courses, trained a great number of research workers and professors (twenty percent of the post-1930 classes, compared with eight percent earlier), reduced the proportion of those going into industry (the percentage of production engineers fell from fifty-seven to thirty-seven percent), and, curiously, lost much of its democratic recruitment.[26]

Conclusions

Thus, after a period of high expectations and extensive structural changes, the move for managerial reform had met with disappointment and social frustration. Of course, negative attitudes had prevailed in the past among engineers, but they were stronger in the 1930s: first, because of the fear, expressed by the unions at the time of the depression, of an anticipated loss of status – an antimodernist, antiegalitarian protest came from first-line engineers and middle management – and, second, because of the feeling of work alienation that had been spreading through the hierarchy in scientifically managed factories. The protest movement could have been transient, as it had been in the past. There had been expressions of resentment among French and German state engineers in the 1850s and 1860s, when parts of the public domain under their control were transferred to private interests; and again in the 1880s, when the whole profession had to submit to wage cuts, unemployment and a major fall in prestige, all of which were made up for in the next upswing. But in the interwar years, with large-scale corporations covering whole sectors of the economy, the factor of work alienation came to last. A more permanent conflict was developing between the modern organizational constraints that business hierarchies were imposing upon their staff and the desire for autonomy. In order to live a truly professional life, engineers had to go through longer courses and to undertake more research-oriented studied in the changed schools. Job dissatisfaction and perhaps some prejudice – for the coming generations – were thus being institutionalized both in industry and in the educational system.

Two remarks, though, may still be in order.

1. Organizational theory teaches that business hierarchies are not identical in all sectors: they tend to be vertical and centralized, with few layers of responsibility, in industries in which single products are manufactured using fairly simple, stable technologies (such as those for coal, steel, automobiles, etc.); but they are horizontal and flexible in the modern sectors, in which firms have to extend processes and to experiment with new ones, to diversify products and to build up markets, and thus have to stimulate initiative in their staff. In the first type, social unrest was only to be expected when falling output and prices called for drastic cost reductions. But peace and loyalty were preserved or quickly restored, during the 1930s, by small firms in advanced sectors, which were able to limit themselves to short series and could produce items that did not displace skilled

hands, and by large-scale corporations using stable technologies, in which labour had been upgraded from the 1890s and large engineering staffs had steadily been built to diversify production and to develop research as a hedge against a possible setback.[27]

2. Career mobility had been severely restricted in the past by the recruitment of directors from outside the firms, and by the building up of new business elites based on academic training and kinship as a substitute for older forms of discrimination. But, in the 1930s, with the near exhaustion of financial markets, industrial corporations became more autonomous and were thus able to take new initiatives to create more participative structures. Much of the power that had formerly been vested with boards of directors, as representative of capital, became centralized in the hands of a general manager, assisted by an executive committee recruited from within which had to approve all policy decisions, including those dealing with appointments and promotions. This brought a new balance between outside and inside groups: bankers and politicians, who in 1911–1913 had supplied one-third of the presidents and vice-presidents in the electrical industries, were reduced to thirteen percent in 1937–1939 (from twenty-four to five percent in the forty largest corporations); while career men, who often had had twenty years' experience in the same firm or group, formed a total of fifty-five percent of all business executives in the same sample, as compared with thirty-two percent in the 1920s. By the same token, the administrative experience of former civil servants lost its value, and the engineers as a group had to retreat: in 1929–1939, seventy-three percent of French business leaders had come from engineering schools; in 1959–1973 only fifty-one percent did so, and graduates of the *École polytechnique* had fallen to thirty-two percent.

Notes

1. Haber, L. F. (1958) *The Chemical Industry during the Twentieth Century*, Oxford; Morsel, H. (1970) *Les Industries électro-techniques dans les Alpes françaises du Nord de 1869 à 1921*. In: CNRS, *L'Industrialisation en Europe au XIXe siècle*, Lyon, pp. 557–595; Beaud, M. (1975) *Une Multinationale française: P.U.K.*, Paris.
2. Alström, G. (1978) Higher technical education and the engineering profession in France and Germany during the twentieth century. *Econ. Hist.*, **21**, 51–88. Cf. Torstendahl, R. (1975) *Dispersion of Engineers in a Transitional Society: Swedish Technicians 1860–1940*, Uppsala; Day, C. R. (1978) The making of mechanical engineers in France: the Ecoles d'arts et métiers 1803–1914. *Fr. hist. Stud.*, **10**, 439–460.
3. Kocka, J. (1978) *Entrepreneurs and managers in German industrialization*. In: Mathias, P. & Postan, M. M., eds, *The Cambridge Economic History of Europe*, Vol. 7, part 1, pp. 534, 582, Lévy-Leboyer, M., ed. (1979) *Le Patronat de la seconde industrialisation*, Paris; Kaelble, H. (1980) Long-term change in the recruitment of the business elite: Germany compared to the US, Great Britain and France since the industrial revolution. *J. soc. Hist.*, **13**, 403–423.
4. Lévy-Leboyer, M. (1979) *Hierarchical structure and incentives in a large corporation: the early managerial experience of Saint-Gobain 1872–1912*. In: Horn, N. & Kocka, J., eds, *Recht und Entwicklung der Grossunternehmen im 19. und frühen 20. Jahrhundert*, Göttingen, p. 465.
5. Kaelble, H. (1978) Social mobility in Germany 1900–1960. *J. mod. Hist.*, **50**, 439–461,

and his *Business Elite*, (note 3), Tables 2 and 3, pp. 409 and 414; see also the contributions of R. Torstendahl (on Swedish business leaders), T. R. Gourvish (on the UK), J. Kocka (on Germany), P. Lanthier (on French executives in the electrical industries) and M. Lévy-Leboyer (on France), in Lévy-Leboyer (1979) (note 3).

6. No systematic social analysis of university staff has been undertaken. However, of the 485 engineers in the nineteenth-century Corps des Mines, seventy-five percent were of the upper classes (business and higher civil service); Thépot, A. (1981) *Les Ingénieurs du Corps des Mines* (in press); and Shinn, T. (1980) *L'Ecole polytechnique 1794–1914,* Paris, pp. 156–170.
7. Anon. (1917) The future of higher technical education. *Bull. Ing. civ. Fr.*, 3 November–27 April; see also Le Chatelier, H. (1911) L'enseignement technique supérieur. *La Technique moderne.*
8. Couriot, M. (1917) *Bull. Ing. civ.*, 20 March; the number of degrees awarded each year was 3400 in the US and 1800 in Germany, according to Robertson, P. L. (1970) *Employers and Engineering Education in Britain and the US 1880–1914,* Edinburgh Economic History Conference.
9. Day, C. R. (1980) *Education for the industrial world: Technical and modern instruction in France under the Third Republic, 1880–1914.* In: Fox, R. & Weisz, G. eds, *The Organization of Science and Technology in France 1808–1914,* Cambridge, p. 153; and Paul, H. W. (1980) *The applied science institutes in twentieth century French science faculties. Ibid.*, pp. 155–181; Lévy-Leboyer, M. (1976) *Innovation and business strategies in nineteenth century France.* In: Carter, E. C., II, Forster, R. & Moody, J. N., eds, *Enterprise and Entrepreneurs in Nineteenth and Twentieth Century France*, Baltimore, pp. 87–135.
10. Kuise, R. F. (1967) *Ernest Mercier, French Technocrat,* Berkeley; Moutet, A. (1975) Les origines du système Taylor en France. Le point de vue patronal 1907–1914. *Mouvement soc.*, **93**, 15–49, and (1978) La politique de rationalisation de l'industrie française au lendemain de la première guerre. *Recherches,* 32–33, 449–492.
11. Detoeuf, A. (1927) La réorganisation industrielle. *Cah. Redressement fr.*, pp. 24, 44; Moutet, A., *Rationalisation* (note 10), pp. 465–467.
12. Fayol, H. (1920) *Administration industrielle et générale,* Paris, and (1921) *L'Incapacité industrielle de l'Etat,* Paris; Brodie, H. B. (1967) *Fayol on Administration,* London; Blancpain, F. (1973) Les carnets inédits de Fayol: présentation et textes. *Bull. Inst. int. Adm. publ.*, pp. 590–622.
13. Lanthier, P., *Le Patronat de l'industrie électrique* (note 5), pp. 117–126.
14. Nusbaumer, C. (1926) De l'idée d'organisation dans son application à l'industrie. *Bull. Com. natl Ind. fr.*, October, pp. 1–7.
15. An ambiguity has existed from the start, since Henri Le Chatelier, who publicized Taylor's method in France, had been first attracted to the latter's invention of high-speed steels (it was presented in Paris as early as 1901) and more specifically by the series of experiments that led to their discovery. Thus, he always referred to the method as a way of solving unit tasks, human or other. French foremen and engineers, who had used clocks to measure performances from the 1890s (from 1904 at Peugeot), were more concerned with segmented tasks or with the definition of new piece rates when an invention like high-speed steel had reduced work in a major way. See, among others, Le Chatelier, H. (1923) La méthode scientifique dans l'industrie. *Trans. Soc. Glass Tech.*, 7. And for the opposite view, Citroën, A. (1919) L'organisation scientifique du travail. *Inf. ouvrière soc.*, 16 January; Mattern, E. (1927) *Etude de l'amélioration des conditions de travail,* quoted in Schweitzer, S. (1980) *Organisation du travail, politique patronale et pratiques ouvrières aux usines Citroën, 1915–1935,* Thesis, Université de Paris VIII, p. 225; Cohen, Y. (1981) *Les Usines Peugeot et le pays de Montbéliard, 1897–1917,* Thesis, Université de Besançon.
16. Schweitzer, S. (1980) (note 15), pp. 306–313; Fridenson, P. (1978) *The coming of the assembly line to Europe.* In: Layton, T., ed., *The Dynamics of Science and Technology,* New York.
17. Fridenson, P. (1972) *Histoire des usines Renault,* Paris; Baudant, A. (1979) *Pont-à-*

Mousson, 1918–1939, Thesis, Université de Paris I; Moutet, A. (1980) Ingénieurs et rationalisation en France de la guerre à la crise. *Coll. Ing. Soc.*, Le Creusot.

18. Le Chatelier, H. (1929) *Bull. C.N.O.F.*, October.
19. Legouez, R. (1918) L'organisation industrielle. *Rev. gén. Electr.*, 23 March, pp. 447–456; de Fréminville, C. (1926) La collaboration des patrons et des ouvriers. *Bull. C.N.O.F.*, November; Moutet, A. *Ingénieurs et rationalisation* (note 17), pp. 29 ff.
20. Montgomery, D. (1973) Immigrant workers and scientific management. In: *Immigrants in Industry Conference* (2 November 1973), and (1978) Quels standards? Les ouvriers et la réorganisation de la production aux Etats-Unis 1900–1920. *Mouvement soc.* The experience of *worksgemeinschaft* at the Vereinische Stahlwerke in Geselkirchen, under K. Arnhold, and that of the Institut fur soziale Betrieslehre at Borsig by G. Briefs, are described by Roche, G. (1979) *Idéologie politique et syndicale des ingénieurs allemands de la période weimarienne*, Thesis, Université de Paris X, pp. 199 ff.
21. Baudant, A. (1979) (note 17), p. 87; Schweitzer, S. (1980) (note 15), pp. 544–551.
22. Hardy-Hémery, O. (1970) Rationalisation aux mines d'Anzin, 1927–1938. *Mouvement soc.*, 72, 3–48.
23. For the positive side of the debate, see Lederer, E. (1910) Die Privatbeamtenstellung. *Arch. soz. Wiss. Sozialpolitik*, **31**, quoted in Roche, G. (note 20), pp. 30–32; Lamirand, G. (1933) *Le rôle social de l'ingénieur*, Paris; Dautry, R. (1937) *Métier d'homme*, Paris, etc. And for the critical side, Veblen, T. (1921) *The Engineers and the Price System*, New York; Layton, E. T. (1973) *Engineers in revolt.* In: Layton, E. T., ed., *Technology and Social Change in America*, New York.
24. Feelings of frustration in a new bureaucratic environment are reflected in the journal of the *Bund des Ingenieures*, the German Union of Chemical Engineers (15 June and 15 October 1930), quoted in Roche, G. (note 20), pp. 58, 81. See also Schweitzer, S. (note 15), p. 404 (on the attempt by the firm to foster a new loyalty in its staff); and Dejonghe, E. (1980) *Ingénieurs et société dans les houillères du Nord.* In: *Coll. Ing. Soc.*, Le Creusot.
25. A reform of the Berlin Technische Hochschule similar to those then being discussed in France in 1930 was contemplated by E. Spranger: Roche, G. (note 20), p. 226.
26. Shinn, T. (1981) Des sciences industrielles aux sciences fondamentales: la mutation de l'Ecole supérieure de physique et chimie 1882–1970. *Rev. fr. Sociol.*, **22**, 167–182.
27. Beaud, C. (1980) *Les Ingénieurs du Creusot du milieu du dix-neuvième au milieu du vingtième siècle*, and Parize, R. (1980) L'ingénieur au Creusot, 1919–1939. In: *Coll. Ing. Soc.*, Le Creusot; Shinn, T. (1980) The genesis of French industrial research, 1880–1940. *Soc. Sci. Inf.*, **19**, (3).

DISCUSSION: AN ENGINEER IS AN ENGINEER IS AN ENGINEER?

SVANTE LINDQVIST

Department for the History of Technology, Royal Institute of Technology Library, Stockholm, Sweden

The three papers in this session span a wide variety of subjects. They are highly specialized, and yet, I think, all raise the same essential question, namely, current definitions of 'engineer', 'science' and 'technology' in comparison with the definitions prevailing at the time of Alfred Nobel.

In 1971, Dr Haber published a comprehensive monograph on the history of the chemical industry between 1900 and 1930;[1] and he has recently written a general survey of the chemical industry during the twentieth century.[2] This paper on Anglo-German industrial rivalry is an important addition to his other works and focuses on many fundamental issues. It is, among other things, an interesting comparison of governmental intervention and technological change. (Dr Haber discussed the British side of this question in detail in an article in *Minerva* in 1973.[3]) Furthermore, it stresses the relationship between war and technological growth — a correlation often neglected by historians of technology.[4] But even more interesting is that it brings out the influence of general cultural issues in the shaping of technology.

Dr Haber described the chemical industry at the opening of the century as being technologically *static* in England, and as being technologically *dynamic* in Germany. He said that the differences between the two countries were not due solely to technical or economical factors, but that the most significant forces in accelerating or retarding technological change were *attitudes*. In late-Victorian Britain, the pursuit of science *per se* was held in higher public esteem than was engineering; whereas the Wilhelmine Empire held a different attitude to the practice of chemistry and to the role of the professional chemist. There were, basically, different attitudes to the concept of science. Dr Haber showed that these attitudes were manifested in differences in managerial structure, in scientific education, in the pursuit of academic *versus* industrial research, and in the relationship between universities and chemical enterprises; he also showed how these differences in turn affected receptivity to technological change, risk-taking, rate of innovation and marketing. This broad study of how attitudes to science in two societies affected technological development is an important contribution to the theme of this Symposium — the interaction of science, technology and society.

The difference between the British and the German chemical industries, as described in Dr Haber's paper, is strikingly illustrated by the library in the Royal Institute of Technology in Stockholm. There, the books have been put on shelves in each subsection in the order that they were acquired, since the foundation of the library in 1825.[5] This practice is historically useful, since it provides a simple way of following the major chronological developments within any one field of technology. The subsection for industrial chemistry (*Signum Dd: Teknisk kemi*) contains some seventeen shelf metres of books, or more than 600 volumes, for the period in question, 1900–1925. The interesting thing is that these books are almost exclusively German handbooks on industrial chemistry, including *Handbücher der chemischen Technologie, Monografien über chemisch-technische Fabrikationsmetoden* and *Laborationsbücher für die chemische und der verwandte Industrien.*

The pattern is first broken in the early 1920s (i.e. by books acquired at that time), and then by the presence of a few British works on 'applied chemistry'. One of the first British books to break the German supremacy is a small volume entitled *What Industry Owes to Chemical Science*, published in 1918.[6] It is a collection of popular articles that had been published in the journal *The Engineer* in 1916–1917, and was an attempt to bridge the distance between teachers of science and leaders of industry. It aimed at informing the former of the kinds of scientific education that would be useful in industry, and the latter of 'the means whereby scientific discovery, scientific methods and scientifically trained men can be used'.[7] In comparison with the solid mass of German handbooks on the shelves above, this little book looks almost pathetic. Even its title echoes a simple belief in a one-way-only relationship between science and technology, and it does little more than stress the late-Victorian view of technology as applied science and of the scientist as the legitimate tenant of the ivory tower. In his introduction, the British chemist Sir George Beilby wrote:

> 'The deeper and more prolonged search into the phenomena and laws of Nature must, of necessity, be left to those who by natural endowment and by opportunity can pursue this search apart from the distractions of the work-a-day world. The workers of the applications of science may well realise their debt to these seekers after knowledge, for the seeds of achievement of the practical side must, in many cases, be planted and watered under their fostering care.'[8]

This, if any, is a clear-cut statement of an attitude that shaped industrial development in England. I have quoted it since I would like to suggest that it would be fruitful to look for similar statements of belief in other contemporary works designed to be read by 'the general educated public', in order to examine in detail the attitudes that Dr Haber has brought to our attention.

My second point concerns the 600 German handbooks on industrial chemistry on the shelves above. Dr Haber stressed the importance of know-how in the German industry, and I would like to suggest that an investigation of the exact nature of the information in these and other contemporary German handbooks might tell us more about the range and diffusion of the know-how of chemical technology on a manufacturing scale. Even the mere existence of such a huge amount of technical information is remarkable. Who published it, and why? To what extent was it diffused and available? Did it, in any way, influence British industry? If not, why not? The fact of publication of a technical handbook is fundamentally different from that of a scientific paper: It is the proof that a technology is already well established, that its machines and processes have been tested and perfected under practical working conditions for years.

A study of such a knowledge does indeed require, as Professor Rosenberg pointed out, a very different approach to the study of the history of science 'from that which dominates the specialist historian of science . . . who is concerned with important enlargements of the stock of scientific knowledge'.[9] Such a study might also further clarify German attitudes to the practice of chemistry. It may prove that our present definitions of science and technology are too rigid, too blunt a tool to form the basic concepts of a true understanding of the reasons behind the German industrial success.[10]

In 1975, Professor Torstendahl published two books about Swedish engineers. One, *Teknologins nytta*, was a study of the reasons given by Swedish parliamentarians and administrators in favour of technological education between 1810 and 1870;[11] and the other was a study of social mobility among Swedish technicians between 1860 and 1914.[12] His present paper is a comparative study of the sociological literature on engineers in industry before 1914, which places the Swedish case in a broad perspective. Professor Lévy-Leboyer has treated the French case in his book, *Le Patronat de la seconde industrialisation*, and more recently in an article on the French modern corporation in the book *Managerial Hierarchies*.[13] His present paper stresses the national context that shaped the development of managerial structures in France. His results differ from those of Professor Torstendahl, and bring out the complexity of comparing the roles of 'engineers' in different cultures and over a period of time. My comments deal with the concepts of 'research and development' and 'engineer' used in the literature surveyed by Professor Torstendahl.

'Research and development' is today not only a frequent entry in the accounts of industrial enterprises but also a common concept in research. As such, I would like to question its validity. Its abbreviation, 'R & D', in particular, seems to exercise a strong charismatic attraction on scholars. It is the kind of concept that is so commonly used that no one dares ask exactly what it stands for, since they think that everybody else knows; but this, too, implies the assumption of a simple, one-way causal relationship between science and technology and the taking for granted of scientific advancements as prerequisites for technological change. Recent figures for Swedish industry show that as much as eighty-eight percent of the costs for 'research and development' are actually spent on product and process development, and only three percent on basic scientific research.[14] This implies that major socially and economically important technical developments are due to the systematic accumulation of practical knowledge and workable solutions, i.e. day-to-day improvements in machines and process plants, by means of trial and error, parametric variation and optimization: A thousand small steps by the men in grey coats, rather than one giant leap by the men in white coats. This kind of technological work is epistemologically distinct from basic scientific research. The fact that the two are often lumped together under the entry 'R & D' in the statements of accounts does *not* mean that they should be treated as one concept in research, which says nothing about their ratio, individual significance or causal relationship.

Use of this concept also entails the risk of confusing what *was* labelled and what we today *would* label 'research and development'. Professor Torstendahl said that 'occasional monographs on big business suggest that organized research and development departments were sporadic before 1914'.[15] But that is a question of denomination, with the impending risk of exaggerating the importance of research and development from the time when the phrase was coined in the early twentieth century. Surely, there was highly advanced research and development in the large industrial enterprises long before then, even if it does not appear under that entry in the accounts and organization lists of those industries. In the Swedish mining industry, for example, there was a constant search for, and experimenting with, new processes from the early seventeenth century, when the industry became firmly organized. This is reflected in the continuous technical changes in industries directed towards less energy-demanding processes and a higher quality of end products. Such work was undertaken by iron-masters, skilled engineers in executive positions trained according to the customer practice of their time: They often had a university education, had worked as apprentices in various mining industries, had served on the Board of Mining, and had travelled abroad extensively to study technical innovations in other European mining industries.

This brings me to my second point. Those men should be labelled 'engineers', if by 'engineer' we mean a trained person who is responsible for the maintenance and/or development of a technological system. 'Engineer' is both a denomination and a title, and its meaning has changed considerably over time.[16] The 'engineer' taken as a unit of current sociological analysis is a member of the professional group that appeared during the nineteenth century, who gained the title 'engineer' by graduating from a technical school or technical university. But they are only a subset of those people who were engaged in capacities that we, today, denominate 'engineering'. It goes without saying that there have always been engineers in industry in the denominative sense of the word, although the distinction between denomination and title has not always been clear.

The engineers in the literature referred to by Professor Torstendahl are defined by their formal training in engineering schools. But the process of social selection of individual careers is influenced by additional factors. Professor Torstendahl has discussed this problem in his book *Dispersion of Engineers in a Transitional Society*, in which he presented a model of individual occupational role-change which included external social factors and personal motives.[17] I would like to add that training in engineering school was a relatively minor part of the educational process of an engineer: His future choice of occupation was also influenced by his work experience before engineering school, apprenticeships during summer holidays, the industry in which he carried out his graduate work, and the industry in which he worked during the first years after graduating. These factors were probably more influential than the engineering school he chose.

On another level, there is an additional risk in relating the social status acquired through formal training to the role-behaviour of an engineer. Everyone outside the engineering community seems to think that they know what an engineer is, and what he stands for. What we often have in mind is a vague idea of the archetypal engineer at the turn of the century, a caricature of Weber's ideal type: A person with a rather single-track mind, awkward social manners, a bowler hat, side-whiskers, a slide rule, and a slightly dirty collar. We assume that he shared certain values with his colleagues – values that had been fostered by his training in one of the new engineering schools and further nourished in his professional society. We assume that the articles in the journal of that society expressed his opinions, and that the society worked to promote his demands that engineers attain social recognition as a professional group.

I also am an engineer, and I dare say that I and my friends in engineering school may have appeared to be a uniform group to any outside observer: people with rather single-track minds, awkward social manners, side-whiskers, pocket calculators, and slightly dirty T-shirts. Yet I know that we share few values in common, and that those that we share were by no means fostered by our training in engineering school. We are divided on all fundamental questions, such as, for example, the safety of nuclear power or the moral justification of contributing to military technical development. Our professional society is nothing more to us than a trade union with more or less compulsory membership, and we know that its journal expresses nothing other than the opinions of the journalists on the editorial staff. Still, the records are there for future historians: number of students, year of graduation, professional affiliation, as well as the subject matter of the editorials. This material will probably be used by future historians who are studying Swedish engineers of the second half of the twentieth century. They will count us, place us in columns, draw diagrams, and quote the editorials as expressions of our opinions. But this material will tell them nothing about our individual motives and aspirations, our views on the social role and ethics of engineering. Our individual values were fostered by ideas, structures and processes in our own time – by far more subtle factors than the fact that we at one time happened to take a number of exams together and all have diplomas to prove it. It will be a history of available data, rather than our history.

In his paper, Dr Haber quoted from *Through the Looking Glass*; I would like to quote a passage that occurs in the same book a few pages later:

> '"What's the use of their having names," the Gnat said, "if they won't answer to them?"
>
> '"No use to *them*," said Alice; "but it's useful to the people that name them, I suppose." '[18]

Notes

1. Haber, L. F. (1971) *The Chemical Industry 1900–1930. International Growth and Technical Change*, Oxford.
2. Haber, L. F. (1978) *The chemical industry: A general survey*. In: Williams, T. I., ed., *A History of Technology*, Vol. VI, Oxford, pp. 499–513.
3. Haber, L. F. (1973) Government intervention at the frontiers of science. British dyestuffs and synthetic organic chemicals 1914–1939. *Minerva*, **11** (1), 79–94.
4. Dr Haber shows that the importance of technological supremacy in warfare became more clearly understood as a result of the War. It is also worth pointing out that the effect of war as an accelerator of technological growth became evident due to the development of chemical technology; this was fully understood at the time and was a permanent result. The British chemist Sir Edward Thorpe, for example, wrote in 1921 that 'there can be no doubt that the war will be found to have affected for good the progress of manufacturing chemistry in this country'. [Thorpe, E., ed. (1921) *A Dictionary of Applied Chemistry*, Vol. 1, London, p. v.]
5. For a bibliometric study of the collection of the Royal Institute of Technology Library, relating its acquisition to the development of technology, see: Sahlholm, B. (1979) *Tillväxt och gränser. Studie kring accessionen år 1825–1923 till bokbeståndet i Kungl. Tekniska Högskolans Bibliotek (Stockholm Papers in History and Philosophy of Technology TRITA-HOT-3002)*, Stockholm.
6. Pilcher, R. B. & Butler-Jones, F. (1918) *What Industry Owes to Chemical Science*, London.
7. *Ibid.*, p. ix.
8. *Ibid.*, p. xi.
9. Rosenberg, this volume.
10. *Ibid.*
11. Torstendahl, R. (1975) *Teknologins nytta. Motiveringar för det svenska tekniska utbildningsväsendets framväxt framförda av riksdagsmän och utbildningsadministratörer 1810–1870 (Studia Historica Upsaliensia, No. 66)*, Uppsala.
12. Torstendahl, R. (1975) *Dispersion of Engineers in a Transitional Society. Swedish Technicians 1860–1914 (Studia Historica Upsaliensia, No. 73)*, Uppsala.
13. Lévy-Leboyer, M. (1979) *Le Patronat de la seconde industrialisation* (*Cahier du Mouvement social, No. 4*), Paris: Chandler, D., Jr & Daems, H., eds (1980) *Managerial Hierarchies: Comparative Perspectives on the Rise of Modern Industrial Enterprise*, Cambridge, Mass.
14. *Industriföretagens forsknings- och utvecklingsverksamhet 1977–1981* (*Statistiska meddelanden serie U, Svenska Statistiska Centralbyrån*), Stockholm.
15. Torstendahl, this volume.
16. For a historical survey for Sweden, see: Berner, B. (1981) *Teknikens värld. Teknisk*

förändring och ingenjörsarbete i svensk industri (*Diss., Arkiv avhandlingsserie, No. 11*), Lund, pp. 116–130.
17. Torstendahl (note 12), pp. 55–68; see especially Figure 1 on p. 59.
18. Carroll, L. (1981) *Through the Looking Glass* [1872], Harmondsworth, p. 225.

Round Table: The use of archival materials concerning the Nobel Prizes in Science and Medicine for research in the history of science: Problems and methods

INTRODUCTION

ELISABETH CRAWFORD

Groupe d'Études et de Recherches sur la Science, École des Hautes Études en Sciences Sociales – Centre National de la Recherche Scientifique, Paris, France

In 1974, a new section was added to the provision in the statutes of the Nobel Foundation (§10) that deliberations concerning the prizes be kept secret. This addition stated that a prize-awarding institution could after due consideration in each individual case, permit access to the material that had formed the basis for evaluation and for the decision concerning a prize, for purposes of historical research. Such permission may be given for prize decisions dating back fifty years or more.

At the Nobel Symposium, some of the individuals who had been granted permission to use archival materials for research on the prizes awarded in science and medicine between 1901 and 1930 were invited to present papers. The organizers were cognizant of the fact that the findings presented would be preliminary, since the studies from which they were drawn were still in early stages or at mid-course at the time the Symposium was held. Nevertheless, it was felt that the Symposium provided a unique opportunity for discussion of the early Nobel prizes in science and medicine among a group of historians of late nineteenth- and early twentieth-century science. Among the many lively exchanges that took place in connection with this Round Table, it was possible to include only the comments of the discussant, Professor Armin Hermann. Others are reflected, however, in the revisions that the papers underwent before publication.

The papers that follow require some background information concerning the structure of the Nobel institution and the procedures for awarding the prizes. Since this information is common to the four papers, it seemed most appropriate to include it in an introduction.

The Nobel prizes in science and medicine

The prizes in science and medicine* were (and still are) awarded by two separate institutions: the *Royal Academy of Sciences,* for the prizes in physics and chemistry; and the *Karolinska Institute* (the medical school in Stockholm), for that in physiology or medicine. To carry out evaluations and make recom-

*For a list of Nobel prizewinners in physics, chemistry and physiology or medicine, 1901–1930, see Appendix I.

mendations concerning each year's prizewinners, the Academy of Sciences appointed two committees, each composed of five members: The *Nobel Committee for Physics* and the *Nobel Committee for Chemistry*. At the Karolinska Institute, the same functions were performed by the *Nobel Committee for Physiology or Medicine*. The members of the Nobel committees for physics, chemistry and physiology or medicine elected between 1900 and 1930 are listed in Appendix II. Final adjudication rested with the Royal Academy of Sciences, meeting in plenary session, and with the professorial staff of the Karolinska Institute. At that time (as now), the Nobel Foundation was responsible for the economic administration of the fund but was in no way involved in the selection of prizewinners.

Nomination and choice of prizewinners

To be eligible for a prize, a scientist had to be proposed by specially invited nominators. Two broad categories of nominator were involved: *those with permanent nominating rights* and *those invited to submit proposals for the prizes of a given year*. At the Academy of Sciences, the former category was composed of:

(1) Swedish and foreign members of the Academy of Sciences;
(2) members of the Nobel committees for physics and chemistry;
(3) scientists who had already been awarded the Nobel prize by the Academy of Sciences; and
(4) permanent and acting professors of physics and chemistry at universities and analogous institutions in Sweden and the other Nordic countries.

Each year, the Academy, following suggestions from its Nobel committees, would also invite nominations from people in the following categories:

(5) chairholders in physics and chemistry at foreign universities (at least six); and
(6) an unspecified number of scientists invited to nominate in their individual capacities.

At the Karolinska Institute, there was a similar distinction between permanent and specially invited nominators. The former category was composed of:

(1) members of the professorial staff of the Karolinska;
(2) members of the medical section of the Academy of Sciences;
(3) Nobel prizewinners in medicine or physiology; and
(4) members of the medical faculties at the universities of Uppsala, Lund, Oslo, Copenhagen and Helsinki.

The specially invited nominators were:

(5) members of at least six medical faculties other than those listed above; and
(6) other scientists.

In actual practice, the two prize-awarding institutions made very different use of categories 5 and 6. In general, the Academy of Sciences tended to restrict the number of chairholders and individual scientists invited to nominate, whereas the Karolinska, as shown in the article by Salomon-Bayet, cast the net much more widely. This was partly a function of the statutory provision that *members of medical faculties* rather than just chairholders be invited to nominate.

Each year, invitations to nominate candidates for the prizes were sent to both permanent and *ad hoc* nominators. Nominations had to reach the committees before 1 February. The committees started their work by verifying that the nominations met the requirements of the statutes, e.g. that they had not arrived too late, that they did not represent 'self-nominations', and that they specified the work for which the candidate was being proposed. At several meetings held during the spring, the committees would first consider the merits of the overall group of candidates proposed for the year's prizes and then select a certain number for more detailed evaluations. These took the form of reports written by members of the committees. The evaluations for the medical prize differed somewhat from those for the prizes in physics and chemistry in that the candidates were grouped by fields of physiology or medicine. During most of the period studied here, the deliberations for the medical prize covered six groups: anatomy-histology; general biology, physiology and physiological chemistry; pathology; medicine, surgery and ophthalmology; bacteriology, hygiene and aetiology; and immunology.

Having reached a final decision on which candidate or candidates they would recommend, the committees submitted their recommendations to the respective prize-awarding bodies. The recommendations were accompanied by (1) a *general report*, in which most of the candidates for the year's prizes were reviewed, and (2) *special reports*, which comprised, first and foremost, the report on the scientist who was being recommended for the prize, but also, depending on the years and the committees, reports on leading contenders for the prizes.

At the Academy of Sciences, committee recommendations were examined by the sections (*klasser*) of physics and chemistry (4th and 5th, but after 1904, 3rd and 4th sections, respectively). When the sections had elaborated their recommendations, the Academy was ready to reach final decisions on the year's awards at a plenary meeting, which generally took place in early or mid-November. At the Karolinska, permanent members of the professorial staff deliberated and voted on the recommendation of the Committee, and the final decision was reached in November. The statutory provision for secrecy mentioned earlier meant that neither the opinions expressed during the final prize adjudication nor the votes taken were recorded.

During the first ten years, the decisions were not made public until the prizes were presented at a ceremony held on 10 December, the anniversary of Nobel's death. This arrangement proved to be impractical, and prize decisions have subsequently been announced shortly after they were reached. Finally, it should be mentioned that it was, and still is, incumbent upon the prizewinner to give a lecture on the work for which he was awarded the prize.

THE PRIZES IN PHYSICS AND CHEMISTRY IN THE CONTEXT OF SWEDISH SCIENCE: A WORKING PAPER

ELISABETH CRAWFORD

Groupe d'Etudes et de Recherches sur la Science, Ecole des Hautes Etudes en Sciences Sociales – Centre National de la Recherche Scientifique, Paris, France

and

ROBERT MARC FRIEDMAN

Institute for Studies in Research and Higher Education, Norwegian Research Council for Science and the Humanities (NAVF), Oslo, Norway

Introduction

It is a curious fact that, with everything that has been written about the Nobel prizes, so little has been said about the Swedish scientists who, year after year, have decided who is to be the laureate in each area covered by the prizes. Yet, from the beginning, the prize decisions, and eventually the Nobel prize institution itself, were shaped by the Swedish scientists who served on the prize-awarding bodies, and particularly those who were members of the Nobel committees. Upon further reflection, it seems reasonable to assume that their judgements, both of what specialities should receive consideration within the general fields designated for the awards and what specific works should be rewarded, were influenced by what they themselves considered important in those fields. Not surprisingly, Nobel prize deliberations became enmeshed in the process by which different factions and personalities within the Swedish scientific community attempted to define their science and to legitimize particular sub-fields within their disciplines.

Committee members played critical roles in the selection of prizewinners both by evaluating nominees and by selecting nominators. During the period for which committee archives are accessible (1901–1931), the physics and chemistry committees each year judged the nominations received and then proposed to the physics and chemistry sections (*klasser*) of the Royal Swedish Academy of Sciences how the prize should be allocated. Following approval from the *klass* or its alternative proposals, the full Academy then voted on the award. On some occasions, the *klasser*, and even the Academy, disagreed strongly with the original choice; thus, in their reports the committees had to justify their actions carefully, especially when these were controversial or not unanimous. The committee members also drew up each year a list of institutions and individuals who would be invited to send nominations for the following year. Thus, the committee influenced to some extent which nominators would supplement the fixed group entitled to make nominations – members of the Academy, former prizewinners, and professors of physics and chemistry at Scandinavian universities and technical colleges.

Uncertainty and disagreement could arise during these proceedings, since the statutes of the Nobel Foundation did not attempt to define the general fields of physics and chemistry. Many other issues entered into the process of prize selection; but, generally, treatment of these questions would require detailed historical analysis of the backgrounds of the Swedish participants and of their scientific contexts. In this paper, the influence of the Swedish scientific community on Nobel prize decisions is illustrated by three cases, which treat, respectively: (1) Arrhenius and physical chemistry; (2) the experimental bias of the Physics Committee and the prize for Einstein; and (3) restriction of the scope of the Nobel Prize in Physics (especially in relation to astrophysics and geophysics). At the present stage of our work, this analysis is of necessity a preliminary one.

Arrhenius and Physical Chemistry

The discussions that took place before the awarding of prizes to J. H. van't Hoff, S. Arrhenius and W. Ostwald give insights into the significance of Arrhenius' membership on the Physics Committee and his role in the Chemistry Committee (which amounted to an unofficial membership) as well as raise broader issues concerning the rewarding of work in interdisciplinary specialities. Particularly in view of Arrhenius' presence on the scene, the question was not *if* the men who had helped to found physical chemistry should be rewarded, but *when* and *how* this should be done.

Arrhenius' progress from the theory of electrolytic dissociation to his more speculative work in immunology and cosmical physics is sufficiently well known that it does not require elaboration. More important in the context of Swedish science are the problems he encountered early on in his scientific endeavours. Most of Arrhenius' biographers agree that the poor reception given his doctoral dissertation, which he presented in Uppsala in 1884 and which contained the first version of the theory of electrolytic dissociation, opened a wound that never healed.[1] His bitterness towards the members of the jury was directed particularly at Robert Thalén, professor of physics and Arrhenius' future colleague on the Nobel Committee, who had already refused him access to the Uppsala physics laboratory and was now instrumental in giving his dissertation the low grade of *non sine laude*, which quashed his hopes for a scientific career in Uppsala.[2] From then on, he showed an excessive preoccupation with the regard in which his work was held by the Uppsala physicists. Invited to lecture at the Royal Society after he had received the Nobel prize, he wrote to Vilhelm Bjerknes about the accolade he had received from British physicists, adding that 'it looks as if in the area of cosmical physics as well, Hasselberg, Hildebrandsson and Ångström [his colleagues on the Physics Committee] shall have the pleasure of first leering and then "heartily" congratulating me on my success'.[3]

The questions of 'when' and 'how' the group known as the 'Ionists' should be rewarded were posed in the first year the prizes were awarded, with respect to van't Hoff and Arrhenius, who were both nominated in that year. Specifically, the questions concerned, on the one hand, in which *order* the two should receive the prize and, on the other, to which *field* – physics or chemistry – the work of Arrhenius should be considered as having contributed the most. As for the question of order, the nominations did not provide much guidance, for, although van't Hoff received the largest number of nominations in chemistry in 1901, with one exception he was being commended for his overall contribution to theoretical chemistry, organic as well as inorganic. Since, according to Nobel's testament, the award should be given for a specific 'chemical discovery or improvement', it fell upon the Committee to single out the achievement that would form the basis for an award. This process had already begun at the nominating stage when P. T. Cleve and Otto Pettersson – the two most influential members of the Committee – proposed Arrhenius and van't Hoff for a joint prize, citing both Arrhenius' papers on the electrolytic theory of dissociation published in the *Proceedings* of the Swedish Academy of Sciences (1884 and 1887) and the memoire by van't Hoff (*Lois de l'équilibre chimique dans l'état dilué, gaseux ou dissous*) published in the same journal in 1886.[4]

By proposing van't Hoff and Arrhenius for a divided prize, Cleve and Pettersson were advancing the most judicious solution to the problem of who should be rewarded first; however, it was not the solution adopted by the Committee or by the Academy. Although Arrhenius would later claim that he had postulated his theory of ionic dissociation in his ill-fated doctoral dissertation, published in the *Proceedings* (1884), historians of chemistry who have studied the matter closely agree that his theory emerged only in his paper of 1887. By that time, van't Hoff's paper had already appeared in the *Proceedings* (1886) and had, furthermore, been instrumental in setting Arrhenius on the path that would lead to his theory.[5] These points of chronology would obviously have been of importance had the issue at hand been one of simultaneous discovery.[6] Instead, it was the significance of the works of the two men in terms of their *complementarity* that was stressed in the Pettersson-Cleve proposal and that is therefore examined here.

The chief innovation in the paper of van't Hoff was the analogy he established between dilute solutions and perfect gases through the use of osmotic pressure (which he found corresponded to gaseous pressure). This made it possible for him to apply thermodynamics to the calculation of the chemical equilibrium of dilute solutions; and his general results were expressed as $PV = iRT$, where P is osmotic pressure, i an empirical factor characteristic of a given solution and R a constant which is the same for solutions and gases (V and T being, of course, volume and temperature). The term i remained the puzzle in the equation, for it was an *ad hoc* variable that van't Hoff had had to introduce to correct for the deviations found with electrolytes from the characteristics of ideal solutions. It was this puzzle that sparked Arrhenius' imagination and led him to postulate that the dissociation observed with electrolytes did not result in chemically 'active' or 'inactive' molecules, as he had stated in his thesis, but in the formation of ions. This idea was buttressed by his finding that the number of ions produced by dissociation corresponded to the coefficient of i. Whatever the specific merits of the work of each man – and they were undeniably great – when viewed in terms of their complementarity, it was Arrhenius' idea that made for the generality of van't Hoff's earlier results. This point, which was made clearly in the Pettersson-Cleve proposal, was perhaps most explicitly stated by van't Hoff in his lecture on receiving the prize, when he said: 'I would not have had the pleasure of giving this lecture had it not been for Professor Arrhenius pointing out the reasons for these deviations [i.e. of strong electrolytes from the law of ideal solutions].'[7]

Why, then, did Arrhenius not receive the prize before or together with van't Hoff? There may well have been a feeling among members of the Academy that to attribute all or a share of the first year's prize to a Swede would be to diminish the international standing of the prizes. Subsequent events also point to the importance of the second question posed above, that is, in which *field* should Arrhenius receive the prize?

The most likely explanation for the 'passing over' of Arrhenius is that he himself made it known that he did not want to receive the prize in that year. For the reasons cited above, he may not have been sure of having sufficient support in the Academy. He may also have been holding off in the hope of receiving the prize in physics. The latter explanation seems likely in view of the strategy that Arrhenius' friend, Otto Pettersson, member of the Chemistry Committee, devised to bring this about (as discussed below). It should also be mentioned here that in 1902 and 1903 Arrhenius withdrew from the prize discussions in the Physics Committee, giving his own candidature as the reason.[8]

Arrhenius wanted to be awarded the prize in physics for several reasons, among which the following seem to have been the most important. First, although Arrhenius and his colleagues had set out to revolutionize chemical theory, this had been largely accomplished by applying the methods and reasoning of physics, chiefly thermodynamics, to the problem of chemical activity in solutions. Second, starting in the 1890s, Arrhenius' own research had been devoted increasingly to physics, particularly cosmical physics. Third, given his feelings of resentment towards the Uppsala physicists, there could have been no sweeter victory, or revenge, than to be attributed the Nobel prize in spite of his old adversaries.

In 1902, van't Hoff, in his capacity as a former prizewinner, nominated Arrhenius for the physics prize, in a proposal that emphasized the physics components of his theory of electrolytic dissociation. In that year, the Physics Committee was helped out of potential embarrassment by a campaign launched in favour of H. A. Lorentz by Gösta Mittag-Leffler, a prominent Swedish mathematician and member of the Academy who took an active interest in the prizes. In the field of chemistry, after the first year's celebration of physical chemistry, it was not surprising that the prize was awarded to an organic chemist, Emil Fischer, who was honoured for his work on sugar and purine synthesis. By late 1902, the intransigence of the Physics Committee to proposals for an undivided prize for Arrhenius must have been sufficiently clear for his supporters to shift strategy and to propose, instead, that he receive a share of the prizes in both physics and chemistry. For the prizes of 1903, then, van't Hoff renewed his proposal in physics but also nominated Arrhenius for that in chemistry. In that year, Arrhenius received seven nominations for the prize in physics and eleven for that in chemistry.[9]

Early in 1903, the Chemistry Committee, probably at the instigation of Otto Pettersson, addressed a letter to the Physics Committee suggesting that Arrhenius be attributed half of the prize in physics and half of that in chemistry. The attribution of a share of the physics prize was justified, the letter stated, by the fact that Arrhenius' theory of electrolytic dissociation had its origin in a physical problem, that of the electrical conductivity of solutions, and also by its ability to explain a wide range of phenomena in physics, particularly those associated with the ionization of gases and related radiation phenomena (cathode rays, X-rays and Becquerel rays). The letter suggested that the other part of the chemistry prize be attributed to William Ramsay for his discovery of the inert gases, argon and terrestrial helium. This would require, however, that the Physics Committee likewise entertain the possibility of awarding the remaining share of *its* prize to Lord Rayleigh, whose discovery of argon had been made more or less simultaneously with that of Ramsay and which, furthermore, had been announced in a jointly authored paper (1895).[10]

The letter received a cool reception from the physicists, whose reply stated that Arrhenius' theory of electrolytic dissociation 'indisputably . . . has been most consequential for chemistry'. Although the Committee did not deny that, when viewed from this standpoint, it had been of significance for physics as well, it did not feel that Arrhenius' works 'were of greater, or even as great significance as others proposed for the physics prize'. Since the latter could not be passed over, it would not be possible for the Committee to propose Arrhenius for the prize. The proposal that the two committees join forces in rewarding Rayleigh and Ramsay would be acceptable only if each committee decided to recommend them for an undivided prize, that is, Rayleigh for the physics prize and Ramsay for that in chemistry.[11]

The reply of the Physics Committee quashed Arrhenius' hopes that it would recommend him for any part of the physics prize; moreover, it spelled trouble for the chemists. The reply indicated, in fact, that if they were to recommend to the Academy that Arrhenius and Ramsay share the prize, the proposal would not only appear to be disloyal to the physicists, it would also have little chance of succeeding over the physicists' opposition. The choice for the chemists, then, was a stark one: either Ramsay or Arrhenius. There was more at stake, though, than a simple choice between the two, for the candidates favoured by the Physics Committee – those 'others' whose work was of greater significance than that of Arrhenius' – were Becquerel and the two Curies. Looking beyond the matter of rewarding Arrhenius, which could always be postponed, some members of the Chemistry Committee – O. Widman and H. G. Söderbaum, in particular – saw attribution of the physics prize for pioneering work in radiation as a dangerous precedent, since it could mean that in the future this prestigious field would be appropriated by the physicists for *their prize*. For Widman (writing to Söderbaum), an important reason for supporting Ramsay in combination with Rayleigh (with or without Arrhenius) was that 'in this manner, the physicists and the Academy would be saved from the foolishness of pulling in the *chemistry* couple Curie already this year'.[12]

In the first voting on the prize in 1903 in the Chemistry Committee, which took place in the spring, Widman, Söderbaum and Pettersson supported Ramsay for the prize, whereas Cleve and P. Klason favoured Arrhenius. Given this split, the Committee could do no more than await the decision of the physicists (which would not come until the autumn) and commission reports on the leading contenders. The task of arguing for the importance for physics of Arrhenius' theories fell to Otto Pettersson, who must still have hoped that the Academy could be persuaded to attribute a share of the physics prize to Arrhenius. Cleve, who strongly favoured that Arrhenius be attributed the prize in chemistry, took on the task of reporting on the significance of his theory for chemistry, whereas Söderbaum was assigned the report on Ramsay.[13]

When the Physics Committee met in the autumn, it decided unanimously to recommend Becquerel and the two Curies for the prize. It adopted an alternative recommendation for Rayleigh, but only on the condition that the chemistry prize be awarded to Ramsay alone. Shortly thereafter, this decision was endorsed by the fourth *klass* (physics) of the Academy.[14] The final vote in the Chemistry Committee still showed a three-to-two split in favour of Ramsay; in the *klass*, however, Arrhenius gained an absolute majority of seven votes, as against three for Ramsay.[15] At the plenary meeting of the Academy, it was decided, at the request of the Physics Committee, that, exceptionally, the Academy would vote on the chemistry prize before moving on to that for physics. This was necessary in view of the condition that the physicists had attached to their alternative recommendation for Rayleigh.

The draft of the speech that Widman made in the Academy, preserved among his papers, represents a concise statement of the two main considerations that had motivated the majority of the Committee to propose Ramsay for the prize. First, it shows clearly that Widman, and probably also Söderbaum and Pettersson, had not abandoned their earlier proposal for coupling the prizes in physics and chemistry by awarding them jointly to Arrhenius, Rayleigh and Ramsay. Widman felt obliged to appeal directly to the Academy, since the initial refusal of the Physics Committee to consider the proposal had meant that it could not be discussed either in the committees or in the *klasser*. Second, it shows how important it was to himself and to his colleagues that the Academy delay its decision to attribute the prize to the Curies, since, in his opinion, even if their work had not yet been the object of proposals for the prize in chemistry, the situation was bound to change in the years to come. He argued that the isolation of radium (accomplished by Marie Curie in 1902) was of greatest significance to chemistry, since it was the most important new chemical element discovered in recent times. Furthermore, the discovery was one that was likely to change chemists' notions of the basic and invariable nature of the elements. Widman's speech notwithstanding, and he was no doubt supported in his stand by Söderbaum and Pettersson, the Academy decided to attribute the prize in chemistry to Arrhenius.[16]

Now that both van't Hoff and Arrhenius had received their prizes, there remained only one of the original triumvirate of Ionists, Wilhelm Ostwald. In 1904, it appeared that he would be rewarded imminently, since in that year he had advanced to the select group of candidates whose work is made the subject of special reports in the Chemistry Committee. The report written by Otto Pettersson, mainly served, however, to point up the difficulties raised by Ostwald's candidature. Although Ostwald had done research, lectured and written textbooks on most aspects of modern chemistry and had also been instrumental in winning acceptance for the theory of ionic dissociation, it was difficult to pinpoint one particular achievement that could be held up as an important discovery. This was stated in the general report on the candidates prepared for the Academy in 1904: '. . . were the Nobel prize . . . to be awarded for scholarly work promoting the general development of chemistry, there can be no doubt that Ostwald should have been put ahead of other candidates this year.'[17] This statement was reiterated in the general reports of 1907 and 1908. As far as the Committee was concerned, then, in 1904, Ostwald's candidature was put in abeyance and was to remain so until the Committee reversed its position in 1909, the year he received the prize. In the intervening years, his chances of being awarded the prize appeared to be diminishing, since a decision would have to be based on achievements that were receding into the past. In fact, by 1906 Ostwald was no longer active in chemical research, having left his chair at the University of Leipzig to set himself up as an independent scholar and philosopher at his country property, where he devoted his time to a host of scientific, educational and cultural causes – including energetics, monism and the creation of a new world language (Ido).[18]

Since Arrhenius' support eventually proved critical in the decision to award Ostwald the prize, it is interesting to reflect on some of the reasons that he did not act until 1909. In the first instance, Ostwald's name probably did not figure at the top of the list of candidates that Arrhenius sought to promote for the chemistry prize. He gave priority, instead to the rapidly advancing field of radiation, in which his own nominations were instrumental in securing the chemistry prize for Rutherford in 1908. Second, for Arrhenius, who was avidly following the evidence accumulating for an atomistic view of matter (in particular through the research programmes of Thomson and Rutherford), Ostwald's advocacy of energetics, which rejected such a view in favour of one based on transformations of energy, must have been something of an embarrassment. It is probably no accident that Arrhenius threw his support behind Ostwald only after the latter had confessed to his 'mistake' and adopted the idea of the physical existence of atoms.[19] Third, Arrhenius was enough of a strategist when it came to Nobel prize matters to know when the moment was ripe for action.

In 1909, the Chemistry Committee was in considerable disarray as to which candidate or candidates it should recommend to the Academy. The consensus that had reigned among the nominators in the early years had dissolved, and there was now a host of candidates, none of whom had received more than a handful of nominations. For that year's prize, Ostwald was nominated by Arrhenius, van't Hoff and G. Bredig, professor of physical chemistry in Heidelberg and a former assistant to Ostwald.[20] Shortly after the end of the nominating period, Arrhenius wrote to Widman, suggesting that a prize for Ostwald would be analogous to the one attributed to von Baeyer in 1905, since 'just as the organic chemists at that time wanted to give recognition to their great master and leader, despite the fact that his best works were somewhat removed in time, now the physical and inorganic chemists want to do the same'. Having suggested to Widman the wording of the award citation for Ostwald, Arrhenius ended his letter with an argument that carried particular weight for him: in contrast to many other candidates, he stated, Ostwald, who was not independently wealthy, could well use the Nobel prize money.[21]

Widman must have taken Arrhenius' arguments to heart, for, shortly after the spring meeting of the Committee, he wrote back with an encouraging report, which also provides a rare insight into the workings of the Committee. After the other members had talked for a couple of hours, 'murdering each others' candidates', he threw in the names of Nernst and Ostwald. Having no problem in eliminating Nernst's candidature, he then argued for an award to Ostwald on the basis of the latter's investigations of catalytic processes. Although these had figured in one of the nominations for Ostwald in 1904, as well as in the above-mentioned proposal by Bredig, to quote Widman, 'this was a viewpoint that the other members of the Committee had not considered until now, but it evidently impressed them'. The chances of success were good, but it was necessary to give particular care to the special report on Ostwald, which Widman had agreed to write only because Arrhenius had promised to give a hand.[22] When the report came due, however, Widman found to 'his surprise and horror' that Arrhenius was in Manchester doing radiation research in Rutherford's laboratory. To Widman this was not a sufficient excuse for delay, since 'you are probably not doing anything in particular over there and since it is important that you do not bother him [Rutherford] in *his* work, you might as well write the report'. If Arrhenius did not comply, Widman would give up Ostwald and, instead, write a report on Victor Grignard, the French organic chemist and future Nobel prizewinner (1912), with whose work he was intimately familiar.[23]

The report on Ostwald that Arrhenius wrote after his return from England and which Widman handed in under his own name at the autumn meeting of the Committee focused on catalytic phenomena, which were described as the thread running through Ostwald's experimental work. The report shows the utility of this device, for it permitted the Committee to hold up Ostwald's work on a phenomenon, or rather a conceptualization of chemical processes, which in itself was specific enough, but which also, through its applications to different branches of chemistry, was sufficiently general to subsume a significant body of Ostwald's writings. It was thus that, in addition to work specifically on catalysis – Ostwald's *Über Katalyse* (1901) – which, furthermore, had the advantage of recency, Arrhenius could point to his own and Ostwald's investigations on the conductivity of acids in solutions as another example of catalysis studies. Likewise, the rapidly developing field of enzyme research could be brought in under the catalytic 'umbrella', making Ostwald's award a logical follow-up of that to Buchner (1907), who had been the first to demonstrate experimentally the role of enzymes in cell-free alcoholic fermentation. Hence, in its report to the Academy recommending Ostwald for the prize, the Committee could explain its sudden acceptance of Ostwald's catalysis research by referring to 'the development of science proper, especially investigations into the catalytic action of enzymes and colloids, which have highlighted the importance of a group of works by Ostwald, that is, those dealing with catalytic reactions'.[24] Although the arguments presented were sufficiently persuasive for Ostwald to be awarded the prize, J. R. Partington, in his *History of Chemistry*, states that 'since Ostwald had no theory of catalysis, he proposed superficial analogies'; E. Hiebert and H.-G. Körber note that on several basic questions concerning catalytic reactions 'no convincing answers were supplied by Ostwald and his collaborators'.[25]

The foregoing shows that Arrhenius' influence was instrumental in the rewarding of the original triumvirate of Ionists. He also used his influence, in a series of actions too complex to elucidate here, to delay until 1921 the awarding of the prize to W. Nernst.[26] This hiatus doubtless had the effect of aggrandizing the contribution of the pioneers. The case of Arrhenius and physical chemistry raises broader issues, though, since the failure of the physics and chemistry committees to cooperate in awarding Arrhenius a share of the prize in each discipline set a precedent which precluded any such actions being undertaken in the future. In the opinion of A. Westgren, a former member of the Chemistry Committee, if this precedent had been created, it 'would have rendered the treatment of similar questions unnecessarily complicated', particularly in cases involving the overlap of chemistry and medicine, which are the domains of separate prize-awarding institutions.[27] Another view might be that, by creating such a precedent, the recognition of interdisciplinary specialities would have been facilitated and the scope of the areas in which prizes were awarded would also have been broadened. Given the extent to which this question was enmeshed in conflicts peculiar to the Swedish scientific community at the time, however, the outcome was probably a foregone conclusion.

The Experimental Bias of the Physics Committee and the Prize for Einstein

At the time the first prizes were awarded, three of the five members of the Physics Committee belonged to the Uppsala tradition of experimental physics: B. Hasselberg, R. Thalén and K. Ångström. This predominance carried over until after the First World War. Thus, an experimentalist bias in the Committee proved significant from the start in Nobel prize decisions.

A quick glance at the list of early prizewinners testifies to the experimentalist predilection of the Committee: Röntgen (1901), Zeeman (1902), Becquerel and the Curies (1903), Rayleigh (1904), Lenard (1905), J. J. Thomson (1906), Michelson (1907), Lippmann (1908), Kamerlingh-Onnes (1913). In many cases the work honoured lay in the rapidly developing area of 'ray' physics (X-rays, cathode rays and radiation). It was undoubtedly useful that such breakthroughs easily met the stipulation that they be 'discoveries and inventions' made during 'the preceding year', the latter having been modified in the statutes to mean 'the most recent achievements in the fields of culture referred to in the will'. Moreover, the universal excitement that these discoveries created among the public would naturally enhance the prestige of the new institution. However, upon closer examination, the preponderance of prizes in experimental physics seems to reflect more a conscious design from within the Committee than merely the wishes of the nominators.

The prize awarded to Michelson is an example of how support from the experimentalist majority, led in this instance by Hasselberg, played a major role. As stated in a letter to G. E. Hale, who had proposed Michelson for the prize, Hasselberg set as his goal 'to do all in my power to procure the prize for him [Michelson]'. In a subsequent letter to Hale, Hasselberg allowed that his high regard for Michelson's work was 'in some way an opinion of sympathy for an area closely connected with my own speciality' and went on to state: 'I cannot but prefer works of *high precision*.'[28] The specific elements of Michelson's achievements that evoked Hasselberg's strongest admiration were the applications of his major instrumental innovation – the interferometer – to spectroscopy and metrology. These specialities were especially close to Hasselberg's heart, both because of his own investigations and because of his work as the Swedish representative on the International Committee of Weights and Measures. Michelson's work with the interferometer to determine experimentally the length of the international meter introduced the notion that metrology could be based on natural constants. Furthermore, Michelson's interferometer had been used to recalculate wave-length tables for spectroscopy. It is significant that these two studies were in Hasselberg's opinion the most important aspects of Michelson's work to be presented to the Nobel Committee for Physics. The lengthy treatment of these achievements stands in marked contrast to the one-sentence reference to the negative results of the Michelson-Morley experiment (1887), for which Michelson is best known in the history of science.[29]

The admiration of Hasselberg and (although less well documented) of other members of the Uppsala school for Michelson's work in precision physics was sufficiently strong to overcome the fact that Michelson was proposed by only two people for the year in which he received the prize and had been proposed only once before then (by E. C. Pickering, the Harvard astronomer). Moreover, Michelson's work did not contain any specific elements that could be held up as a discovery. That Hasselberg was aware of this possible conflict with the statutes is borne out in one of his letters to Hale, in which he asks the opinion of the latter as to which of Michelson's works might constitute a discovery.[30] Thus, the desire to reward precision measurement outweighed these otherwise negative considerations. Attempts to reward mathematical and theoretical innovations did not generally receive comparable treatment.

Debates on the role of mathematics and theory in physics carried over into deliberations on the awarding of the prize, so that the experimentalist bias also seems to have acted on occasion to oppose the rewarding and legitimizing of certain forms of scientific contribution. Attempts during the early years to place a mathematical physicist on the Committee, the lack of which had been characterized as a 'serious deficiency'[31] in the evaluation of H. A. Lorentz repeatedly met with failure. Uppsala physicists acknowledged that the Committee needed to be strengthened in the area of mathematical physics, but it was also pointed out that this need was not very urgent in view of the statutory provision that the prize be awarded for 'discoveries or inventions'. 'An investigation in mathematical physics would only rarely be considered for the prize', it was stated, 'unless, as was the case, for instance, for Lorentz's electron theory, it leads to "discoveries or inventions" in the area of experimental physics'.[32] Even in the case of the world-wide campaign that Mittag-Leffler launched in 1910 to procure the prize for Henri Poincaré, in which his mathematical physics *qua* physics was separated from Poincare's purely mathematical acomplishments, the majority did not see fit to recommend him for the prize.[33]

In a protest note of 1911, the Committee's newly elected supporter of mathematical and cosmical physics, V. Carlheim-Gyllensköld, compared the standing of mathematical physics in the scientific world, on the one hand, and the extent to which it had been represented in Nobel prize decisions on the other. He pointed out that, with the exception of H. A. Lorentz (1902) and, to some extent, J. J. Thomson (1906), 'the Nobel prizes have up to now been attributed to experimental physicists'. He stressed that the neglect of mathematical physics was not due to an absence of nominations for, since the beginning, prominent representatives of this speciality (e.g. Boltzmann, Duhem, Heaviside, Lord Kelvin, Planck, Wien) had, in fact, been put forward for the prize. In most instances, such nominations had been made by experimental physicists: Lenard, Becquerel, Röntgen, Zeeman, J. J. Thomson, to mention only the first handful of names in the long list that he presented. 'These numerous votes', he writes, 'merit attention.'[34] Nevertheless, Committee members continued to rely more on their own opinions than on those of the nominators in selecting prizewinners and prizewinning work.

The influence of the Swedish physics discipline on Nobel prize decisions is well illustrated by the discussions leading up to the awarding of Albert Einstein's prize. In 1922, the Academy voted to award Einstein the previously reserved Nobel prize of 1921 'for his services to theoretical physics, and especially for his discovery of the law of the photoelectric effect'. Both the timing and the citation have long raised questions among Nobel prize watchers. If the opinions of the scientific community at large, or more specifically of the nominators, had been decisive, Einstein would certainly have received a prize earlier, and for his theories of special and general relativity.

As early as 1910, Ostwald began nominating Einstein for his work in special relativity. During the First World War and immediately thereafter, many nominations for Einstein arrived, specifying his research on relativity, Brownian motion and specific heats of solids and/or quantum theory in general. Following the famous 1919 solar eclipse experiment, which, according to many, confirmed Einstein's prediction of the bending of light by the gravitation of a massive body, a virtual flood of nominations for Einstein ensued.

Yet in 1920, the Committee and the Academy agreed to award the prize to Charles Guillaume, director of the International Bureau of Weights and Measures, citing 'the services he has rendered to precision measurements in physics by his discovery of anomalies in nickel steel alloys'. One year later, claiming that they could find no grounds for awarding a prize to Einstein, the Committee, followed by the Academy, voted to reserve the prize for 1921.[35] Finally, in 1922, Einstein received a prize, specifically for the law of the photoelectric effect – not for his quantum theory to explain the law, nor for relativity in any form. In the approximately fifty nominations of Einstein over the years, the law of the photoelectric effect was specified only in 1921 and 1922 and only by one nominator, C. W. Oseen, professor of mechanics and mathematical physics at the University of Uppsala.[36] It is evident that knowledge about the Committee members is necessary to begin understanding these events.

Immediately after the First World War, the Committee was still dominated by physicists who adhered to an experimentalist philosophy of scientific advance. Naturally, the majority's demands that theory be completely verified by experience virtually precluded the awarding of a prize to Einstein for his various theoretical endeavours. Yet, did the 1919 eclipse experiment have any impact on the committee's evaluation? Although it is often hailed as a 'crucial experiment' which proved Einstein's general theory of relativity, the test left many reasonable physicists unconvinced, among them the members of the Nobel Committee for Physics. In 1920, Arrhenius prepared a special report on the subject for the Committee, in which he noted that since objections could be made against the accuracy of the observations they were not a proof of the prediction.[37]

Nevertheless, the Einstein campaign continued to grow, with strong support for the relativity theories. To ensure a sound evaluation, the Committee assigned the task of writing a special report focusing on Einstein's theories of relativity and gravitation to Gullstrand, who would have received the 1911 prize in physics for his contributions to geometrical optics had he not first been named for a prize in medicine or physiology.[38] In his fifty-page report, Gullstrand concluded that neither the general nor the special theory of relativity warranted a Nobel prize.[39] In 1922, he brought his report up to date and came to the same conclusion: Acceptance of these theories remained simply 'a matter of faith (*trossak*)'.[40] He stated privately that Einstein should never receive a prize.[41]

In general, the Committee appears to have supported Gullstrand. Hasselberg asserted from his sick bed that 'it is highly improbable that Nobel considered speculations such as these to be the object of his prizes'.[42] Nor was there much sympathy for honouring Einstein's other theoretical contributions. Regarding Oseen's proposal in 1921 specifying the law of the photoelectric effect, Arrhenius concluded his special report by noting that it would be 'strange [*egendomligt*]' to reward that work while passing over Einstein's contributions on Brownian motion and on relativity, with which the discovery of the law could not favourably 'compete [*tävla*]'.[43] When the *klass* met to vote on the Committee's proposal to reserve the prize, Oseen – who did not become a member of the Committee until 1923 – suggested that they consider the significance of Einstein's discovery, which provided a basis for the recent remarkable advances in atomic physics. The *klass* then voted to acknowledge the significance of the Einstein discovery, but stopped short of proposing a Nobel prize.[44]

Oseen did not intend to let the issue lie there. True, he wanted Einstein to receive a prize, but not for relativity; he seemed equally, if not more, concerned with Niels Bohr, who had been nominated repeatedly since 1917.[45] For him, 'of all the beautiful . . . Bohr's theory of the atom [was] the most beautiful', and he was thus determined that this work should receive a prize.[46] Moreover, he had already begun a campaign to strengthen Swedish physics, particularly by introducing theoretical physics. For Oseen, atomic physics offered the challenge of theoretical investigation while remaining tightly disciplined by experiment. Previously, the experimentalists on the Committee had balked at approving Bohr's atom model, which, they claimed, stood 'in conflict with physical laws' and hence with reality.[47]

In 1922, Oseen again proposed Einstein for the prize on the basis of the discovery of the law of the photoelectric effect, and then joined the Committee as a special extra member. Oseen wrote special reports on Einstein's law and on Bohr's atomic theories. He carefully argued that experimental investigations had confirmed Einstein's law so thoroughly that it must be considered among the 'soundest propositions physics now possesses'.[48] Hence, Bohr's models, which, he argued, were based on this law, must be regarded as being in solid agreement with physical reality.[49] Finally, not without some difficulty, Oseen prevailed. Although further study is needed, this episode demonstrates the importance of the interests of the Committee members and their philosophies of science for Nobel prize decisions.

Defining the Scope of the Nobel Prize in Physics

During the early years of the prize, the Committee understood that physics should be defined as broadly as possible. At the first Committee meeting, held at the Meteorological Institute of the University of Uppsala, the Committee agreed that astrophysics was eligible while astronomy was not.[50] Moreover, the inclusion of meteorology raised little, if any, objection; as well as hosting this meeting, the director of the Meteorological Institute, Professor H. H. Hildebrandsson, served on the Committee. In fact, over the next thirty years, nominations to honour various astrophysicists, geophysicists and meteorologists (e.g. G. E. Hale, H. Deslandres, K. Birkeland, C. Störmer, N. Shaw, J. Hann, V. Bjerknes) were not and could not have been formally disqualified as being for scientists outside 'physics'. In short, neither Nobel's will nor the statutes restricted the definition of physics to any particular group of specialities. Yet, as noted above, those members of the Committee who belonged to a nineteenth-century tradition of experimental physics generally interpreted restrictively the statute that prizes be awarded to a 'discovery or invention' as excluding innovations and discoveries of a theoretical nature. After the First World War, these interpretations had to be adapted to new conditions in the discipline of physics, both in Sweden and abroad.

The deliberations concerning Einstein came at a turning-point in the development of Swedish physics and of the Committee. In 1923, after the deaths of Hasselberg and Granqvist, Oseen was elected to the Committee and was joined by his Uppsala colleague, atomic physicist Manne Siegbahn. Together with

Oseen's friend, Gullstrand, the three Uppsala physicists commanded a majority on the Committee and could plan and act together on Nobel prize matters. Oseen's broad knowledge, international reputation and aggressive determination allowed him to assume a dominating position on the Committee. His vision called for an overhaul of Swedish physics: stronger links with major foreign research centres, and greater visibility and prestige to ensure adequate research funds and university positions.[51] By promoting atomic physics, in which experiment and theory progress together intimately, he hoped to overcome traditional prejudices and to institutionalize a theoretical physics that stood apart from abstract mathematical physics and mechanics. Yet, to be successful in the 'new epoque'[52] that he declared after the election of Siegbahn to the Committee, they had not only to promote and to legitimize their preferred research programmes, but also to eliminate those that they considered insignificant specialities which could obstruct their plans. Starting in the 1920s, the Uppsala group, often together with Arrhenius, began a determined campaign to restrict the definition of physics within the Academy and with regard to the Nobel prize. To pursue this strategy they attempted to control membership in the *klass* and Committee and to establish traditions for interpreting the statutes by honouring particular specialities and blocking others.

This change in attitude became clear when, in 1923, the new Uppsala group and Arrhenius attempted to eliminate astrophysics from consideration for the prize. This discussion arose in connection with the candidature of George Ellery Hale and Henri Deslandres for their work related to solar physics. Prior to the War, they had been considered almost certain recipients of the prize. In the general report for 1913, the Committee noted that their endeavours should be followed very closely since these 'without doubt sooner or later very rightly will probably come to be regarded as deserving the physics prize'.[53] Again in 1914, they were considered among the very top possibilities; however, because only Hale had been nominated that year, the Committee decided to wait for a detailed evaluation of their work. When, in 1915, nominations for both were received, an unfortunate circumstance virtually precluded any chance of their success: Carlheim-Gyllensköld, who supported the consideration of cosmical physics and whose vote would therefore be crucial, had that year proposed his own candidates, Kristian Birkeland and Carl Störmer. Of course, a more detailed analysis of the proceedings in 1915 will be necessary before it is possible to understand how the Committee came to its unanimous vote to award the prize to W. H. and W. L. Bragg; and why, in the draft of the 1916 general report, the remarks that Hale's and Deslandres' work was worthy of recognition with a Nobel prize were crossed out.[54]

In 1923, after the Committee had finished with the difficult nominations of Planck, Einstein and Bohr, Hale and Deslandres again appeared to be possible contenders. In a letter to Arrhenius, Gullstrand expressed concern that Carlheim-Gyllensköld (who, after the death of Hasselberg, had become the main champion of the astrophysicists) would argue that Hale and Deslandres were the strongest candidates. Gullstrand noted that Oseen doubted whether this work could be considered as physics, but that he would be willing to follow Arrhenius' authority on the matter.[55] Indeed, Carlheim-Gyllensköld concluded his special report with highly flattering praise of the results and methods of Hale and Deslandres, which, he felt, must be considered among 'science's most beautiful achievements in recent times [*vetenskapens vackraste landvinningar under senaste tiden*]'.[56] Arrhenius presented a note to the Committee in which he first attempted to dissociate himself from earlier apparent support of Hale and Deslandres, and then expressed concern that a prize for their work would establish a dangerous precedent. He foresaw a considerable increase in the number of nominations for work in astrophysics and astronomy. In an attempt to reverse the earlier agreement (1901) on the inclusion of astrophysics, which had been pressed for by Hasselberg, Arrhenius argued that astrophysics had progressed so rapidly in recent years that it now encompassed all physical astronomy; hence, it could not be considered physics. Special studies would be required for each nomination received to determine whether the work in question was truly part of physics – a situation that would swamp the Committee and the Academy with work.[57] In the general report, written by Gullstrand, Carlheim-Gyllensköld's superlatives were eliminated; and, instead, comparable praise was bestowed on Hale's American colleague, Robert A. Millikan. Although Carlheim-Gyllensköld sent a rebuttal to Arrhenius, in which he showed that the feared increase in astrophysical nominations was nonsense and that Hale and Deslandres had been repeatedly nominated by physicists,[58] the question of formally declaring astrophysics outside the scope of physics did not arise. As long as the Committee and the *klass* shared a common notion of the boundaries of physics, nominations in border areas, such as astro- and geophysics, could be dismissed as 'not being significant for physics'. During the 1920s, the Uppsala group and to some extent Arrhenius worked to eliminate opposition to their notions of the proper direction for physics.

Although repeated efforts to eliminate Carlheim-Gyllensköld from the Committee failed because of his support in the Academy,[59] the Uppsala group had greater success in reducing the influence of meteorologists in the *klass*. When the physics *klass* of the Academy expanded from six to ten members in 1904, it received the title 'Physics and Meteorology'. Having established strong research traditions in meteorology during the nineteenth century, Swedish meteorologists received a number of places in the *klass*. By 1919, however, the promise of Swedish meteorology seemed to have diminished. Moreover, a long, bitter feud concerning atmospheric thermodynamics raised doubts about the personal and/or scientific reliability of some of the meteorologists. Oseen was prompted to express concern over whether any branch of 'terrestrial physics' could become rigorous, given the scale of geophysical phenomena and the lack of possibilities for laboratory testing; he sensed a 'relativity of all knowledge' within these sciences.[60] The Uppsala physicists shared his belief; they attempted to reduce the influence of the meteorologists in the *klass* and to prevent the awarding of a Nobel prize to workers in this science, since that would legitimize once and for all meteorology as part of physics in the Academy.

During the early 1920s, the majority of the Committee worked to elect 'physicists' to the *klass* as replacements for dead members who had been 'meteorologists'. Finally, the elderly Hildebrandsson protested to the Academy that 'should this happen yet again, then it must be regarded as a wish of the physicists to get rid of meteorology within the third *klass*'.[61] Although following this challenge a meteorologist was elected, almost immediately thereafter Gullstrand warned the Committee members that 'under all circumstances we must act [*ställa oss*], so that we do not get in yet another weak meteorologist'.[62]

Comparable strategies were used at this time within the Committee to avoid the awarding of Nobel prizes for meteorological research. Occasional nominations of meteorologists were dismissed with a simple 'not significant', or 'not confirmed by experience'; however, one nominee – Vilhelm Bjerknes – appeared repeatedly. Irrespective of the merits of this candidate, observers on the scene seemed to believe that the Uppsala-dominated Committee did not want to consider meteorology, or geophysics in general, as part of physics.[63] Although, in the 1930s, arguments were presented that developments in dynamic meteorology and in weather forecasting that were grounded in a solid foundation of physics could benefit mankind in a manner that perfectly fulfilled Nobel's intentions, Oseen declared that meteorology was not part of physics and thus meteorologists were not eligible for the prize.[64] This opinion was not shared by all of the members of the *klass*. To what extent the moves of Oseen and his colleagues were directed against meteorology as a science and a branch of physics, or against their meteorologist colleagues in Sweden, whom they regarded as unwelcome members of the *klass* and a hindrance to their plans for introducing and developing theoretical physics as a vital subdiscipline, remains unclear. Yet, it does seem clear that the commonly accepted, restricted definition of the scope of the Nobel prize in physics has its origins in past controversies and contingencies. Of course, the issues involved in the relation between the fragmentation of 'physics' into subdisciplines and the awarding of the Nobel prize in physics must be studied in greater detail.

Conclusions

The three cases described above show first and foremost that there are no simple answers to the questions of how the early Nobel prizes were awarded and how the Nobel prize institution has evolved. Indeed, official documents in the Nobel archives pose rather than resolve questions concerning the prize decisions. True, the general development of disciplines and the standing of scientists in international communities to a large extent set the parameters for the selection process. Nevertheless, as should be clear from the cases presented here, the specific choices of specialities, discoveries and people to be honoured by the prizes were determined by the Swedish participants in the process. It is only when documents from the Nobel archives are made the object of historical analyses and the dynamics of Swedish science are taken into account that their significance can be understood. In this manner, the opening of the Nobel archives to scholars may well prove an incentive to developing the study of the history of modern Swedish science.

Acknowledgements

We want to thank Professor Bengt Nagel for providing helpful criticisms and suggestions in the revision of this article. Our work was supported by grants from the Centre National de la Recherche Scientifique (A.T.P. 3682 'Recherche sur la recherche') and the Norwegian Research Council for Science and the Humanities.

Notes

Abbreviations and terms used: *KVA*, Kungl. Vetenskapsakademien (Royal Swedish Academy of Sciences). *Förslag, fysik alt. kemi,* Förslag till utdelning av Nobel-priset i fysik alt. kemi (Letters of nomination concerning the awarding of the Nobel prize in physics or chemistry). *KU, fysik alt. kemi,* Kommittéutlåtande, Nobelkommittén för fysik alt. kemi (Committee report, Nobel committee for physics or chemistry); refers to general reports handed in to the Academy. *Utkast, KU, fysik alt. kemi,* Utkast, Kommittéutlåtande, Nobelkommittén för fysik alt. kemi (Draft committee report, Nobel committee for physics or chemistry); refers to drafts of general or special reports prepared for the committees, the date being that of the meeting at which the report was handed in to the committee. *Protokoll, NK, fysik alt. kemi,* Protokoll vid Kungl. Vetenskapsakademiens Nobelkommittés för fysik alt. kemi sammanträde (Minutes of the meetings of the Nobel committee for physics or chemistry of the Royal Swedish Academy of Science). (a) *Nobelprotokoll, KVA* and (b) *Nobelprotokoll, KVA, 3dje alt. 4de klassen,* Protokoll vid Kungl. Vetenskapsakademiens sammankomster för behandling av ärenden rörande Nobelstiftelsen (Minutes of meetings of the Royal Swedish Academy of Sciences, for discussion of matters concerning the Nobel Foundation); (a) refers to plenary meetings of the Academy; (b) to meetings of its sections for physics and chemistry.

1. Arrhenius, O. (1959) *Svante Arrhenius – det första kvartsseklet.* In: *Svante Arrhenius till 100-årsminnet av hans födelse (K. Vetenskapsakademiens årsbok för år 1959 Bilaga),* Stockholm, Almqvist & Wicksell, p. 63; Root-Bernstein, R. (1980) 'The Ionists: Founding Physical Chemistry, 1872–1890'. Doctoral thesis, Department of History, Program in History and Philosophy of Science, Princeton University, p. 114.
2. It is not correct, as stated, for example, in the biographical notice drawn up at the time of Arrhenius' Nobel prize award (*Les prix Nobel en 1903*, Stockholm, Norstedts, 1906, p. 67), that 'the dissertation received only one grade above a failing one', *non sine laude* being, in fact, two grades above a fail. That this version of the event, which Arrhenius himself probably propounded, remained a thorn in the side of the Uppsala physicists is illustrated by the reply of K. Ångström to G. Mittag-Leffler who had inquired about the verity of the statement quoted above. Ångström ended his letter as follows: 'Poor Cleve

[member of the jury and chairman of the Nobel Committee for Chemistry]! He tried to repair his unforgivable mistake through a Nobel prize – and nobody championed this more than him – but for many more years to come the story of the idiotic Uppsala faculty will be told.' (K. Ångström to G. Mittag-Leffler, 2 March 1909, Mittag-Leffler Collection, Mittag-Leffler Institute).

3. S. Arrhenius to V. Bjerknes, 11 June 1904, Bjerknes Collection, Oslo Universitets-bibiliotek.
4. Förslag, kemi, 1901.
5. Partington, J. R. (1972) *A History of Chemistry*, Vol. 4, London, Macmillan, pp. 675–677; Root-Bernstein (note 1), pp. 114–119, 375–386.
6. There were also problems of simultaneous discovery, in that Arrhenius and Max Planck had announced similar versions of the dissociation theory in 1887; but these were not touched on in any of the Committee reports [Partington (note 5), p. 679; Root-Bernstein (note 1), pp. 460–473].
7. (1904) *Les prix Nobel en 1901 (Les conférences Nobel en 1901)*, Stockholm, Norstedts, p. 6; Partington (note 5), p. 655; Root-Bernstein (note 1), pp. 482–485.
8. Protokoll, NK, fysik, 16 August 1902, 5 August 1903; Nobelprotokoll, KVA, 9 September 1903.
9. Förslag, fysik, kemi, 1903.
10. Protokoll, NK, kemi, 5 April 1903. For the text of the letter, see Protokoll, NK, fysik, 9 May 1903 (Appendix A). Hiebert, E. (1963) *Historical remarks on the discovery of argon: the first noble gas.* In: Hylan, H. H., ed., *Noble-gas Compounds*, Chicago, University of Chicago Press, pp. 3–20.
11. For the text of the letter, see Protokoll, NK, kemi, 20 May 1903 (Appendix).
12. O. Widman to H. Söderbaum, 26 August 1903, Söderbaum Collection, KVA.
13. Protokoll, NK, kemi, 20 May 1903.
14. Protokoll, NK, fysik, 5 August 1903; Nobelprotokoll, KVA, 4de klassen, 31 October 1903.
15. Protokoll, NK, kemi, 19 September 1903; Nobelprotokoll, KVA, 5te klassen, 31 October 1903.
16. 'Motiverade vota för Nobelpris, m.m.,' Widman Collection, Uppsala Universitetsbibliotek; Nobelprotokoll, KVA, 12 November 1903.
17. KU, kemi, 1904.
18. Hiebert, E. & Körber, H.-G. (1970) *Wilhelm Ostwald.* In: *Dictionary of Scientific Biography*, Vol. 15, Suppl. 1, New York, Scribner, p. 456.
19. Ostwald's conversion was first recorded in his review of T. Svedberg's *Studien zur Lehre von den kolloiden Lösungen* in *Z. phys. Chem.*, **64**, 508–509 (1908).
20. Förslag, kemi, 1909.
21. S. Arrhenius to O. Widman, 1 April 1909, Widman Collection, Uppsala Universitets-bibliotek.
22. O. Widman to S. Arrhenius, 28 April 1909, Arrhenius Collection, KVA.
23. O. Widman to S. Arrhenius, 9 June 1909, Arrhenius Collection, KVA.
24. Utkast, KU; protokoll, NK, kemi, 18 August 1909. See also S. Arrhenius to O. Widman, 7, 13 July 1909, Widman Collection, Uppsala Universitetsbibliotek; O. Widman to S. Arrhenius, 12 July 1909, Arrhenius Collection, KVA.
25. Partington (note 5), p. 600; Hiebert & Körber (note 19), p. 462.
26. Nernst was awarded the prize for 1920 in 1921.
27. Westgren, A. (1962) *The chemistry prize.* In Schück, H. *et al.*, eds, *Nobel: The Man and his Prizes*, Amsterdam, Elsevier, p. 359.
28. B. Hasselberg to G. E. Hale, 5 July 1907, 29 December 1907, G. E. Hale Papers, American Institute of Physics.
29. Protokoll, NK, fysik, 2 September 1907; (1909) *Les Prix Nobel en 1907*, Stockholm, P. A. Norstedt & Söner, pp. 13–18 (Comptes-rendus des travaux couronnés). See also Hasselberg, B. (1908) Om det metriska mått och viktsystemets uppkomst och utveckling. *K. Vetenskapsakademiens årsbok för år 1908*, pp. 199–219; Holton, G. (1969) Einstein, Michelson and the crucial experiment. *Isis*, **60**, 133–197.
30. B. Hasselberg to G. E. Hale, 5 July 1907, G. E. Hale Papers, American Institute of Physics.

31. Nobelprotokoll, KVA 4de klassen, 29 November 1902.
32. Nobelprotokoll, KVA 4de klassen, 28 November 1903.
33. It is of interest to note that Hasselberg – together with Carlheim-Gyllensköld – formed the Committee minority which entered a dissenting opinion in favour of Poincaré in 1910 (Protokoll, NK, fysik, 26 September 1910). Hasselberg's support for Poincaré was based on an argument analogous to that he had successfully developed in the case of Michelson, i.e. that instrumental innovations in physics could qualify as discoveries. In assuring Mittag-Leffler that he would support Poincaré, he stated his opinion that 'mathematics was an instrument of investigation although not in the same category as telescopes and scales. If an important discovery has been made in the use or development of this instrument for physics, I consider it as a "discovery or invention", therefore, I can see no statutory objections to its being rewarded with a prize.' (B. Hasselberg to G. Mittag-Leffler, 24 January 1910, Mittag-Leffler Collection, Mittag-Leffler Institute.)
34. Protokoll, NK, fysik, 2 September 1911.
35. Protokoll, NK, fysik, 7 September 1921; Nobelprotokoll, KVA 3de klassen, 29 October 1921.
36. Although the law of the photoelectric effect was considered to have been thoroughly confirmed by experiment, the theoretical/conceptual basis from which Einstein derived and explained the law was much more problematic. As several historians of science (e.g., G. Holton, R. Kargon, M. Klein and R. Stener) have shown, it was not until the Compton effect (1922) was generally accepted in the mid-1920s that large numbers of physicists began accepting Einstein's light quanta. Robert Millikan, whose experimental work played the most significant role in confirming the law, thoroughly distanced himself from Einstein's theory of the photoelectric effect. In 1916, he stated that 'despite . . . the apparently complete success of the Einstein equation, the physical theory of which it was designed to be the symbolic expression is found so untenable that Einstein himself, I believe, no longer holds it'. Of course, Einstein did not retract his theory; and Millikan in his Nobel lecture (1923) insisted on the same separation of the law and the theory: 'At the present time it is not too much to say that . . . Einstein's equation is of exact validity . . . [it] is perhaps the most conspicuous achievement of experimental physics during the past decade. . . . [But,] the conception of *localized* light-quanta out of which Einstein got his equation must still be regarded as far from being established . . . '. Similarly, the motivation for awarding a Nobel prize to Einstein for the law also attempted to maintain a distance from his theory. In the general report for 1921 and 1922, Oseen's nominations 'solely' for the discovery of the law are kept separate from 'Einstein's quantum theory for the photoelectric effect', listed with Einstein's other theoretical innovations. The deliberate divorcing of Einstein's law from his theory can be seen in changes in the draft of the 1922 general report, in which the link originally suggested is broken, and, instead, the law of the photoelectric effect is recorded as having contributed to the development of quantum theory *in general.*
37. Protokoll, NK, fysik, 8 September 1920.
38. Protokoll, NK, fysik, 16 March 1921; 21 September 1911.
39. Utkast, KU, protokoll, NK, fysik, 7 September 1921.
40. Utkast, KU, protokoll, NK, fysik, 6 September 1922.
41. Mittag-Leffler, G. (1922) 'Dagbok: Resan sommaren 1922', 27 July 1922, Kungliga Biblioteket.
42. B. Hasselberg to H. H. Hildebrandsson, 9 September 1921, Hildebrandsson papers, Uppsala Universitetsbibliotek.
43. Protokoll, NK, fysik, 7 September 1921.
44. Nobelprotokoll, KVA, 3de klassen, 29 October 1921.
45. Oseen's reasoning, as revealed in his letters, special reports and in his lecture (1922) (*Den Einsteinska lagen.* In: *Kosmos. Fysiska uppsatser utgivna år 1922*, Stockholm, Norstedts, pp. 105–131), strongly suggests his interest in Bohr's investigations. His feelings about Einstein are expressed in a letter to Bjerknes dated 14 December 1920 (Bjerknes Collection, Oslo Universitetsbibliotek).
46. Oseen, C. W. (1919) *Atomistiska föreställningar i nutidens fysik. Tid, rum och materia,* Uppsala, Almqvist & Wiksell, p. 15.

47. Protokoll, NK, fysik, 15 September 1919, 7 September 1921.
48. Protokoll, NK, fysik, 6 September 1922.
49. *Ibid.*
50. Protokoll, NK, fysik, 8 January 1901. The decision with respect to astrophysics was made in response to an inquiry which S. Newcomb had addressed to B. Hasselberg (S. Newcomb to B. Hasselberg, 15 October 1900; B. Hasselberg to S. Newcomb, 26 October 1900, Newcomb Collection, Library of Congress).
51. C. W. Oseen to G. Mittag-Leffler, 17 October and 15 November 1918, Mittag-Leffler Collection, Mittag-Leffler Institute; C. W. Oseen to V. Bjerknes, 12 November 1920, 3 February 1923, Bjerknes Collection, Oslo Universitetsbibliotek; C. W. Oseen to S. Arrhenius, 29 September 1920, Arrhenius Collection, KVA.
52. C. W. Oseen to V. Bjerknes, 3 February 1923, Bjerknes Collection, Oslo Universitetsbibliotek.
53. KU, fysik, 1913.
54. Utkast, KU, Protokoll, NK, fysik, 13 September 1916.
55. A. Gullstrand to S. Arrhenius, June 1923, Gullstrand Collection, Uppsala Universitetsbibliotek.
56. Utkast, KU, Protokoll, NK, fysik, 5 September 1923.
57. Protokoll, NK, fysik, 5 September 1923 (Appendix F).
58. Protokoll, NK, fysik, 31 January 1924 (Appendix A).
59. C. W. Oseen to A. Gullstrand, 1 November 1928 and 5 November 1928, Gullstrand Collection, Uppsala Universitetsbibliotek; Nobelprotokoll, KVA, 3dje & 4de klassen, 30 November 1929.
60. C. W. Oseen to V. Bjerknes, 18 June 1919, Bjerknes Collection, Oslo Universitetsbibliotek.
61. Letters, H. H. Hildebrandsson to KVA, 6 September 1923; H. H. Hildebrandsson to KVA, 3dje *klass*, 6 September 1923.
62. A. Gullstrand to S. Arrhenius, 25 December 1923, Arrhenius Collection, KVA.
63. O. Pettersson to S. Arrhenius, Midsummer 1926, Arrhenius Collection, KVA.
64. W. V. Ekman to C. W. Oseen, 7 November 1937, 15 November 1937, Oseen Collection, KVA.

THE AWARDING OF THE NOBEL PRIZE: DECISIONS ABOUT SIGNIFICANCE IN SCIENCE

GÜNTER KÜPPERS, NORBERT ULITZKA & PETER WEINGART

Universität Bielefeld, Forschungsschwerpunkt Wissenschaftsforschung, Bielefeld, Federal Republic of Germany

Summary

The awarding of the Nobel prizes in physics and chemistry is seen as a process in which the crucial problems for the Royal Swedish Academy of Sciences are to reach decisions about significance in science and to have those decisions accepted by the scientific community and by the public. These problems are met by developing a procedure for making decisions. Two complexes are analysed: the structure and pattern of the nominating process and its interrelationship with the prize decisions, as well as the operation of implicit evaluative criteria in the handling of two cases of revolutionary ideas in physics: those of quantum theory and of relativity theory.

Introduction

For the sociologist, the process of awarding the Nobel prizes, the influences which shape this process, the structure it assumes and its interaction with the development of the disciplines in question are important objects of analysis. The will of Alfred Nobel provided that prizes be awarded in several domains, including those 'for the most important discovery or invention within the field of physics' and 'for the most important chemical discovery or improvement'. No criteria for choice were given, other than stating that the prizes should be given 'to those who during the preceding year shall have conferred the greatest benefit on mankind'. This phrase left the problem of decision-making wide open and made it incumbent upon the prize-awarding institution to establish its own criteria and a procedure for making such decisions. Discoveries, inventions and improvements in each of the fields of science under consideration can be widely diverse in nature and function, and cannot always be compared.

A further complication was that the *type* of decision required was fundamentally different from the continuous process of evaluation in science: The latter process is informal and also incremental; it is the result of complex communication, by discussions, publications and citations. As the outwardly visible indices of peer-group evaluation, these lead to the emergence of a scientist's reputation. In this process, no one is obliged to make a decision fixed in time as to the 'most important discovery'. On the contrary, the process is fraught with irrationalities, such as resistance to innovative ideas and revolutionary concepts, disbelief of evidence, adherence to established paradigms, etc.

From the start, then, the concept of Nobel prizes implied some basic problems. First, the difference between the prerequisites of the awarding of the prize, i.e. an *annual decision* about the most important discovery, and those of the evaluation process in the scientific community required that the prize-awarding institution set up its own rules and define its own criteria for choosing. Second, since the awarding of the prize implies that the decision will be highly visible, because it is claimed to be an authoritative judgement of achievements in the respective scientific fields (a claim which is substantiated in no small measure by the amount of prize money), a burden of legitimation is put on the prize-awarding institution: the awarding body has to legitimate its decisions in the view, primarily, of the scientific community but also of the public.

The two problems are interdependent. The first concerns the ability to make decisions, and the second of how those decisions gain acceptance. The question, then, is how the Academy copes with these problems. The approach that seems most appropriate conceives the prize-awarding as a decision-making procedure.[1] In the overall structure of this process, one can differentiate between an 'input' (the nominations to the committees), the processing of the nominations (the evaluations and recommendations by the committees), and the final decision by the Academy.

With respect to the nominating process, we analyse the profile of the nominations from the scientific community in each of the two fields under investigation, the role played by national affiliations, the degree of consensus among the nominators, its development over time, and whether the nominations are connected in some way with *preceding* prize decisions.

The second stage of the decision-making process involves the processing of the nominations before the final decision. It is assumed to obey the same basic principles of any decision-making process in society, and is thus governed by formal, explicit rules (in this case, the statutes of the Nobel Foundation) and by implicit rules and criteria developed over time. While the formal rules represent fixed premises of the decision, the informal rules and criteria are crucial to qualitative evaluations and judgements. The use of the formal preselection criteria (such as those which rule out unauthorized nominations, self-nomination and nomination of a life's work) was not problematic but had little differentiating power. The use of informal criteria (such as those for distinguishing which of several works, for scientific as well as for practical and humanitarian reasons, was the best choice for a Nobel prize) was implicit, but led to a final decision. We hypothesize that the development of such informal criteria of decision-making can be assumed to follow a certain pattern, which is determined by the need to establish a consistent, defensible and thus 'legitimate' procedure. In other words, the application of such criteria and their representation in the final decisions establish a 'history' of prize awards. This pattern can be assumed to consist of:

(i) '*Continuation*'. By this is meant various strategies to establish the continuity and consistency of prize decisions (and selections of candidates). One example is the principle of 'completion', i.e. the rewarding of candidates in relatively clearly demarcated and well-developed research areas, such as that of gas-discharge physics during the first decade in which the prizes were given.

(ii) '*Relating*'. This is the choice of highly reputed scientists who later serve as 'standards' by which other candidates are measured. This strategy was used especially for scientists within the same research area.

(iii) '*Definition and structuring of frames of reference*'. Any decision implies a selection among different choices and thus requires adequate criteria. The disciplines and the specialities within the disciplines become the important frames of reference for the development of such criteria, and the committees developed (although in slightly different ways and at different times) procedures to delineate such frames of reference. Early on, the Academy requested extensive reports on candidates to be prepared by members of the committees. The latent function of these reports was to place the achievements of the nominees in a research context, making it possible to establish an 'internal connection' among the awarded achievements, a reference to earlier decisions and a justification for rejecting nominations. Implicitly or explicitly, thereby, a *structuring* of the frame of reference for the pertinent evaluations took place.

(iv) '*Evaluation*'. This is the development or implicit use of criteria for evaluation. Such criteria were used to differentiate between disparate phenomena (different scientific achievements): they had to be generalizable over at least an entire discipline; and they had to be acceptable as objective and just. Examples of such criteria developed by the Nobel committees are their insistence on 'experimental proof' and on 'correspondence to experience'.

Identification of a pattern of procedural strategies during the course of prize decisions is still in its preliminary stages. Questions remain regarding the implicit criteria that governed the evaluations by the Nobel committees and by the Academy, of whether the hypothesized mechanisms for establishing consistency and continuity of decisions can be corroborated, and of how they functioned with respect to revolutionary discoveries.

The history of the awarding of these prizes is of interest because of its relation to the history of the disciplines, physics and chemistry. The awarding of the prizes as described above intervenes in the continuous process of evaluation in science; it ascribes reputation which is otherwise conferred by informal and incremental rewards. Thus, the prize-awarding institution could concur with this process — which we term the 'disciplinary history' — or it could compete with or diverge from it. In the latter case, it ran the risk of losing its legitimacy in the eyes of the scientific community. In such a case, the 'prize history' does not coincide with the 'disciplinary history'. We focus on cases in which such a divergence of the two histories occurred, and on the ways in which the Nobel committees and the Academy coped with them.

Thus far, we have presented a sketch of the interpretive framework for a sociological analysis of the process of awarding the Nobel prizes. Given that this is a report on research in progress, in the following we present data and findings with respect to only two aspects of the entire process: the structure of the *nominating process*, and problems arising in the *decision-making process* in two cases of revolutionary achievements, namely, the development of quantum theory by Planck and of relativity theory by Einstein.

The Nominating Process

Questions of structures and patterns of nominations, the 'input' into the decision-making process within the context of the entire process of prize awarding, are approached by quantitative analysis, since this approach is best suited to understanding the overall relation between the nominating process and the decision-making process within the Nobel committees and the Academy.

Between 1901 and 1929, an annual average of thirty-three nominators in each field proposed candidates for the prizes in physics and in chemistry. The nominators made an average of forty-six nominations per year in physics and thirty-eight nominations in chemistry, proposing twenty-one candidates in physics and nineteen candidates in chemistry.[2] In this manner, each nominator proposed 1.6 candidates in physics and 1.7 candidates in chemistry per year. These figures provide the background for the following quantitative analysis. Since the nominating process is considered in its entirety, the analysis includes those nominations which in the course of the decision-making process were excluded on the basis of statutory provisions (self-nomination, belated nominations, etc.).

The impact of national affiliation

The Nobel prize frequently serves as an indicator of national achievements in science; and writings about the Nobel prize have often made comparisons of the nationalities of Nobel prizewinners. Use of the prize for international competition contrasts with the statutes of Nobel's will, which, reflecting the values of universalism in science, explicitly ruled out any consideration of nationality in awarding the prizes. It is obvious that national preferences, i.e. the fact that scientists tend to nominate colleagues of their own nationality, can compromise the internationality of the scientific community. Nationalist tendencies are due not only to chauvinist trends in the scientific community (particularly at the time of the First World War and several years after), but also to the existence of national schools and communication structures, and to other similar factors. These patterns, should they occur, present a problem to the committees and to the Academy who would have to neutralize them in order to comply with Nobel's will and to reach an unbiased decision.

As a starting-point in analysing this factor, we look at the distribution of nominators and nominees by country.[3] Not surprisingly, the number of Germans was relatively high over the entire period considered; of the nominators they represented 25.9 percent in physics and 36.2 percent in chemistry; and of the nominees, 27.9 percent in physics and 34.1 percent in chemistry. The second largest group of nominees were French, followed by the English, the Americans and the Scandinavians. Among the nominators, the Scandinavians constituted the second largest contingent in chemistry; while in physics they were only slightly fewer than the French. This distribution is shown in more detail in Table 1. Figures 1 and 2 portray the trends in the national distributions of nominees and nominators between 1901 and 1929. Strong increases in the numbers of German nominees and nominators in physics as well as in chemistry are obvious.[4]

TABLE 1. *Distribution of nominees and nominators by country*

Country	Nominees (%)		Nominators (%)	
	Physics	Chemistry	Physics	Chemistry
Germany	27.9	34.1	25.9	36.2
France	19.8	12.4	13.9	10.2
Great Britain	16.9	10.3	7.6	5.7
United States	13.9	10.8	11.0	9.5
Scandinavia	11.0	10.3	13.4	11.2
Benelux	2.3	4.9	6.1	4.0
Austria, Hungary, Czechoslovakia	3.5	3.2	7.0	8.5
Russia, Poland, Baltic States	1.2	3.8	3.5	3.0
Switzerland	—	4.3	1.7	6.5
Italy	2.3	2.7	7.0	5.0
Others	1.2	3.2	2.9	0.2
Total	100.0	100.0	100.0	100.0
No.	172	185	344	401

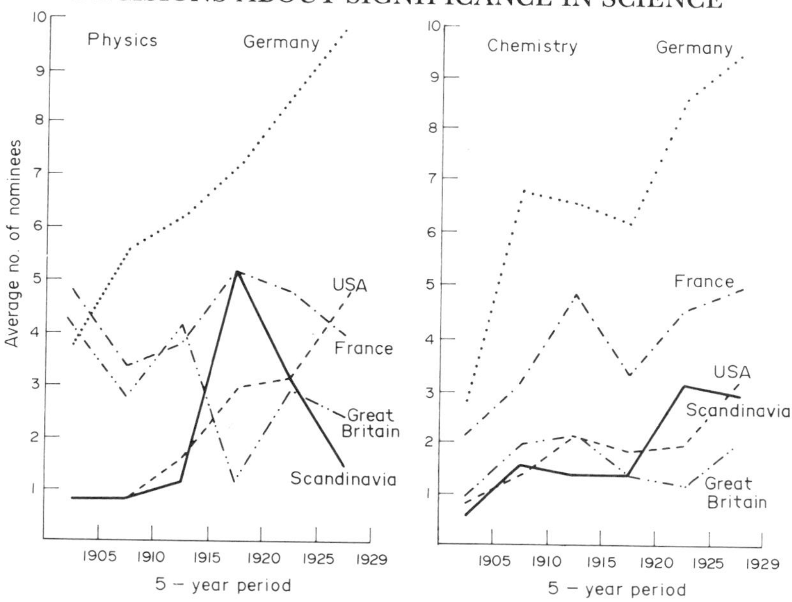

FIGURE 1. Numbers of nominees by nationality, 1901–1929.

The question now is whether the national distributions of nominees and nominators are also reflected in the nominating pattern, i.e. how the 'votes' of nominators from one country were distributed among candidates from other countries (national nominating pattern) and how the 'votes' that candidates from a particular country received were distributed among nominators from the various countries (pattern of reception of 'votes').

Tables 2 and 3 show how the nominations were distributed among the various nations and indicate the share of nominations received by candidates of the same nationality as the nominators; they show, in addition, the distribution of received nominations among various nations and the share of nominations from nominators of their own country for particular candidates. It is apparent that nominators clearly favoured candidates from their own nation. Almost eighty percent of the English nominators in physics 'voted' for English candidates, seventy-five percent of the French 'voted' for French candidates, and over sixty percent of the Germans did so. Both German and French candidates received more than fifty percent of their 'votes' from nominators of their own country.

Given this relatively strong pattern of national affiliations in the nomination process, the question arises of what its impact was on the decisions of the Nobel committees. Table 4 shows that the relationship of prizewinners to nominees for the leading nations in science remained relatively homogeneous in physics, although an aberration is seen with regard to Great Britain in chemistry. The United States is clearly underrepresented; and the relatively low figures for the Scandinavian countries may be due to their 'holding back'

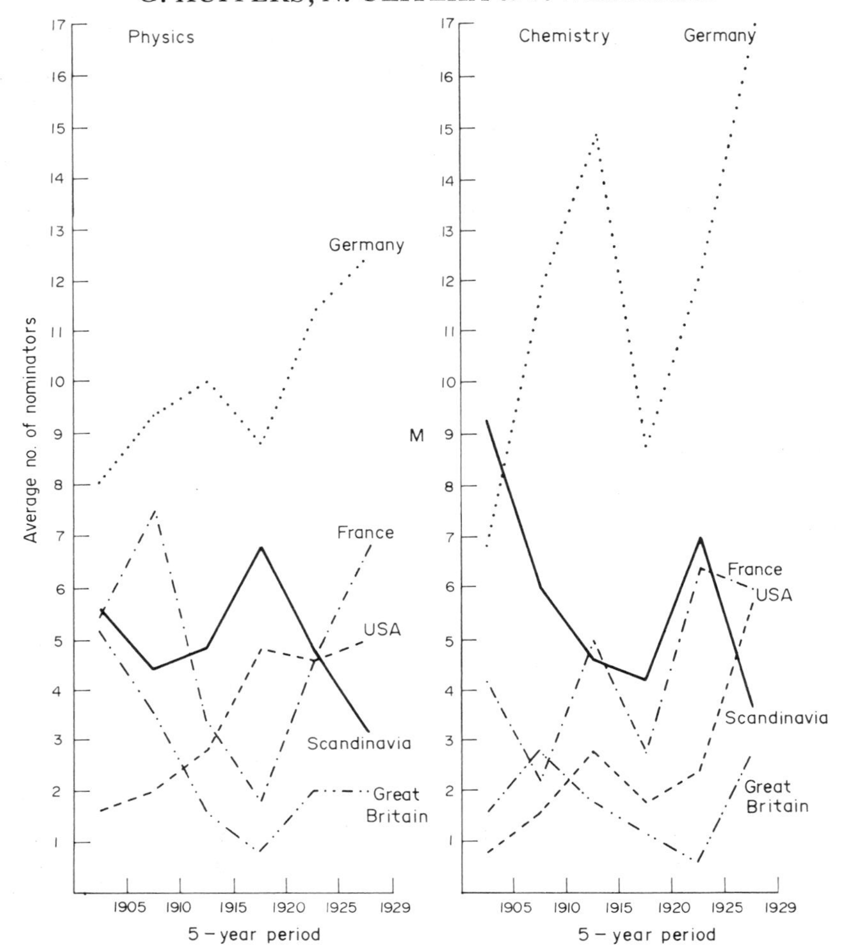

FIGURE 2. Numbers of nominators by nationality, 1901–1929.

in the nominating and awarding processes. The Nobel committees and the Academy in their prize decisions therefore did not level the national differences that exist in the nominating procedure, with the exception of the preeminence of Great Britain in chemistry and (to a lesser extent) in physics.

With respect to the impact of national affiliations on the nominating process, the question arises of what happened during and immediately after the First World War, when German science was boycotted by the leading scientific nations, and whether this was reflected in the nominating process. Figures 3 and 4 provide no conclusive evidence that either the War or the boycott afterwards had a marked impact on the nominations for the prizes: In physics, although the lowest percentage of German scientists nominated for the prize, 26.7 percent, occurred in 1915, the highest percentage of all was reached in 1921, with

TABLE 2. *Distribution of nominations for the prize in physics by country, 1901–1929*

Country	All nominations cast (%)	Nominations given for nominees of own country (%)	All nominations received (%)	Nominations received from nominators of own country (%)
Germany	30.5	68.6	35.8	58.4
France	15.7	74.5	20.7	56.4
Great Britain	5.8	77.9	12.4	36.4
United States	10.0	63.9	11.2	57.0
Scandinavia	17.3	32.2	8.5	65.5
Benelux	5.0	23.9	4.1	29.1
Austria Hungary Czechoslovakia	5.6	16.0	1.9	48.0
Russia Poland Baltic States	2.9	15.8	0.5	85.7
Switzerland	2.0	0.0	–	–
Italy	4.5	60.0	4.1	65.4
Others	0.7	0.0	0.7	0.0
Total	100.0	–	100.0	–
No.	1329		1329	

TABLE 3. *Distribution of nominations for the prize in chemistry by country, 1901–1929*

Country	All nominations cast (%)	Nominations given for nominees of own country (%)	All nominations received (%)	Nominations received from nominators of own country (%)
Germany	37.1	65.4	37.7	64.3
France	12.1	84.2	20.3	50.4
Great Britain	5.0	50.9	7.8	32.6
United States	7.0	67.5	9.0	52.4
Scandinavia	18.7	29.4	10.0	54.8
Benelux	3.2	10.8	1.8	19.0
Austria Hungary Czechoslovakia	6.5	21.6	2.7	51.6
Russia Poland Baltic States	2.2	12.0	2.1	13.0
Switzerland	4.4	43.1	5.0	38.6
Italy	3.7	45.2	3.1	52.8
Others	0.1	100.0	0.5	16.7
Total	100.0	–	100.0	–
No.	1145		1145	

TABLE 4. *Distribution of nominators, nominees and Nobel prizewinners by selected countries (1901–1929)*

Country	PHYSICS Nominators No. (%)	PHYSICS Nominees No. (%)	PHYSICS Prizewinners in relation to nominees (%)	CHEMISTRY Nominators No. (%)	CHEMISTRY Nominees No. (%)	CHEMISTRY Prizewinners in relation to nominees (%)
Germany	89 (25.9)	48 (27.9)	20.8	145 (36.2)	63 (34.1)	19.0
France	48 (13.9)	34 (19.8)	20.6	41 (10.2)	23 (12.4)	17.4
Great Britain	26 (7.6)	29 (16.9)	24.1	23 (5.7)	19 (10.3)	26.3
United States	38 (11.0)	24 (13.9)	12.5	38 (9.5)	20 (10.8)	5.0
Scandinavia	46 (13.4)	19 (11.0)	15.8	45 (11.2)	19 (10.3)	15.8
Others	97 (28.2)	18 (10.5)	–	109 (27.2)	41 (22.1)	–
Total	344 (100)	172 (100)		401 (100)	185 (100)	

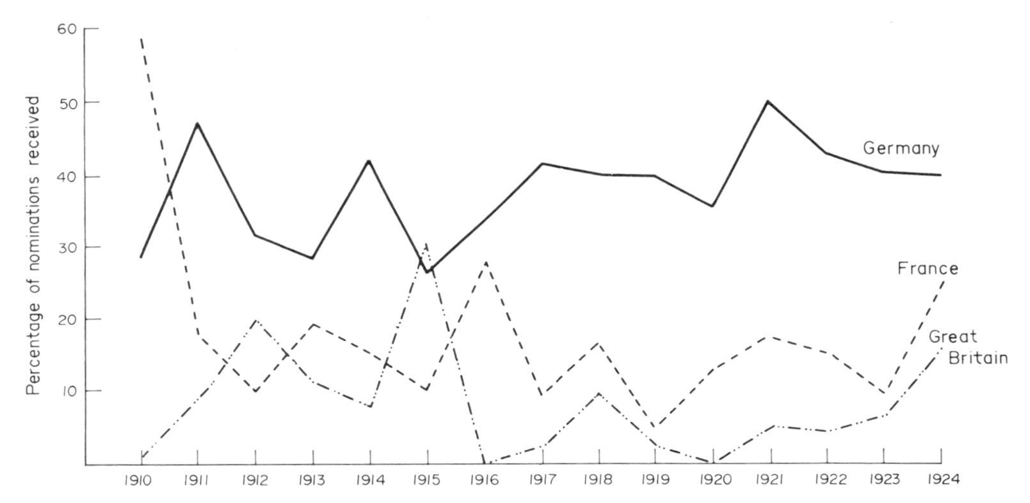

FIGURE 3. Distribution of nominations for the prize in physics, 1910–1924: The impact of the First World War and the boycott of German science.

50 percent. In chemistry, a peak of 64.1 percent was reached in 1916, but the lowest share, 22.6 percent, occurred in 1919. The continued dominating role of German scientists throughout those years cannot be attributed to greater support from German nominators: The largest share of nominations that German candidates in both physics and chemistry received from nominators of their own country, 89.5 percent, occurred in 1910 in physics and in 1920 in chemistry, outside the period of the War; and the smallest shares, 37.5 percent in physics and 42.8 percent in chemistry, were seen in 1915 and 1919, respectively. The values for the period 1910–1924 do not differ greatly from those for years outside that period.

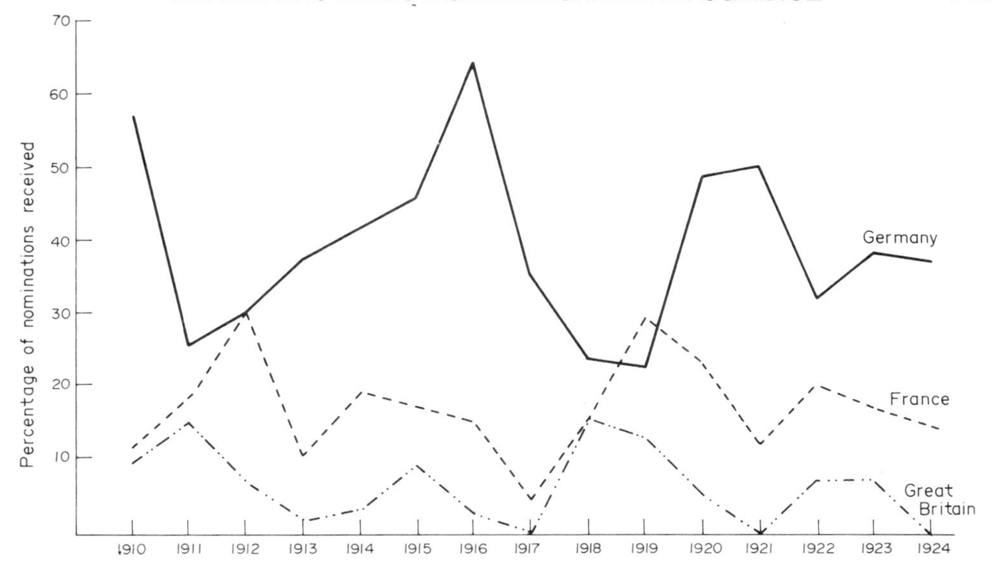

FIGURE 4. Distribution of nominations for the prize in chemistry, 1910–1924: The impact of the First World War and the boycott of German science.

The nominators as a group: consensus

The nominating process may have been structured by allegiances other than nationality. One likely tendency would have been to nominate the elder statesmen of science. This did, in fact, occur quite frequently in the case of scientists nominated for their life's work – a motive that the statutes explicitly rule out. A significant feature of the nominating process is revealed, therefore, by calculating the number of years in which a candidate was nominated: 54.1 percent (in physics) and 62.2 percent (in chemistry) of all candidates were nominated up to two times and only 26.2 percent (in physics) and 20.5 percent (in chemistry) were nominated more than four times. Of the group who never received a Nobel prize, 58.9 percent (in physics) and 70.6 percent (in chemistry) were nominated at most twice.

One reason for this relatively large fluctuation may be the fact that most nominators were asked to serve in that function only a few times, so that repeated nominations were made less likely. (The role of the groups with permanent nominating rights will be determined quantitatively at a later stage.) Another possible explanation is that the nominating process remained in close connection to the dynamics of the research process and its evaluation by the scientific community. The few cases in which candidates were nominated over a longer period of time thus acquire special significance: If their nominations were not merely indicators of collegial or institutional loyalty and favouritism by the nominators, these cases may reflect a response to earlier prize decisions. We come back to this point later.

The central question is whether the nominations in each year demonstrated a significant consensus, i.e if the nominators favoured a particular candidate. If one looks at the annual degree of consensus, measured in terms of the number of candidates who received a certain percentage of nominations (Figures 5 and 6), it becomes evident that, with few exceptions, there was no meaningful unanimity. As a first approximation, one may consider the most successful nominees to be those that received an exceptionally high percentage of nominations. Only a few cases stand out. In physics, Poincaré, in 1910, received by himself more than a half of the nominations; in chemistry, Ramsay alone received sixty-six percent in 1904, and Moissan alone had fifty-two percent in 1905.

A numerical measure of consensus, and thus a more systematic picture, is provided by determining how many candidates shared in fifty percent of the nominations each year (the fifty percent level, being, of course, arbitrary). The figures indicate a clear trend, namely, that between 1901 and 1906 the level of consensus was notably higher than in the years after and that this was more

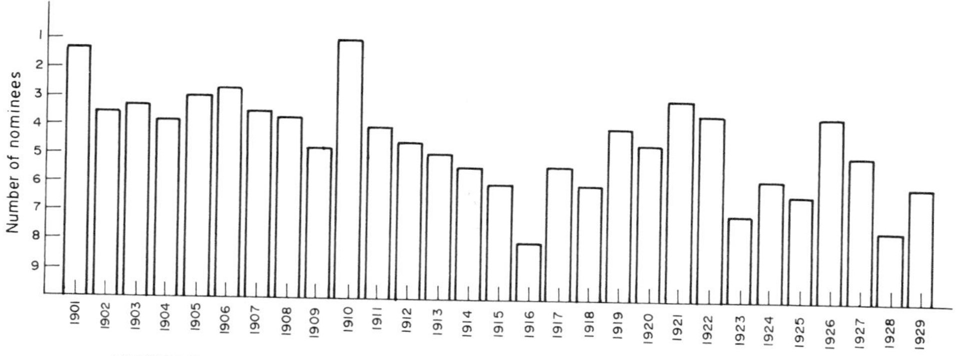

FIGURE 5. Consensus: Number of nominees sharing 50% of all nominations for the prize in physics.

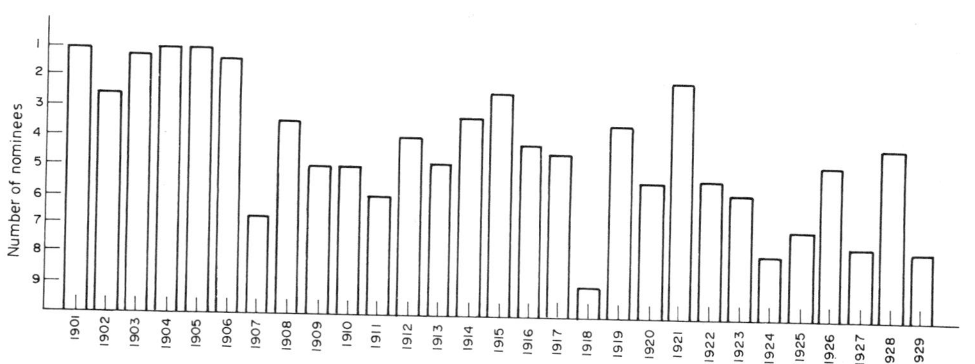

FIGURE 6. Consensus: Number of nominees sharing 50% of all nominations for the prize in chemistry.

pronounced among nominators to the Chemistry Committee than among those to the Physics Committee. In fact, the overall trend over the first twenty-nine years of the awarding of the Nobel prize was a gradual decrease in the level of consensus among the nominators. The concentration of nominations on particular candidates in the early years of the prize awarding may be due to the fact that the prize provided a new opportunity to reward very eminent scientists. It is characteristic of such a situation that a backlog of some five or six eminent scholars can be identified. As the prize became institutionalized, the search for prizeworthy candidates became more difficult. It is, of course, tempting to hold developments in the disciplines (such as the alleged 'crisis' in physics that occurred at the beginning of the twentieth century) responsible for the disappearance of the relative consensus in the community 'vote'. But the striking similarity of the receding consensus in *both* physics *and* chemistry suggests that the observed patterns were created in the nomination process itself.

Interrelationship between the nomination process and the prize decisions

The relatively low level of consensus appears to leave the Nobel committee with little guidance from the nomination process. It is necessary, however, to look more closely at the interrelationship between the nominations and the prize decisions.

One approach is to determine the role of those nominees who received the majority of nominations in one year. The nominee who was most 'successful' in this respect was not necessarily the laureate of the year: This was true in only seven cases out of twenty in physics and in eight out of twenty-five in chemistry, although in five and four cases, respectively, the Academy chose the leading candidate of the preceding year; however, only four out of twenty in physics and eleven out of twenty-five in chemistry who had at one time received the majority of 'votes' never received the prize. In the early years of the awarding of the Nobel prize, when the consensus was relatively good, the majority of the candidates with the largest share of nominations actually received the prize. (During the first ten years, this was true for seven out of eight candidates in physics and for six out of eight in chemistry.) Thus, it is plausible to assume that in that early phase, the Nobel committees and the Academy, lacking a 'history' of prize decisions, relied more heavily on the community 'vote'.

There is also an indication that nominators reacted to preceding prize decisions. Looking once more at the degree of consensus over time, as measured by the concentration of 'votes' on one candidate, it is noteworthy that the relatively good consensus dissipated after 1906 but recurred in the year 1910 for Poincaré, and in 1921 and 1922 for Einstein. In chemistry, the situation was similar, although not quite so pronounced; in 1910 and 1921 Nernst received an unusually high share of nominations; in 1915, Willstätter did so and in 1926,

Urbain. Although it cannot be proven that these peaks were due to concerted actions on the part of the nominators, they appear to reflect a reaction to the fact that the committees had 'waited too long' to make what the community considered to be obvious awards.

The numerical analysis, though still not complete, has revealed characteristics of the nominating process that would have eluded the study of individual cases. The questions related to the decision-making process proper, however, are of a different nature.

The Decision-making Process – The Establishment of a Procedure

Analysis of the nomination process has shown that there were complex interrelationships between the nominating process and the prize decisions. However, the nomination process served only as a preselection, and the actual burden of decision-making and of its legitimation rested with the Nobel committees. Consequently, it is the evaluations by referees in the committees that link the decision-making process to the contemporary history of the discipline. An inherent problem of these evaluations is that although they are designed to provide authoritative information about the validity and relevance of the research nominated for the prize, they must remain to some extent inconclusive, since they intervene in the ongoing research process. The Nobel committees were thus caught in the dilemma either of having to decide in the face of some degree of uncertainty, or waiting so as to award past achievements, thereby losing touch with the development of research frontiers, thus, not only contradicting the statutes but also putting at risk general acceptance of the prize. Our interest is focused on this relation between 'disciplinary' history and the prize decisions, and the procedures the committees developed in order to achieve an optimal overlap of the two.

New physics – Planck and Einstein: Two problematic cases for the Nobel Committee

To elucidate the relation between the prize decisions and the disciplinary history, we have analysed cases in which the decision-making procedures and evaluation criteria could be expected to lead to extraordinary difficulties and in which the risk was especially high that the prize decisions diverged from or even contradicted the continuous assessment of research within the discipline. Such cases are revolutionary discoveries that redefine the fundamental principles of a discipline and therefore defy evaluation, since they cannot be related to other achievements. Early quantum physics and the theory of relativity are well-known cases of this kind.

Apart from the specialities into which physics is broken down, such as thermodynamics, optics, electrodynamics, etc., research achievements can also be grouped as follows:

- new observations that extend existing knowledge (X-rays, radioactivity, the Zeeman and Stark effects);
- explanations of observations which allow their integration into theoretical concepts which, in turn, are often changed in the process (classical theoretical physics, with van der Waals, Boltzmann and Heaviside, but also Planck's radiation law);
- new methods and instruments of measurement which open up new possibilities for observation and extend the realm of experience (Wilson, Hale and others);
- precision measurements and experiments that make it possible to decide between competing theories (von Laue, Michelson and others);
- new theoretical concepts which allow an integration of experimental anomalies and revolutionize traditional principles (Planck, Einstein).

The first three groups relate to the three traditional areas of cognitive action in science: experiment, construction of theories and development of methods. Precision measurement is a sub-category of the first but is mentioned here specifically because its relevance to the development of science was discussed extensively by the Nobel committees. The last group, the work of Planck and Einstein, is the one we shall concern ourselves with here.

Achievements that fall into the last group are more difficult to evaluate than those in the other. First of all, their relevance for the progress of the respective field cannot be evaluated unequivocally because that progress has become unpredictable as a result of the revolution. Since the importance of revolutionary theories cannot be estimated, the discussion on early quantum physics and relativity theory was limited to whether they corresponded to reality or, conversely, contradicted experience. An evaluation of Einstein that the Physics Committee commissioned from one of its members, Allvar Gullstrand, in 1921 underscores the importance of this point; it also reveals aspects of how that referee conceived science. Gullstrand wrote that a theory is the 'summing up of a number of descriptions of related phenomena'.[5] The categories and concepts being used are derived from experience. 'In its efforts to find the connections between phenomena on which knowledge has been gained by objective observation, science, wherever possible, usually illustrated this rapport by describing the actual phenomena with categories and concepts which are well-known from earlier experience.'

If, in those cases in which a causal connection between different phenomena has been found empirically, a theory makes it possible to deduce those phenomena from earlier experience, it is said that 'the theory explains the phenomena'. This explanation is more valid 'the more general the experience is from which the causal connection seems to appear'.

Experience plays a crucial role. 'Already in the earliest stages of the development of the human psyche many experiences based on man's own body were gathered about force as the cause of motion, about velocity and by and by also about mass.' The work of Planck and Einstein, however, contradicted existing experience. Nonetheless, the question of their experimental corroboration had to be answered differently. Planck was nominated for the first time in 1907 for his radiation law, work which at that time fell into the category of classical theoretical physics. He had succeeded in 1900 in formulating the correct radiation law on the basis of certified experimental results. Many attempts, from those of Kirchhoff to those of Wien, had failed. Planck's radiation law comprised all the others that had hitherto been formulated and integrated the experimental data produced so far. Thus, it is not surprising that already in 1908 the Nobel Committee suggested to the Academy that Planck be awarded the Nobel prize. At that time, the Committee had not yet realized the far-reaching theoretical impact of Planck's formula.

The Academy decided against Planck and in favour of Lippmann, who had also been proposed by the Committee as a possible candidate for the prize. The reasons for the Academy's divergence from the Committee's recommendations are not known, but in 1909 Planck was rejected as a possible candidate in view of that decision.

Up to 1912, the majority of those who nominated Planck did not realize the revolutionary nature of the quantum approach and nominated Planck repeatedly for his radiation law. Hasenöhrl, from Vienna, constituted an exception; he summed up his nomination of 1910 as follows:

> 'An exceptionally remarkable conclusion also follows from Planck's viewpoint: a resonator cannot receive energy from the surrounding radiation in whatever quantity but only the multiple of a specific quantum. The assumption of such an "energy-atom" is at first sight quite astonishing and considered by some as the weak point of Planck's theory. . . . Of course the "quantum hypothesis" of energy is not evident to us now; but is that not characteristic of innovative original thoughts? Has not every significant progress in science presupposed new forms of thinking?'[6]

In 1912, several nominators stressed the novelty of Planck's theory. Lummer nominated Planck, for among other things, the 'discovery of the elementary quantum'.[7] Waldeyer spoke in his nomination of the 'Umgestaltung der Physik'[8] that could be foreseen due to Planck's work on the theory of heat radiation; and T. W. Richards from Cambridge, Massachusetts, mentioned in his nomination Planck's 'unusual contribution to the scientific discussion of energy'.[9]

Now the Nobel Committee, too, in its discussion of Planck's work recognized it as a theory of 'heat quanta'.[10] The Committee's reason for not recommending Planck for the prize was stated as follows: 'This theory has just recently become the subject of a lively debate and Planck himself has modified his theory.' The

Committee decided to wait for further developments 'in order to gain as much clarity as possible on the importance of this new theory', although it already conceded that Planck's work 'has had an immense influence on the development of modern science'.[11]

In the ensuing years, the influence of the quantum theory on science was emphasized again and again. It was pointed out at the same time, however, that the theory, despite continuing work, was not internally consistent and contradicted certain firmly established principles of physics. Awarding of a prize was thus postponed until an explanation for the contradictions could be found.

In an extensive discussion in 1917, the important role of Planck's work in explaining the spectrum was pointed out, and an application of his thought was seen in Bohr's atomic model. The precise wording was that the quantum theory was now confirmed by the newly discovered combination law of the spectra. However, 'there are still great theoretical difficulties, as Planck points out, and he has looked for a solution of this issue in his latest work (1917). . . .'[12]

The Nobel Committee found itself in a dilemma. On the one hand, quantum theory 'has not reached the satisfactory completion which seems desirable'. This at least was Arrhenius' judgement in his report of 1918, in which he cited W. Wien's remark made in a *Festschrift* honouring Planck's sixtieth birthday:

> 'The insight that the hitherto held basic principles of theoretical physics are unsatisfactory and that, on the contrary, premises that have been made up to now have to be abandoned according to which only continuous processes exist in nature, "which make no jumps", this insight is certainly one of the greatest achievements of science. . . . A logically complete presentation of the quantum theory is not yet possible. However, through its manifold application it has had a more inspiring effect than most other theories. As a newly opened area of research it will command the work of physicists for a long time, but it also promises rich fruit.'[13]

On the other hand, Planck's approach revealed a high degree of correspondence to reality: From it, the combination law of spectra could be deduced; this amounted to a correspondence to reality in contradiction to experience.

The dilemma was compounded by the fact that the problem of Planck blocked the awarding of the prize in the area of the new atomic physics: The Committee agreed with von Laue that the prize could not be given to Sommerfeld or Bohr before Planck himself had received it. Thus, a solution had to be found.

The Committee's way out was to stress the universal importance of Planck's constant. In 1919, the Committee stated: 'As the importance of the theory has been more and more clearly revealed, it has become increasingly probable that the Planck constant, the discovery of which summarizes the theory, corresponds to some reality in the constitution of matter, and there is no longer any doubt but that this constant gives the measure of an atomistic property, albeit of yet unknown nature.'[14]

Arrhenius was again asked to prepare an evaluation. He stressed the fundamental nature of Planck's *constant* but dealt primarily with the fruitfulness of Planck's *theory*: 'All these circumstances make it evident that the quantum theory of Planck is the leading theory in the modern development of physics; although it has not yet obtained a final, satisfactory formulation, quantum theory plays a role making it fully comparable even with the foremost theories so far invented by the pioneers of this science.'[15]

Planck received the prize for 1918 in 1919, for 'the services he rendered to the advancement of Physics by his discovery of energy quanta'. The vast majority of the nominators, however, had nominated Planck for the discovery of a *natural law* in atomic physics, namely, the discovery that the exchange of energy between radiation and matter did not occur continuously but in discrete packets of energy, the quanta. But he was not nominated for the discovery of a *natural constant*: Evidently the Nobel Committee was not in a position to reward his theoretical approach, although its fruitfulness and its quantitative correspondence to reality were undisputed. Perhaps it was still its contradiction of experience that caused the Committee to act so cautiously.

In the case of Planck, the revolutionary content of his discovery emerged slowly. In the case of Einstein, it was visible from the very beginning. He was nominated for the first time by Ostwald in 1910 for his theory of relativity. This theory, which was in total contradiction of experience and had no experimental support, was labelled purest speculation by the Nobel Committee. In contrast, Lorentz, who was nominated in the same year for his work on relativity theory, received praise for the care with which he had formulated his hypothesis. Einstein was criticized for his 'views', and at this point his relativity theory was considered not important enough for scientific progress 'as to support the view that it would be of greatest benefit for mankind'.[16]

In 1913, Ostwald portrayed the theory of relativity as a Copernican revolution and compared its significance to that of the Darwinian theory of evolution. The Nobel Committee doubted this judgement. It stated: 'Even though Einstein's and Lorentz' work may have its own value independent of the principle of relativity, this new knowledge does not have so much generality or so much importance for scientific development that it would merit a Nobel prize.'[17] Einstein, who was now being nominated each year, was criticized for his speculations, which were supposedly not in accordance with experience.

Between 1917 and 1919, the lack of agreement between the theory of relativity and observation was pointed out repeatedly. The red shift in the spectrum and the deviation of light beams by the sun, which were called for by the theory, could not be proved. The proof of the latter effect during an eclipse of the sun on 29 May 1919 was not considered sufficient by the Nobel Committee.

Given the bias of the members of the Committee against relativity theory, which was termed unimportant and an act of faith, the stress on experimental proof in the Committee's reports is easily understood: 'The relativity theory provides more a mathematical formulation than an explanation of natural phenomena.'[18]

In 1920, Einstein was nominated for the first time for his work on the photoelectric effect. It was stated that, in general, his work had begun to occupy a prominent place in more recent research and had several times led to important progress. However, Einstein was nominated primarily for his theory of relativity, and on this the Committee's opinion had not changed.

The year 1921 marks the turning-point in the discussion in the Nobel Committee of Einstein's work. A first report on the photoelectric effect was prepared by Arrhenius, who evaluated Einstein's achievements in this area critically: 'It cannot be denied that Einstein's idea was a stroke of genius. But it was a natural and obvious consequence of Lenard's, J. J. Thomson's and Planck's great works. When it was put forward, it was done so only tentatively as a little worked out conjecture based on qualitatively, only approximately correct observations.'[19]

The breakthrough came in 1922, when Oseen, a student of Bohr's who was familiar with recent development in physics, prepared a report on Einstein. In contrast to Arrhenius, Oseen stressed Einstein's achievement in connection with the interpretation of Planck's constant '*h*'.

> 'The internal relevance of Planck's constant subsequently became the central problem in the area of heat radiation. Planck's own views have achieved no lasting impact. They suffer from inner contradictions. . . . The person who led the theory of heat radiation out of this isolation, the person who first showed that the quantity "*h*" is of basic importance for the whole of atomic physics is Einstein . . . an allegation which often appears in the literature, that it was through another more or less parallel hypothesis that Einstein reached his proposition is absolutely wrong. It was, on the contrary, through an analysis the originality and intellectual power of which has few equals in physics.'[20]

The Nobel Committee concurred with Oseen's summarizing evaluation: '. . . when physicists, with few exceptions, rejected the quantum theory of Planck, it was Einstein who through an unusually original and ingenious analysis showed the [quantization of the transfer of energy between matter and ether].'[21] Measurements by Millikan confirmed Einstein's theory, and Bohr constructed his atomic model on the basis of Einstein's ideas. 'Almost all confirmations of Bohr's atomic model are also confirmations of Einstein's principle.'[22] The law of photoelectric effect was considered of greater significance and motivated the awarding of a Nobel prize, which Einstein finally received in 1922 for 1921.

With Planck and Einstein, a new era had begun in physics. Maxwell was perhaps the first theoretical physicist to turn away from the 'old' theoretical physics. He no longer claimed to have deduced equations immediately from experience. While old theoretical physics proceeded from experience and strove for mathematical descriptions of physical realities which were made visible by experiments, the new theoretical physics became, in principle, independent of experience. It proceeded from simple basic assumptions, such as the constancy of the velocity of light and the isotropy of space. The 'thought experiment' became an established methodology for determining when fruitful experiments could be made and what results could be expected. Absence of correspondence between theory and experiment was blamed first on the experiment: The theory was held to represent the pure, idealized reality.

The discussions over Planck and Einstein clearly show the sort of decision-making problems the Nobel committees faced when dealing with new, revolutionary discoveries. In both cases described above, not only the content of the theories was revolutionary but also their methodology. Thus, an evaluation of the new theoretical physics could not be based on their experimental corroboration alone — the new theories could be refuted only on theoretical grounds. As long as the Nobel Committee adhered to the principles of traditional physics, it was using an epistemologically inadequate criterion; and not until Oseen became a member of the Committee did it have a representative of the new theoretical physics. Only then could coherence and consistency of theories become a new criterion for evaluation. Thus, the decision-making process in the first twenty years of the awarding of the prize was determined by a specific concept of science which prevailed among the members of the Nobel Committee for Physics. When that concept became invalidated by new developments in physics, the Committee was in danger of losing touch with those developments and thus also of losing its legitimacy as an evaluating institution by becoming party to a scientific conflict.

After the founders of modern atomic physics — Planck, Bohr and Einstein — had received the prize, all after a long period of waiting, other scientists in the area were awarded the prize with little delay; and thus from 1919 on, after Planck had received the prize, atomic physics acquired a place in the prize awarding which adequately reflected its role in the development of physics.

Acknowledgement

We thank Barbara Schöber and Ursula Ruschhaupt for preparing the numerical material.

Notes

1. The concept of 'procedure' is part of an analysis of decision-making processes in terms of sociological systems theory. For an elaboration, cf. Luhmann, N. (1975) *Legitimation durch Verfahren*, Darmstadt und Neuwied.
2. Since many nominators proposed several candidates, sometimes within the same year, the number of nominations is considerably greater than the number of nominators. Thus, the basis for the numerical analysis is always the number of nominations and not the number of nominators.

3. The data available made it necessary to attribute nominators and candidates to their *native* country.
4. Average numbers for five-year intervals were used in order to simplify the presentation.
5. Minutes of the Nobel Committee for Physics, Gullstrand's report on Einstein, 1921; the following citations are also from that report.
6. Letter of nomination, Hasenöhrl 'voting' for Planck, 1910.
7. Letter of nomination, Lummer 'voting' for Planck, 1912.
8. Letter of nomination, Waldeyer 'voting' for Planck, 1912.
9. Letter of nomination, T. W. Richards 'voting' for Planck, 1912.
10. Minutes of the Nobel Committee for Physics, general report, 1912.
11. *Idem.*
12. Minutes of the Nobel Committee for Physics, general report, 1917.
13. Minutes of the Nobel Committee for Physics, Arrhenius' report on Planck, 1918.
14. Minutes of the Nobel Committee for Physics, general report, 1919.
15. Minutes of the Nobel Committee for Physics, Arrhenius' report on Planck, 1919.
16. Minutes of the Nobel Committee for Physics, general report, 1912.
17. Minutes of the Nobel Committee for Physics, general report, 1913.
18. Minutes of the Nobel Committee for Physics, Gullstrand's report on Einstein, 1921.
19. Minutes of the Nobel Committee for Physics, Arrhenius' report on Einstein, 1921.
20. Minutes of the Nobel Committee for Physics, Oseen's report on Einstein, 1922.
21. Minutes of the Nobel Committee for Physics, general report, 1922.
22. Minutes of the Nobel Committee for Physics, general report, 1922.

THE DISCUSSION CONCERNING THE NOBEL PRIZE FOR MAX PLANCK

BENGT NAGEL

Department of Theoretical Physics, Royal Institute of Technology, Stockholm, Sweden

Introduction

Planck's discovery of the quantum in 1900 marks the beginning of quantum theory, and hence of modern physics. The story of the origin and development of this theory, or rather embryo of a theory, through the first decades of its existence is one of the most fascinating tales in the history of physics. Planck was a 'reluctant revolutionary', who, though well aware of the importance of his discovery, struggled tenaciously to incorporate his new idea into the framework of classical physics, fighting the 'younger revolutionaries', represented above all by Einstein, who realized quite early on that a break with the old concepts was irrevocable.

The story of how Planck got the Nobel prize in physics is also the story of one of the most dramatic and extended episodes in the history of Nobel prizes in physics, covering the period from 1907, when Planck was first nominated, to the final prize award in 1919. As in the history of Planck's involvement in quantum theory, the most dramatic moment came right at the beginning. In 1908, on the basis of a single nomination that year and only two previous nominations, a very hesitant Nobel Committee for Physics recommended Planck as prizewinner 'for his work on the laws of heat radiation'. In the decisive voting in the Royal Swedish Academy of Sciences, this suggestion met a devastating defeat, through the eloquence of a powerful mathematician. The arguments that ensured the defeat were probably supplied by another famous Swedish mathematician/mathematical physicist, who also happened to be the only nominator of Planck that year. The crucial mistake of the Committee may have been that they did not want Wien to share the prize with Planck, as was actually proposed in the nomination.

In 1908, the clash between the new quantum idea and seemingly well-established classical physics, and, hence, the inconsistency of using them in combination in arguments involving 'energy quanta', was not generally known in the physics community. But, in 1911, the picture had changed: at the same time as the quantum idea turned out to be of importance in more and more domains of micro-physics, its inconsistencies and contradictions became more obvious. This motivated the Committee to cite only Wien for the prize of 1911, 'for his discoveries regarding the laws governing the radiation of heat'. By then, Planck had been the first or second name in the 'nomination league' for three years.

In spite of a consistently high nomination pressure, the Committee, while carefully following developments in quantum physics and acknowledging the importance of the quantum idea, postponed proposing Planck for the prize, in the hope that the inconsistencies would be resolved. At last, in 1919, the Committee lost hope, and stated that such resolution might occur only after 'a lapse of time exceeding the present range of vision'; they suggested that Planck be given the prize — actually the previous year's reserved prize — 'in recognition of the services he rendered to the advancement of Physics by his discovery of energy quanta'. In retrospect, one can readily agree that the developments of the subsequent six or seven years, up to the birth of quantum mechanics, could not have been foreseen in 1919.

When a prize had been given to the 'father of the quantum', the gate was open, and in 1922 prizes were given to Einstein and Bohr, representatives of the younger generation, for their contributions to the development of quantum theory — contributions that had actually been crucial in the argumentation for giving the prize to Planck.

The purpose of this paper is to present and discuss, in a preliminary way, a selection of the Nobel archival material pertaining to the discussion concerning Planck. The main body of the material consists of nominations of Planck and of the reports of the Nobel Committee sent to the Academy, including in some cases, as enclosures, separate reports on Planck written by Committee members. Some information was taken from minutes of the Committee meetings and of the Nobel prize meetings of the *klass* (or section) of physics of the Academy. According to the statutes of the Nobel Foundation, no record is to be made of discussions or of voting figures at the final plenary meeting of the Academy to award the prize. However, for the most interesting cases — the prize decisions of 1908 and 1919 — there is available some information in private letters from Academy members.

The interest of doing a study such as this one is at least twofold:

Primarily, the material is useful for the study of the history of quantum theory, in the sense that it reflects mainly the knowledge and opinions of that part of the physics community that was not directly involved in research in the area in question. The material usually available for the study of history of science consists of original research papers, conference lectures, letters and textbooks, which are usually produced by the scientists involved in the research. The nominators of candidates for Nobel prizes represent a wider spectrum. The understandable tendency of a nominator not to stress negative aspects of the work of his candidates(s) would be balanced by the more critical attitude of committee and *klass* members, who represent a broader spectrum of the domain of physics and have also acquired experience in evaluating suggestions for prizewinners. Use of such material could hardly be expected to bring to light any revolutionary new facts in the history of physics, but these studies could be used for a sort of independent check on established opinion and as a measure of how quickly new developments in certain fields of physics spread to the general physics community.

Another use of the material is to learn about the workings of the Nobel award system. Possible questions in this connection are: How important is the opinion of nominators? Do different categories of nominators carry different weights? What criteria should a work attain to be considered worthy of a prize? What are the relative influences of different committee members? What influence has the *klass* of physics, or the Academy, on the discussion of the Committee or on the final decision? Although it must be kep in mind that the awarding of the prize to Planck was not typical, some of the conclusions that can be drawn from the relevant material are generally valid.

In the main part of this paper, the reader is usually left to draw his own conclusions on the basis of the summaries and quotations given, although, of course, there may be a subtle or unintentional influence of the author in his selection of material. At the end of the paper, tentative conclusions are drawn concerning the two aspects outlined above.

Some Background

A description of the process of prize awarding is given in the introduction to this 'Round Table' section of the volume. To aid readers with no previous knowledge of the history of quantum theory to follow this presentation of the material, the main points of the development of the idea are outlined. For a more satisfactory description, the reader is referred to the extensive literature on the subject; a selection is given in notes 1–6.

Planck's radiation law, which gives the frequency and temperature-dependence of the intensity of electromagnetic radiation in an enclosed space (cavity – in German, 'Hohlraum' – radiation, also referred to as 'heat' or 'black-body' radiation), the walls of which are kept at a uniform temperature, was first presented in October 1900. This first 'derivation' of the law was actually an ingenious interpolation between Wien's radiation law of 1896 and the measurements of Lummer-Pringsheim and Rubens-Kurlbaum in 1899 and 1900; those measurements deviated systematically from Wien's law and showed a particularly simple behaviour at high temperatures, in agreement with that demonstrated by Rayleigh's radiation law presented in the summer of 1900. To give a theoretical motivation to the law, Planck had to accept Boltzmann's statistical mechanics approach to the treatment of oscillators in equilibrium with radiation, which Planck had introduced. In particular, he used the relationship between the entropy of the oscillator and the probability of a state of the oscillator, which was basic to Boltzmann's theory. Finally, to arrive at the 'October formula', which he knew agreed well with experiments, Planck had to assume that the energy contained in an oscillator is an integral multiple of the smallest unit, which is proportional to the resonance frequency of the oscillator. The proportionality factor was denoted 'h', and was later called 'Planck's constant'. This was the 'quantization of energy' presented in December 1900. From experimental data on radiation, Planck could also deduce a value for Boltzmann's constant k and from this value derive numerical values for Avogadro's number and the elementary electrical charge.

Outside the relatively small circle of physicists who were interested in black-body radiation, Planck's work aroused little interest until seven or eight years later. Planck gave an extended presentation of his and his predecessors' work in this domain in his book *Vorlesungen über die Theorie der Wärmestrahlung*, which appeared only in 1906.

In 1905, Einstein presented his 'light quantum' hypothesis, the idea that the electromagnetic field itself somehow occurs in quantized units. This was an even more revolutionary idea than Planck's assumption of the quantization of energy of material systems, and was generally accepted only at the beginning of the 1920s. In 1907 followed Einstein's use of the quantum hypothesis to explain the behaviour of the specific heat of solids at low temperatures; this was a more 'conventional' use of the idea of energy quanta, and was thus accepted more quickly than his light quanta.

In an important experiment in 1908, Rutherford and Geiger measured the charge of the α particle and determined a value for the elementary charge (and hence also for Avogadro's constant) which agreed very well with the values Planck had calculated in 1900.

The final breakthrough of the quantum idea came in 1913 when Bohr combined it with the Rutherford model of the atom. This led to a partial explanation of the large amount of spectroscopic data available; and it became clear to physicists in general that, in spite of all the difficulties and contradictions inherent in quantum theory, it had come to stay.

I must finally mention three important conferences, which are generally considered to have been of importance for disseminating knowledge about the development of quantum theory in the crucial years 1908–1912: the international mathematics conference in Rome in 1908, where Lorentz gave a rather conservative and (for quantum theory) negative report on radiation theory; the Salzburg conference in 1909, at which Einstein featured as one of the main speakers; and the first Solvay conference in 1911, where quantum theory was the main topic. Only the first of these conferences is mentioned explicitly in the material in the Nobel archives.

Nobel Prize Nominations, Reports and Discussions with Regard to Planck, 1906–1919

The following is a chronological survey of the Nobel archival material concerning Planck, from the time he was first mentioned in connection with a nomination – in 1906 – to 1919, the year in which he was awarded the prize. The data are given in particularly comprehensive form for the years up to 1911–1912, since that was the period of transition, when the importance of Planck's work for physics outside the domain of heat radiation and the recognition of the quantum hypothesis as a definite break with classical physics became widespread in the physics community. As an illustration of this point: the words 'energy quanta', or an equivalent term, appeared in discussions in the physics *klass* and probably also in the Academy in 1908, were mentioned for the first time in a nomination in 1909, but were not used by the Committee until 1911.

When quoting from the material, I have retained the original language, unless it was Swedish. I apologize in advance for my rather heavy-handed translations into English; however, this was not entirely unintentional, since they could serve to convey the somewhat old-fashioned impression the language of reports and minutes from those days makes on a present-day Swede. It would probably have been much easier to translate the excerpts into German.

A list of nominators of Planck, in chronological order, is given in Table 1; it should be noted that a nominator is listed only the first time he nominates. Further information of possible interest concerning the mechanism of prize awarding is given in Table 2.

In the presentation below, the year given refers to the year the prize was awarded; thus, the corresponding nominations could have been written during the last months of the previous year, and no later than at the end of January in the same year.

1906 First nomination for work on radiation laws

As far as I am aware, the first time Planck is mentioned in connection with a nomination is in *Lenard*'s proposal of Boltzmann in 1906 'für seine thermodynamisch-gastheoretischen Arbeiten . . . und so im besonderen auch die Mittel und Wege schafft zur exacten Begründung der wichtigen Strahlungsgesetze des schwarzen Körpers. Dass diese letzteren Gesetze ... zu den hervorragendsten Errungenschaften der neueren Physik gehören, und dass sie also zur Grundlage einer Preisertheilung genommen zu werden verdienen, unterliegt wohl keinem Zweifel, und darauf gründet sich mein Vorschlag.' Among the then living physicists who had contributed to this domain, Lenard cites as the most important Boltzmann, Wien and Planck, on the theoretical side, and Lummer, Rubens and Paschen, on the experimental side. But Lenard is of the opinion that no other single prizewinner, or combination of two or three winners, is

TABLE 1. *Nominators of Planck, 1907–1919*

Name	Address	Nomination years
Bjerknes, V.	Stockholm; Kristiania from 1909	1907*, 1909*, 1912*
Wassmuth, A.	Graz	1907
Fredholm, I.	Stockholm	1908*, 1909*
Brunner, H.	Berlin	1909, 1910, 1911, 1912, 1913
Chowlson, O.	St Petersburg	1909
Fischer, E.	Berlin	1909, 1910
Goldhammer, D. A.	Kasan	1909*
Riecke, E.	Göttingen	1909, 1910
Uljanin, W.	Kasan	1909
van't Hoff, J. H.	Berlin	1909*, 1910
Eucken, R.	Jena	1910
Hasenöhrl, F.	Vienna	1910
Lenard, P.	Heidelberg	1910
Richards, T. W.	Cambridge, USA	1910*, 1911, 1912, 1914*
Warburg, E.	Berlin	1910*, 1911*, 1916
Wiedemann, E.	Erlangen	1910
Rubens, H.	Berlin	1911, 1916, 1919*
Scheiner, J.	Potsdam	1911
Wehnelt, A.	Berlin	1911
Lummer, O.	Breslau	1912
Neesen, F.	Berlin	1912*
von Baeyer, A.	Munich	1912, 1916
Waldeyer, W.	Berlin	1912, 1913
Millikan, R. A.	Chicago	1913
Prandtl, L.	Göttingen	1913
Runge, C.	Göttingen	1913
Julius, W. H.	Utrecht	1914
Kamerlingh-Onnes, H.	Amsterdam	1914
Lorentz, H. A.	Haarlem	1914
Mittag-Leffler, G.	Stockholm	1914
Simon, H. T.	Göttingen	1914*
van der Waals, J. D.	Amsterdam	1914
von Wilamowitz-Moellendorf, U.	Berlin	1914
Wien, W.	Würzburg	1914, 1918, 1919*
Zeeman, P.	Amsterdam	1914
Volterra, V.	Rome	1914
Burnstead, H. A.	New Haven	1915*
Schulze, F. S.	Berlin	1915
von Laue, M.	Frankfurt	1916, 1917*, 1919*
Marx, E.	Leipzig	1917
Radokavić, M.	Graz	1917
Röntgen, W.	Munich	1917
Wiener, O.	Leipzig	1917
Graetz, L.	Munich	1918
Meyer, E.	Zürich	1918*, 1919*
Meyer, S.	Vienna	1918*
Sommerfeld, A.	Munich	1918
von Linde, C.	Munich	1918
Born, M.	Berlin	1919
Einstein, A.	Berlin	1919

*The nominator suggested more than one candidate.
Total: 74 nominations by 50 nominators.

Division of nominators by country		
	Number of nominators	Number of nominations
Germany	30	47
Holland	5	5
Austria	4	4
Russia	3	3
Scandinavia	3	6
USA	3	6
Switzerland	1	2
Italy	1	1

TABLE 2. *Analysis of the nominations and the winners of the physics prize, 1907–1919*

Year	No. of nominators	No. of nominees	No. of nominations of Planck	Three most nominated candidates	No. of nominations	Prizewinner suggested by Committee	Prizewinner	Work cited	Experimental (E) or theoretical (T)
1907	33	19	2	Lippmann	7	Michelson	Michelson	Optical precision instruments	E
				Rutherford	7				
				van der Waals	5				
1908	26	14	1	Lippmann	5	Planck	Lippmann	Colour photography	E
				Rutherford	5				
				Righi	4				
1909	43	21	9	Planck	9	Marconi, Braun	Marconi, Braun	Wireless telegraphy	E
				Wright *et al.*[a]	7				
				Hale	7				
1910	59	17	10	Poincaré	34	van der Waals	van der Waals	Equation of state of gases and liquids	T/E
				Planck	10				
				Righi	5				
1911	27	17	6	Planck	6	Wien	Wien	Heat radiation	T (E)
				Poincaré	5				
				Eötvös, Elster, Geitel	3				
1912	30	17	7	Planck	7	Kamerlingh-Onnes	Dalén	Regulators for lighthouses	E
				Kamerlingh-Onnes	5				
				Einstein	4				
1913	40	27	5	Kamerlingh-Onnes	7	Kamerlingh-Onnes	Kamerlingh-Onnes	Low-temperature physics	E
				Amagat	6				
				Planck, Righi	5				
1914	44	24	11	Planck	11	von Laue	von Laue	X-Ray diffraction in crystals	T
				Righi, Kamerlingh-Onnes	4				
				Eötvös	3				
1915	17	21	2	Bragg, W. H.	4	Bragg, W. H. and W. L.	Bragg, W. H. and W. L.	Study of crystals with X-rays	E
				Hale, von Laue	3				
				Planck, Bragg, W. L.	2				
1916	29	23	4	Stark	5	Reserved	Reserved		
				Planck	4				
				Hale	4				
1917	34	22	5	Hale	6	Reserved	Reserved		
				Planck	5				
				Einstein, Poulsen, Stark	3				
1918	27	23	6	Planck	6	Barkla	Barkla[b]	Discovery of characteristic X-rays	E
				Einstein	5				
				Perrin	3				
1919	29	20	6	Millikan	6	Planck	Planck[c]	Discovery of energy quanta	T
				Planck	6				
				Einstein	5				
						Stark	Stark	Doppler effect in canal rays and splitting of spectral lines in electric fields	E

[a]O. and W. Wright, Voisin and Farman.
[b]Prize for 1917.
[c]Prize for 1918.

possible other than Boltzmann alone: 'Denn die Experimentatoren für sich allein haben zum Erfolg – dem vollständigen und auch theoretisch standhaltenden Gesetz – nicht kommen können. Von den Theoretikern aber könnten Wien und Planck weder ohne Boltzmann, noch auch ohne die Experimentatoren für das Gesetz prämiiert werden; denn ihre Arbeiten haben einerseits zur wesentlichen Vorbedingung die Boltzmann'schen gehabt, andererseits aber der Correctur (Wien) bez. des Wegweises (Planck) durch die Ergebnisse der Experimentatoren bedurft.'

It might be noted that Boltzmann was also nominated in 1906, the year of his death, by Planck and Czerny, but with no explicit reference to his contribution to radiation theory.

1907 Nominations

V. Bjerknes, Stockholm, suggests as a third alternative that the prize be divided between Wien and Planck; his first choice is Heaviside, his second an undivided prize for Wien. Wien is nominated for his radiation law, 'generally acknowledged as one of the most important achievements in physics in later time'. Planck is cited for his contributions to the theory of radiation, consisting, among other things, of a new basis for and refinement of Wien's law; this work is less 'popular' (probably meaning less well known), but not less important, than Wien's. Reference is given to Planck's papers in *Sitzungsberichte Berlin* from 1895 onwards, and to his book *Vorlesungen über die Theorie die Wärmestrahlung* (1906). Bjerknes also says that although Planck's work is related to Wien's, it is of independent importance, so that it seems best to give undivided prizes, first to Wien and then to Planck. It seems clear from the context that 'Wien's radiation law' refers to Wien's displacement law, rather than to his energy distribution law. It should also be noted that no explicit reference is made to Planck's radiation law, not to mention his 'quanta of energy'.

A. Wassmuth, professor of mathematical physics at Graz, however, concentrates on Planck's radiation law in his nomination. He quotes the preface of the second edition of Drude's *Lehbuch der Optik II* (1906), in which the author states that the greatest advance in optics has been made by Planck in the domain of radiation. Wassmuth continues: '*Planck* war es in der That, der ausgehend von der elektromagnetischen Lichttheorie unter Benutzung einer unsterblichen Idee *Boltzmann's* über den Zusammenhang zwischen Entropie und Wahrscheinlichkeit des Zustandes *auf neuem, einwurfsfreiem Wege* zu dem nach ihm benannten, durch die Versuche vollständig bestätigten Strahlungsgesetze für schwarze Körper gelangte. Dieses von ihm gefundene Gesetz und seine Folgerungen erlaubten ihm überdies nicht nur das *elektrische Elementarquantum zahlenmässig* zu berechnen, sondern auch die *absolute Masse der Gasmoleküle* zu ermitteln.'

It can be concluded from explicit references in Wassmuth's nomination that the 'new, non-objectionable' way of deriving the radiation law refers to the derivation given in §§ 145–152 of Planck's *Vorlesungen*, which is essentially the original derivation presented in December 1900. The 'old way' – by implication objectionable – is Planck's interpolation method, presented in October 1900.

Even if the natural tendency of a nominator to present his candidate's achievements in a positive way is taken into consideration, it seems safe to conclude that at this time, the beginning of 1907, a theoretical physicist who was probably well acquainted with Planck's work and in particular with his *Vorlesungen* could still be unaware of the fact that the derivation of the radiation law introduced an element alien to traditional physics, and that this would bring about difficulties.

1907 Committee report

The Committee mentions the nominations of Wien and Planck but does not discuss them.

1908 Nominations

I. Fredholm, Stockholm, nominates Wien and Planck for a divided prize 'for their work in the theory of temperature radiation'. Fredholm says that it is unnecessary to give any detailed motivation, since, on the one hand, 'the works of the scientists mentioned are well known and highly valued by every physicist', and, on the other, 'the great importance of the problems solved by the works of Wien and Planck should be well known to the Nobel Committee'. For Planck's work, Fredholm referred to his *Vorlesungen* and the list of publications therein. 'Wien's work' referred to his papers published in 1893 and 1894 on the displacement law.

1908 Committee report

The Committee's main candidates for the prize are Lippmann, Wien and Planck. Righi, whose experimental work on short electromagnetic waves had already been declared worthy of a prize in 1907, is also mentioned in the 'top list' but is eventually put well behind the other three candidates.

Lippmann had been discussed several times earlier by the Committee, and in 1907 he is one of the most serious candidates for a prize in the coming years.

It is more surprising that Wien and Planck advanced so quickly to top positions, on the basis of only a couple of nominations. There is no doubt that the nomination by Fredholm, and that in 1907 by Bjerknes, carried some weight with the Committee. However, probably of greater significance in this connection was direct, external influence on members of the Committee. By

1908, the importance of Planck's work for physics outside the more limited domain of black-body radiation began to be realized more generally. As discussed below, the experiment by Rutherford and Geiger, described earlier, played a crucial role in the argumentation for giving Planck the prize.

The discussion in the Committee of the alternatives, Lippmann and Wien/Planck, was evidently a very difficult one. The Committee worked out two alternative reports, the one suggesting Lippmann being attached to the main report. However, as stated in the main report: 'In the final consideration of these two proposals the Committee has, yet with great hesitation, given preference to the proposal of Wien and Planck, however, motivated by reasons given in detail in the special motivation below, in such a way that it has decided to suggest that this year's physics prize should be given undivided to Professor M. Planck in Berlin.' Planck is then formally suggested as the prizewinner 'for his work on the laws of heat radiation'. Then follows the special motivation referred to above, a ten-page report on the history of heat radiation and Planck's work. The Committee report is dated 24 September, although the decisive discussion took place in a meeting on 18 September. The minutes of this meeting indicate that Arrhenius was commissioned to work out the special motivation for proposing Planck. Attached to the minutes is an explanation, one page long, written by the chairman, Knut Ångström, and co-signed by all other Committee members except Arrhenius, giving the reasons why Ångström agreed to the final suggestion 'only with the greatest hesitation'. The wording of the document clearly brings out this hesitation; the main reason for it is that, in the elucidation of the laws of black-body radiation, 'it is very far from being that the theoretical works have guided the experimental ones, but rather that one could justly make a completely contrary statement'. In the view of Ångström, 'the prize ought to be divided between the foremost theoretician and the foremost experimentalist, in this case probably Lummer'. Since this was not possible, as no experimentalist was nominated for the prize, Ångström found it desirable to postpone Planck's award to later, especially since Lippmann was obviously a prizeworthy candidate who did not give rise to any objections. 'However, since I could not deny that the radiation laws constitute a more important advance in physical science than Lippmann's colour photography, I have considered myself obliged to join the suggestion of a prize award to Planck in spite of the great apprehensions I have professed here, apprehensions which are not lessened by the fact that the theoretical treatment of the radiation problem is an as yet unfinished chapter, where new efforts to solve the problem are both necessary and constantly tried.'

It should be noted that Ångström was a well-known expert on experimental aspects of the radiation problem. One of his contributions had been to introduce, at the beginning of the 1890s, the basic method of absolute (quantitative) measurement of heat radiation (see, for example note 4, Summary, p. XIII, and p. 153).

The recommendation of the Committee was whole-heartedly supported only by Arrhenius; the other four members, including the Chairman, would rather have chosen Lippmann. As we shall see in the final discussion of the Academy, the defeat of the Committee was probably due in the end not so much to the evidently weak backing of the Committee itself as to the fact that the Committee – in reality, Arrhenius – had omitted Wien from the suggestion, contrary to the recommendations of both Bjerknes and Fredholm.

1908 Special motivation of the Committee (Arrhenius' report)

After mentioning Fredholm's nomination, the report states that its conclusion will be that the work common to Wien and Planck, 'having its most concise expression in the Wien-Planck radiation law' (i.e. Planck's radiation law, in present terminology), is less important than the separate works of the authors, and that Planck's is superior, so he should get the Nobel prize alone.

The report then gives a very detailed account, in particular of early developments in black-body radiation, from Kirchhoff to Planck's *Vorlesungen* in 1906. The discussion concentrates on the experimental aspects, and Planck's derivation of his law is treated very briefly. No mention is made of 'quanta of energy' or related problems.

In the discussion, a very clear separation is made between Planck's derivation of the radiation law, which is considered to be a correction of Wien's radiation law, and his derivation of relations between the constants k and h and radiation data and the use of k to calculate Avogadro's number. There follows a discussion of the result of the Rutherford-Geiger experiment, mentioned above. 'In this way . . . it has been made extremely plausible that the view that matter consists of molecules and atoms is essentially correct, in spite of recent objections by, e.g., Wald and Ostwald. No doubt this is the most important offspring of Planck's magnificent work.'

In the final comparison of the works of Wien and Planck, it is stressed that the correction of Wien's radiation law was not motivated by Planck's investigations but came as a result of extensive experimental work. Hence, the award should be given not for the derivation of the radiation law itself. 'In this way, the connection between Wien's and Planck's works is broken at its main point, and it remains to be decided which one among the works proposed is most worthy of a Nobel prize. In this connection the Committee does not hesitate to give prominence to Planck's work on the theory of heat radiation. . . .'

Although this wording is rather obscure, to put it mildly, earlier parts of the report indicate that the reference in the last sentence is probably to Planck's use of the radiation law to calculate k, and the consequences thereof.

1908 Discussion in the physics klass

At the Nobel meeting of the physics *klass*, nine members were present, including the five Committee members. The most interesting contribution was the two-page comment by *Carlheim-Gyllensköld.* He makes reference to the lecture by Lorentz at the mathematics conference in Rome in 1908, and says that 'Planck has put forward a new, previously unimagined thought, the thought of the atomistic structure of energy, which like its predecessors, the hypotheses of the atomistic structure of matter and of electricity, seems fit to be fruitful for different parts of physics'. He also mentions that Planck's idea consists in assuming that a resonator cannot gain or lose energy in arbitrarily small amounts, and points out that Stark has made use of Planck's hypothesis in the study of the energy relations between X-rays and electrons. This is the first time 'energy quanta' or an equivalent term was used in discussions of awarding the Nobel prize to Planck, and also the first time a connection was made with phenomena outside the domain of heat radiation. *Ekholm* stresses the meteorological importance of Planck's formula, and gives a reference to a paper he had writtein in 1901, in which he used what he called Planck's '*verbesserte Wien'sche Gesetz*'. *Bäcklund*, Lund, who was not present, sends a written vote for Marconi; although he thinks highly of Planck's work, and 'finds it rather probable that among all now living authors in the domain of mathematical physics Planck is in the first place', he believes that Marconi's invention of wireless telegraphy is probably more popular and useful.

The result was that all the *klass* except Bäcklund voted for Planck.

1908 Discussion in the Academy: The prize goes to Lippmann

My source for what happened in the Academy is a letter from *G. Mittag-Leffler*, Stockholm, to P. Painlevé in Paris, dated 9 December. The main purpose of the letter was to persuade Painlevé and Poincaré to propose the invention of the aeroplane for the physics prize in 1909. After detailed instructions as to how Painlevé should proceed, Mittag-Leffler writes (Although the citation is not all of direct relevance for the Planck prize, it gives a good picture of the character of Mittag-Leffler, which was certainly influential in the decision on the Nobel prize of 1908.): 'On ne peut pas proposer Poincaré pour le prix immédiatement après Lippmann, mais je pense que cette proposition pourra être faite avec succès pour l'année 1910 si nous nous organisons de bon temps. Poincaré une fois couronné on viendra à d'autres hommes de la théorie et j'espère bien de vivre assez longtemps pour vous faire donner le prix. C'est moi qui ensemble avec Phragmén a fait donner le prix à Lippmann. Arrhenius voulait le donner à Planck à Berlin, mais son rapport qu'il était pourtant arrivé à faire accepter d'une manière unanime par la commission était tellement bête que je pouvais facilement l'écraser. Il n'a obtenu à la fin que 13 voix (Retzius inclus évidemment) quand j'avais pour moi 46 voix. Deux membres de la commission ont même déclaré qu'après m'avoir entendu ils changeaient d'opinion et votaient pour Lippmann. Je n'aurais rien eu contre à partager le prix entre Wien et

Planck, mais le donner à Planck seul aurait été de récompenser des idées encore très obscures et qui demandent d'être contrôlées par les mathématiques et par l'expérience.'

The argumentation Mittag-Leffler used against the Committee report for Planck was probably supplied by Fredholm: Among the letters from Fredholm to Mittag-Leffler is an unsigned manuscript, dated 10 November, in Fredholm's handwriting, in which he criticizes the Committee report on several accounts. His main objection is that while in Wien's derivation the displacement law is a strictly logical consequence of established laws in physics, Planck has based the derivation of his radiation law on 'a completely new hypothesis, which can hardly be considered plausible, namely the hypothesis of the elementary quanta of energy'. Although this is not a compelling reason for doubting the value of Planck's investigation, it might be wiser to postpone a definitive judgement. Fredholm also argues against the importance that Arrhenius places on Planck's calculation of Boltzmann's constant k and hence of Avogadro's number from the new radiation law and radiation data. Quite correctly, Fredholm points out that k also appears in the classical limiting case of Rayleigh–Jeans' law (although Fredholm, who quotes Planck's *Vorlesungen*, refers to it as 'Lorentz' formula'), and that it should therefore be possible to use this formula with experimental radiation data to calculate k. One could rejoin that the value of k calculated in that way would probably be much less accurate than the one calculated by Planck; uncertain data on the behaviour at large wavelengths would have to be used instead of the much better known constants of the total radiation intensity formula (Stefan–Boltzmann's law) and of Wien's displacement law.

This action of Fredholm's, effectively destroying the chances of his own candidate, might be explained by assuming that he wanted also to get his other candidate, Wien, onto the ticket. It is even possible that Mittag-Leffler argued in the Academy for a division of the prize between Planck and Wien, as the last sentence quoted from his letter above may indicate. From the point of view of the Committee, such a division would have brought back into focus Planck's radiation law itself and, hence, the neglected contributions of the experimentalists stressed in the statement by Ångström; the majority of the Committee could probably not have accepted this.

1909 Nominations

Of the nine nominators in 1909, the most notable ones were Planck's colleagues in Berlin, *J. H. van't Hoff* and *E. Fischer*, the chemistry prizewinners of 1901 and 1902. van't Hoff gives F. Kohlrausch as his first candidate. *Fredholm* repeats his 1908 nomination of Wien and Planck. Only *Riecke*, Göttingen, and *Goldhammer*, Kasan, who suggest a division of the prize between van der Waals and Planck, give somewhat longer motivations.

Riecke's nomination is remarkable because it is the first in which energy quanta are mentioned. Riecke refers to Planck's *Vorlesungen*, and continues, 'Das werk ist allerdings nicht erst im vergangenen jahre erschienen; es dürfte aber den bestimmungen der Nobelstiftung doch in sofern entsprechen, als die weittragende bedeutung der darin niedergelegten neuen und originellen forschungen im laufe der letzten jahre mehr und mehr hervorgetreten ist. Einerseits verbinden sich die darin entwickelten gedanken mit dem princip der relativität, von dem aus eine neue gestaltung unseres ganzen physikalischen weltbildes sich anbahnt; andererseits gewinnt die vorstellung von einer elementaren verteilung der energie wachsende bedeutung für die wissenschaft. Man kann sich zwar des eindruckes nichterwehren, dass diese vorstellung von Planck in etwas willkürlicher weise eingeführt ist; der erfolg rechtfertigt aber die kühnheit der einführung. Die fundamentalen fragen, welches der physikalischen sinn jener atomisierung ist, wie sich der übergang von der annahme einer stetigen ausbreitung der energie in einem continuierlichen äther zu der Planckschen vorstellung von den energieelementen bewerkstelligen lässt, werden noch lange die physiker beschäftigen; ihre bearbeitung wird der physik nach der theoretischen, wie nach der experimentellen seite reiche anregung geben und neue probleme von fundamentaler bedeutung stellen.'

Goldhammer first mentions Planck's radiation law and then also refers to Planck's work from 1897 onwards on irreversible processes. The other nominations generally make short references to Planck's work on thermodynamics and radiation.

1909 Committee report

The Committee notes that Planck has obtained nine nominations, as compared with one in 1908, but, 'in spite of this, and in particular considering the result of last year's prize award, the Committee does not find it possible to take him into consideration as one of the main candidates for a prize award this year'.

1910 Nominations

The year 1910 was the one of Mittag-Leffler's well-organized campaign to give the prize to Poincaré; it resulted in thirty-four nominations, a record that has been beaten only in recent times. Planck was second-best with ten nominations. In this year, three Nobel prizewinners proposed Planck: *Fischer* and *van't Hoff* were joined by *Lenard* (physics prize in 1905).

Lenard writes, 'Nachdem Boltzmann nicht mehr lebt istes meines Erachtens Planck in welchem das Verdienst um die Entdeckung der auch praktisch so wichtigen Gesetze der Strahlung des schwarzen Körpers hauptsächlich sich concentriert.'

Hasenöhrl, Boltzmann's successor in Vienna, gives a motivation comprising five pages. In the beginning, he discusses two different directions of theoretical physics – a more mathematical one, and theoretical physics proper. Of workers in the latter domain, only a few have been able to build up new theories: Maxwell is one example, and Planck is another. Hasenöhrl then goes on to discuss the problem of heat radiation and Planck's theory of resonators in equilibrium with radiation, and finally arrives at the quantum concept: 'Eine ausserordentlich merkwürdige Folgerung ergibt sich noch aus der Planck'schen Betrachtungsweise: Ein Resonator kann von der umgebenden Strahlung nicht Energie in jedem beliebigen Betrage erhalten, sondern nur das ganzzahlige Multiplum eines bestimmten Quantes ($h\nu$). Die Annahme eines derartigen "Energie-Atomes" ist auf den ersten Blick höchst befremdend und wird von mancher Seite geradezu als schwacher Punkt der Planck'schen Theorie angesehen. Doch sind in allerletzter Zeit auch andere Forscher (A. Einstein; J. Stark) von anderen Gesichtspunkten zu demselben Resultate geführt worden.' A little further on, he writes: 'Man kann heute unmöglich überblicken, wie viele Modifikationen die Theorie der Strahlung noch erfahren wird. Eines ist aber, wie ich glaube, sicher: ganz werden die Planck'schen Gedanken nie verschwinden; seine Arbeit wird stets ein Merkstein in der Entwicklung der theoretischen Physik bleiben.'

In 1910 the first American nomination of Planck was made: *T. Richards*, Cambridge, Massachusetts, mentions Planck as an alternative to van der Waals.

Warburg suggests dividing the prize among Lummer, Wien and Planck.

1910 Committee report

The scars of the battle in 1908 are still visible: 'Nominated works of theoretical or mathematical physical contents are those by Planck, Einstein, Poincaré and Gullstrand. The works of Planck, among which the most important ones treat the laws of radiation, were discussed by the Committee in its report of 1908. Planck's investigations of the second law and of the thermodynamics of dilute solutions have been mentioned by some nominees as worthy of a prize. These works are, however, much older and also, in the opinion of the Committee, of less importance than those on heat radiation; hence, the latter should be preferred in a judgement, if any, on the candidature of Planck.

'Some of the works of the Swiss mathematical physicist *Einstein* in the domain of thermodynamics are close to the works of Planck. However, Einstein has not been nominated for these works, but for his introduction of the so-called relativity principle. . . .'

The year 1910 was the one in which Einstein received his first nomination. W. Ostwald – Landhaus Energie, Gross-Bothen, Kgr. Sachsen – nominated Einstein because 'dessen Relativitätsprinzip die weitreichendste Begriffsbildung darstellt, die seit der Entdeckung des Energieprinzipes bewerkstelligt worden ist'.

1911 Nominations

The most notable nominator of Planck in this year is probably *Rubens*, Berlin. In his one-page motivation, he writes, among other things: 'Die Planck'sche Strahlungsformel, welche sowohl das Elementarquantum der elektrischen Ladung als auch die charakteristischen Einheiten für den Energieaustausch aus einfachen Strahlungsmessungen zu berechnen gestattet, muss heute neben den Ergebnissen der radioactiven Forschung und den Schlüssen aus der Brown'schen Molekularbewegung als die wichtigste Stütze des modernen Atomismus bezeichnet werden.'

Scheiner states in his nomination that only in the preceding year had the first extensive practical application of Planck's results been published, in an astrophysical paper by Witling and Scheiner.

1911 Committee report

In 1911, the Committee sent two reports to the Academy. In the first (dated 20 September), there is only a short discussion of the nominations mentioning heat radiation and Planck. The report ends by suggesting that the prize be given to A. Gullstrand, Uppsala, for his work on geometrical optics. Gullstrand, who had already been discussed as a prize candidate in 1910, became a member of the Committee in 1911. Of some interest is the dissenting opinion of Carlheim-Gyllensköld, which is attached to the Committee's report. He favours Poincaré, whose candidature he had investigated for 1911. Carlheim-Gyllensköld's seven-page report on Poincaré opens with a discussion of the importance of theoretical physics; in the view of the author too few prizes have been given for theoretical works (only that of Lorentz in 1902 and, partly, that of Thomson in 1906).

When it became known that the Nobel Committee for Physiology or Medicine had also proposed Gullstrand 'for his work in the dioptrics of the eye', he declined the physics prize, and the Committee had to find a new candidate. This was done in a meeting on 20 October, where the chairman, Granqvist (chosen after Ångström's death in 1910), presented a report on recent work in the domain of radiation phenomena. The Committee decided unanimously to suggest Wien 'for his discoveries regarding the laws governing the radiation of heat', and commissioned Arrhenius to make the necessary adjustments to Granqvist's report. The resulting report was then sent to the Academy.

Like the report of 1908, this one begins with the history of heat radiation. With regard to Planck's contributions, his first interpolation derivation of the radiation law is presented; then the report says that Planck continued trying to construct a theory that would result in the radiation formula. The introduction of energy quanta for the oscillators is discussed: 'It should be pointed out that without a bold assumption, diverging from our ordinary conceptions, it would be impossible to derive Planck's radiation formula.' The report then discusses recent objections by Abraham, Wien and Lorentz against the quantum hypothesis, and mentions in particular Lorentz' objection on the time it would take for an oscillator to absorb a quantum of energy, leading to the conclusion 'that Planck's assumption about the energy quanta of oscillators can hardly be upheld unless one agrees – as do Einstein and Stark – to assume that, also in the free aether, heat and light energy propagate in the form of definite, indivisible quanta, an assumption which both Planck and Lorentz reject . . ..' The report then mentions Planck's 'second radiation theory' in which only the emission of energy from the oscillators is quantized, whereas absorption takes place continuously. It is too early to judge the success of this theory, 'however, one thing is certain, at present we do not have a generally accepted theory explaining the Planck radiation formula, which is the only formula in agreement with all experimental investigations up to now'.

In view of this uncertainty, the Committee finds it best to postpone the award for the discovery of the Wien-Planck radiation law (i.e. Planck's radiation formula), pending an improved theoretical derivation, and to give the prize to Wien for his displacement law, which has a solid theoretical foundation. The Committee also mentions Wien's preparatory work on the Wien-Planck radiation law (i.e. Wien's tentative derivation in 1896 of the Wien distribution law), as well as his work on canal rays.

1911 Discussion in the physics klass (21 October)

As in 1908, all *klass* members, except Bäcklund, voted for the Committee's suggestion. *Bäcklund* this time also sent a written vote, in which he stated that he preferred Planck to Poincaré; there had evidently not been enough time to inform absent *klass* members about the Committee's new choice. Even *Carlheim-Gyllensköld* voted for Wien, stating that Wien and Poincaré were not commensurable.

1912 Nominations

The most interesting nomination in 1912 is that by *Lummer*, Breslau. In his three-page letter, he gives a historical survey of his own involvement in black-body radiation; he also discusses Planck's work, and writes finally: 'Wie befruchtend die Entdeckung des Elementarquantums werden sollte, hat die neueste Zeit gelehrt. Durch diese Entdeckung wurde ein neues Licht geworfen auf die Theorie des photoelektrischen Effekts; auf sie gründet sich ferner die Einsteinsche Theorie der specifischen Wärmen, die mit einem Schlage die längst empfundene Schwierigkeit der Erklärung der Abweichungen vom Dulong-Petitschen Gesets beseitigte.

'Die auf dem Elementarquantum basierende Theorie der specifischen Wärmen ist von Nernst einer eingehenden Prüfung unterzogen worden. Gerade diese glänzenden Untersuchungen bilden das stärkste Fundament für die Annahme der Existenz der diskontinuierlichen Struktur der Energie. Dadurch ist wohl die Anschauung Gemeingut geworden, dass an den Grundlagen unserer Naturauffassung einschneidende Änderungen vorzunehmen sind. Diese Erkenntnisse sind aber direkt aus den Planckschen Untersuchungen hervorgegangen.

'Wollte man nur neueste Leistungen belohnen, so gebührte Einstein der Preis, dem Entdecker des Relativitätsprinzips und Begründer der Theorie von dem specifischen Wärmen. Nachdem W. Wien mit Recht den Nobelpreis erhalten hat, gebührt Planck mit grösserem Rechte der Preis für 1912.'

Typical of the gradual change in stress in the motivations of the nominations is the following quotation from *Waldeyer*'s nomination (actually a nomination for the prize of 1911, which arrived too late). He mentions Planck's investigations in the domain of thermodynamics, 'welche so wesentlich zur Umgestaltung unserer Grundanschauungen in der Physik beigetragen haben'.

1912 Committee report

After discussing Nernst, and Planck's interest in the heat theorem of Nernst, the Committee mentions Planck's quantum theory: 'This theory is at present heatedly discussed, and Planck himself has recently modified it. Therefore the Committee believes that one should await the rapid developments in this domain, so that if possible further clarification would be obtained about this, seemingly, most important contribution of Planck, before a Nobel prize is given to him, who through all his works in the theory of heat has influenced the development of modern science in a fundamental way, as the Committee has pointed out on several occasions.'

1913 Nominations

The term 'Quantentheorie' occurs for the first time in this connection in two of the nominations of this year, first in that by *Waldeyer*, Berlin, and in the joint nomination by *Runge* and *Prandtl*, Göttingen, who propose Planck 'für seine Verdienste um die Strahlungstheorie und speziell für seine Entdeckung der Quantentheorie'. They enumerate a number of different domains in physics and chemistry in which quantum theory is of importance.

Millikan, Chicago, bases his proposal 'upon the demonstrated usefulness of Professor Planck's discovery of the universal constant called by him "Wirkungs-Quantum" *h*'. In spite of possible doubts about the validity of Planck's derivation, 'the constant itself has demonstrated its use in four different domains of Experimental Physics: 1. the domain of specific heats at low temperatures; 2. the domain of radiant heat energy; 3. the domain of photo-electric radiations; 4. the domain of X-radiation.'

1914 Nominations

The most impressive support for Planck's candidature this year is a joint nomination signed by *Lorentz*, Haarlem, *Julius*, Utrecht, *Zeeman*, Amsterdam, and *Kamerlingh-Onnes*, Leiden. They write, among other things, 'Certes, cette théorie . . . ne fait pas encore l'impression d'une doctrine définitivement établie. . . . Mais l'hypothèse des "quanta" a déjà passé victorieusement par bien des discussions — on le voit, par exemple, dans un des derniers mémoires que feu Henri Poincaré — et elle gagne du terrain de plus en plus. . . . Aussi quelle que soit la forme définitive que devra prendre l'hypothèse de M. Planck, on l'honorera sans doute comme un de ceux qui ont puissament contribué au progrès de la Science en lui ouvrant de nouvelles voies.'

Wien is the only one of the nominators who mentions in his nomination the new application of the quantum concept to the theory of spectra (by Bohr, who is, however, not mentioned by name).

The nomination by *Volterra*, Rome, the only one Planck ever received from Latin Europe, should also be noted.

In an interesting and somewhat intriguing letter to the Nobel Committee, *Mittag-Leffler* first suggests that the prize should be reserved, since in his opinion there does not seem to be any discovery left worthy of a Nobel prize award; although, he adds, referring to a 'very notorious case', this fact has not previously prevented the Academy from awarding the prize. In the opinion of Mittag-Leffler, the intention of Nobel was not so much to reward finished and generally acknowledged work as to stimulate the development of science by helping to further important new ideas that could help science to progress. With this in mind, should the Academy not agree to withhold the prize, Mittag-Leffler suggests that the award be given to Planck for his quantum theory, since this theory, although perhaps not a discovery in the sense used in the Nobel statutes, by far supersedes anything else that could be considered for an award (*sic*!).

1915–1918 Nominations

The nominations made during 1915–1917 add no new aspects of Planck's work and quantum theory that are not already contained in earlier nominations or Committee reports. It might be noted that Planck was nominated in this period both by *von Laue* (twice) and *Röntgen*. von Laue nominated A. Sommerfeld for the prize of 1917, provided the (reserved) 1916 prize be given to Planck: 'Sonst wiederhole ich meinen Vorschlag für 1916 in der Erwägung, dass unmöglich ein Nobelpreis für eine Leistung auf dem Gebiet der Quantentheorie verliehen werden kann, bevor Planck ihn erhalten hat.'

In 1918 Planck was nominated by *Sommerfeld*, who wrote a long motivation stressing the application of quantum theory to the study of spectral lines, and discussing Planck's work in 1916 on quantum theory for systems with several degrees of freedom. Other nominations, such as that five pages long by *Graetz*, Munich, discuss applications to the theory of spectra.

1914–1918 Committee reports

Adjoined to the report of 1914 is a fifteen-page special report on Planck's radiation law worked out by Granqvist. To a large extent, it is a summary of earlier reports. The new aspects are, first of all, a short discussion of Planck's 'second theory' of 1910. Granqvist cites a study of Oseen in 1914 showing that the assumption that an oscillator absorbs energy without at the same time emitting energy leads to the conclusion that Maxwell's equations are invalid in all space. This indicates that the inner contradictions of Planck's 'first theory' are present also in his 'second theory'. Granqvist then goes on to discuss the various areas in which Planck's constant is applicable, including Bohr's atomic model and the new investigations by Franck and Hertz on the quantum condition in the ionization of atoms by electron collisions. The contributions of Debye, and also of Born and von Kármán, to Einstein's theory of specific heats, necessary to achieve agreement with Nernst's measurements, are also covered. Granqvist quotes some remarks of Stark, which are in part highly critical of the 'Planck hypothesis'.

Granqvist's conclusion is that 'the quantum theory has so far not been able, in whatever form it has presented itself, to be brought into agreement with other seemingly well-established propositions in physics, or even with itself. Still, it should be stressed that both Planck's formula and the constant h appearing in it are very important in the study of many different parts of physics.'

With regard to Planck, the Committee reports for 1914–1916 contain essentially only this special report by Granqvist and conclude that awarding of the prize to Planck is impossible in view of the inconsistencies of quantum theory.

In its report of 1917, the Committee deals with the 'grand new applications' of quantum theory in the study of spectra. Einstein's new derivation (1916) of Planck's radiation law is also mentioned. The Committee quotes, and concurs with, the statement by von Laue mentioned above, to the effect that the first prize in quantum theory must be given to Planck.

Attached to the Committee report of 1918 is a special report by Arrhenius, two pages long, presenting, among other things, Planck's new work on the rotation spectra of molecules. Arrhenius' conclusion is: 'The theory of quanta has not yet reached a desirable satisfactory completion.'

There is a slight 'softening' in the attitude of the Committee between 1916 and 1917, as reflected in its reports: In 1916, 'a prize award of it [quantum theory] seems also now out of the question'; whereas, in 1917, 'they do not consider themselves able to propose Planck for the Nobel award this year'. The rapid developments, especially in the application of the quantum hypothesis to the analysis of spectra, had evidently made an impression on the Committee. In its report of 1918, the Committee mentions the importance of Planck's constant *h* in various areas of modern physics, but goes on to say: 'However, since on one side these investigations are not yet completed, on the other side they lead to results containing contradictions – e.g. concerning the absorption and emission of light, and the Stark and Zeeman effects – and since the theory hence cannot be considered to have obtained a form corresponding to reality, the members of the Committee cannot recommend Planck as candidate for a Nobel prize this year.'

1919 'The time is ripe'

Nominations

The new nominators of Planck in 1919 were *Born* and *Einstein. Born* gives a long motivation, with a historical review of the development of quantum theory. He mentions Bohr's atomic model (1913) as the greatest success of the theory, and ends by saying, 'So ist der Name Plancks mit fast allen grösseren physikalischen Entdeckungen der Neuzeit auf's engste verknüpft. Hat er auch selbst keine Experimente angestellt, so ist seine Quantenlehre doch der Wegweiser für zahlreiche experimentelle Forscher und das ordnende Prinzip, das die Ergebnisse der Versuche zu einem Weltbilde zusammenfasst. Zwar sind heute erst die gröbsten Züge dieses Bildes entworfen, doch besteht kein Zweifel, dass der Grundgedanke Plancks von der Diskontinuität der Energieverwandlung bestehen bleiben wird.'

Einstein proposes Planck for his work in the domain of heat radiation, explicitly mentioning Planck's 1901 articles in *Drudes Annalen*, 'Über das Gesetz der Energieverteilung im Normalspektrum' and 'Über die Elementarquanta der Materie und der Elektrizität'. Einstein also describes Bohr's theory of spectra as an important application of quantum theory. This is the first nomination that Einstein made. It is also in this year that Planck nominated Einstein for the first time, citing his general theory of relativity.

Committee report

As a basis for the suggestion given in the Committee report of 1919 Arrhenius again wrote a special report, this time comprising fourteen pages. After describing the history of heat radiation and again stressing (cf. his report of 1908) the importance of Planck's calculation of the constants *k* and *h*, Arrhenius discusses the various applications of quantum theory, ending with, 'the most magnificent success of quantum theory [is] . . . in the domain of emission spectra'. He points out that the work of Bohr radically changed the attitude of English physicists towards quantum theory.

Arrhenius' conclusion is, 'All these circumstances make it evident that the quantum theory of Planck is the leading theory in the modern development of physics; although it has not yet obtained a final, satisfactory formulation, quantum theory plays a role making it fully comparable even with the foremost theories so far invented by the pioneers of this science.'

In its discussion of the nominations of Planck, the Committee mentions that no other scientist has been nominated by so many different and such competent nominators. After mentioning the suggestion in 1908 that the prize be given to Planck, and its defeat in the Academy, the Committee states that '. . . it has become increasingly probable that the Planck constant, the discovery of which summarizes the theory, corresponds to some reality in the constitution of matter, and there is no longer any doubt but that this constant gives the measure of an atomistic property, albeit of yet unknown nature'. The report goes on to say that in view of the increasing importance of Planck's constant in a large number of domains, the Committee has found reason to reconsider its previous attitude, which was motivated by the contradictions inherent in quantum theory. There follows a short summary of the later part of Arrhenius' report. After discussing the importance of quantum theory for spectroscopy, the report continues, 'Furthermore, the quantum theory has been of the utmost importance for the theory of specific heats, for the explanation of certain phenomena at low temperatures, and for further domains of physics. Therefore, the Committee does not hesitate to declare that the time is ripe to reward the work of Planck with the Nobel prize for physics. There is now fully sufficient evidence for the importance of his discovery, and no reason exists to await a definitive formulation of quantum theory, much less so since the efforts in this direction, so far futile, seem to indicate that such a final formulation might appear only after a lapse of time exceeding the present range of vision.'

At the end of the report, the Committee suggests that 'last year's reserved Nobel prize for physics be awarded to Professor M. Planck at the University of Berlin in recognition of the services he rendered to the advancement of Physics by his discovery of energy quanta'.

1919 Discussions in the physics klass *and the Academy*

The Committee suggestion that the 1918 prize be awarded to Planck met no opposition either in the physics *klass* or in the Academy. (Information about the deliberations at the Academy meeting comes from a letter sent by E. Phragmén to Mittag-Leffler, who was abroad at the time.) Concerning the awarding of the 1919 prize to Stark, Carlheim-Gyllensköld had already written a reservation; he suggested that the prize be given to Bohr. The main point of discussion in both the *klass* and the Academy seems to have been Arrhenius' proposal that the science prizes be reserved in view of the taxes on large fortunes that were to be introduced in Germany after the War and that would essentially deprive the prizewinners – Haber was the winner of the prize in chemistry – of the award money. This suggestion met little positive response; a counter-argument put forward in the Academy was that if the award were postponed and Planck should die in the meantime, he would be even more effectively deprived of it.

Postcript: 1922: Oseen's view of Planck and Einstein

In the light of the obvious importance of Einstein's contributions to quantum theory in the Committee's evaluation of Planck's work, and in view of recent controversy among historians of science concerning who really introduced the quantum to physics,[5] it might be of interest to give some extracts of Oseen's special report in 1922, which was probably decisive in the awarding of the reserved 1921 prize to Einstein 'for his services to Theoretical Physics, and especially for his discovery of the law of the photoelectric effect'.

After a brief sketch of the history of heat radiation and Planck's introduction of the constant h, Oseen continues, 'The elucidation of the inner importance of this constant of Planck then became the central problem in the domain of heat radiation. The views Planck himself presented have not been of lasting importance. They suffer from inner contradictions. They also contain an even more important error. They throw no light on any other phenomenon except the one that led to their discovery. The person who led the theory of heat radiation out of this isolation, the person who first showed that the quantity h is of basic importance for the whole of atomic physics, is Einstein. The greatest of Einstein's contributions to quantum theory is his first, the proposition that light emission and light absorption take place in such a way that light quanta of energy $h\nu$ are emitted or absorbed.' Oseen's summary begins: 'At a time when physicists, with few exceptions, rejected the quantum theory of Planck, it was Einstein who through an unusually original and ingenious analysis showed that the energy exchange between matter and ether must take place in such a way that an atom emits or absorbs a quantum of energy $h\nu$ where ν is the frequency.'

Some Conclusions

As far as the history of quantum theory is concerned, the nominations and the Committee reports tend to support the established view. Planck's work became generally known only in 1907 or 1908, after the publication of his *Vorlesungen*. The confirmation by the Rutherford-Geiger experiment in 1908 of Planck's calculation (in 1900) of Avogadro's number and of the elementary electric charge played a great role in the discussions of the Committee, in particular since it was considered an important confirmation of the atomic hypothesis. Also in 1908, the discussions about the contradictions inherent in the new quantum hypothesis began, contradictions that would follow the theory up to the advent of quantum mechanics in 1925–1926. The material also shows the importance for its acceptance of the application of quantum ideas to the theory of spectra.

Any conclusions about the mechanisms of the Nobel award system are of necessity more tentative. The crucial point in rewarding a theoretical work – the question of experimental confirmation – which prevented Einstein from getting the prize for his relativity theories, was no problem in the case of Planck: his radiation formula had been beautifully confirmed by experiment. The problem was, instead, the inherent contradiction in the theory itself. With regard to the relationship between the nominators and the Committee members, it is clear that the Committee, given a good reason, could resist constantly high nomination pressure for a considerable time; additionally, it can be seen that the discussions and arguments of the Committee were drawn to a large extent from sources other than the nominations. Although it appears that the nomination by Fredholm was important for what happened in 1908, it is hard to arrive at any reliable conclusion concerning the relative weights of different nominators, or categories of nominators. To make a valid judgement on the relative importance of different Committee members in the decision process, one would need to study outside material that would throw light on the relations among different Swedish physicists; this is contained in the Crawford-Friedman contribution to the round-table discussion. It seems safe to conclude, however, that Arrhenius, who was probably the main supporter of Planck on the Committee, carried more weight in the Committee itself and in the physics *klass* than he did in the Academy. During the period of interest, the Committee, if it had been united, would have had little serious opposition to fear in the *klass*, since it effectively formed a majority there. The reaction of the Committee in the years following the defeat of its proposal in the Academy in 1908 may show a certain sensitivity to the opinion of the Academy; it may be that the cautiousness of the Committee was primarily a consequence of obvious indecision within the Committee itself concerning the awarding of the prize to Planck. On the whole, after its brave attempt in 1908, the Committee followed a relatively consistent and cautious line in relation to Planck's work; in view of the permanence of the Committee, which had the same members from 1911 to 1919 (actually, until 1922), this should not come as a surprise.

Acknowledgement

It is a pleasure to thank Dr Elisabeth Crawford for constructive criticism of an earlier version of this paper. I am also indebted to her for providing me with letters from the Mittag-Leffler collection, giving information on what happened in Academy meetings in 1908 and 1919.

Notes

1. Klein, M. J. (1962) Max Planck and the beginnings of the quantum theory. *Arch. Hist. Exact Sci.*, **1**, 32.
2. Jammer, M. (1967) *The Conceptual Development of Quantum Mechanics*, New York, McGraw Hill.
3. Hermann, A. (1969) *Frühgeschichte der Quantentheorie, 1899–1913*, Baden, Mosbach (English transl.: *The Genesis of Quantum Theory (1899–1913)*, Cambridge, Mass., MIT Press [1971]).
4. Kangro, H. (1970) *Vorgeschichte des Planckschen Strahlungsgesetzes*, Wiesbaden, Stener Verlag (English transl.: *History of Planck's Radiation Law*, London, Taylor & Francis [1976]).
5. Kuhn, T. (1978) *Black-Body Theory and the Quantum Discontinuity 1894–1912*, Oxford, Clarendon.
6. Pais, A. (1979) Einstein and the quantum theory. *Rev. mod. Phys.*, 51 (4), 863.

BACTERIOLOGY AND NOBEL PRIZE SELECTIONS, 1901-1920*

CLAIRE SALOMON-BAYET

Institut d'Histoire des Sciences, CNRS, Paris, France

> 'On the other hand, he knew perfectly well that over the past twenty-five years an extraordinary change had taken place in medicine. As a student, it had seemed to him that medicine would shortly come to suffer the fate of alchemy and metaphysics, but now, in his night reading, medicine moved him, arousing in him astonishment, even enthusiasm. What an unexpected revelation, indeed, what a revolution! Thanks to antisepsis, operations were now being performed which the great Pirogov had considered impossible, even *in spe* . . . .'
>
> Anton Chekhov,
> *Accounts of 1892,*
> 'Room No. 6'

Alfred Nobel's will specified that the third prize – the third equal part of the revenue from the Nobel fund – should be awarded to 'the person who shall have made the most important discovery within the domain of physiology or medicine'. This alternative was already an archaism, for it implied the existence of a separation between the laboratory and the clinic, between experimentation and the care of the sick; it did not take cognizance of that term created in 1802, with all its attendant ambiguities – 'biology' – a term which, by mid-century, was generally accepted[1] to designate sciences which, although making use of the physico-chemical sciences, were concerned with living creatures, whether cells, organs, organisms or populations. Virchow, Darwin and Claude Bernard are the key names in this history, to which should be added the names of Pasteur and Koch, among so many others, who made the microorganism one of the prime actors of life, of its cycle of fermentation and putrefaction, and of its aberration – microbial pathology.

Within the context of the Nobel awards, this archaism underwent a partial then permanent eclipse: From 1900, the standard letters from the Nobel Committee for Medicine in the four official languages (Swedish, French, English and German) invited scientists to submit 'a proposal for the award of the Nobel prize in the group of physiology *and* medicine [my italics]'. At the same time, the terms of the will – 'physiology *or* medicine [my italics]' – were reflected in the official texts, from *The Nobel Foundation Calendar* to H. Schück's reference work *Nobel, The Man and His Prizes* (1962). This is a minor anomaly, on which one may comment without wishing to interpret.

*Translated from the French by Alan Duff.

In 1554, when Jean Fernel created the term 'physiology', he defined it as the study which preceded pathology and made possible its interpretation.[2] In the second half of the nineteenth century, an attempt was made through instrumentation, experimentation and experimental pathology, to understand the manner of functioning of organisms and their disturbances, apart from suffering and inevitable death. The constitution of the living body – cell, organ or organism – as the object of experimental science differed essentially from the Hippocratic injunction to relieve, not to harm, the sick. To the schools of Vienna, Paris or Edinburgh – the great clinical schools which, since the end of the eighteenth century, had continued to define approaches to disease and to refine clinical and nosographical tables – physiology must have appeared to be one of the scientific perversions of the medical art: Not only the anti-vivisectionist leagues challenged the relevance of research by men such as Magendie or Claude Bernard, but also a number of clinicians, who contested the Bernardian idea of an exact, experimental, scientific medicine, which was not in opposition to empirical medicine but which reduced its conjectural aspect.[3]

This reticence on the part of the medical body was similar to that which had long been shown towards the 'same old microbial story'.[4] In 1895, 'physiology *or* medicine' indicated that physiology constituted a basic field of research distinct from the art of medicine and from its advances, exemplified by the science of hygiene since the beginning of the century. This, in a sense, was to be pre-Bernardian. In 1902, to say 'physiology *and* medicine' was to indicate, with the same words, the desire to distinguish no longer between an art and a science, and to recognize *the unity of a field under research*, whatever might be the methods of the approach and its outcome. This, in a sense, was to be post-Pasteurian. From then until the present day, the term used has been the alternative – physiology *or* medicine.

The discussion of the interpretation of the will by the delegates of the Karolinska Institute – analysed by G. Liljestrand – tended towards the proposition that by 'the domain of physiology or medicine is understood all the theoretical as well as the practical medical sciences'.[5] This discussion passed in silence over the technical and polemical context of the development of the medical sciences in the nineteenth century, confronted as they were with the development of the physico-chemical sciences, physiology and experimental pathology, and the systematic use of numbers and statistics, and not merely with the contrast between practice and theory, fundamental or applied. 'Physiology' was not a neutral term in 1895. This must be kept in mind, just as it is necessary to stress that no definition of medicine or of physiology is given in the statutes of the Nobel Foundation: the field, as G. Liljestrand points out, is open, liberal, and allows for cross-relations between chemistry and medicine, whatever the 'professional profile' of the award winners.[6]

In order to cover the field of innovation in physiology or medicine – i.e. the field of what would today be called biomedical research – within the institution of the Nobel prizes between 1901 and 1920 (the first prize awarded and the first awarded after the War), it would be logical to move out of the strict domain of the third prize and to borrow from the chemistry prize whatever concerns the chemistry of surfaces, organic and biochemical chemistry. These areas correspond to fields of research that are common to the disciplines, either in subject matter or in procedure, and to the specific figures who created or extended these areas of overlap.[7] In fact, the territory of those fields is reserved for the third prize, since in the classification of the propositions by group – six as of 1902 – the second group is entitled 'general biology, physiology, physiological chemistry'; the latter term[8] is an archaism dating from the second half of the nineteenth century, deriving from the parallel and occasionally joint development of experimental physiology and organic chemistry.

We shall not be moving out of the conventional framework of the third prize; nonetheless, in the documents relating to those first twenty years, the scientific controversy between S. Arrhenius, T. Madsen and P. Ehrlich on the chemical relation between toxin and antitoxin does deserve more than a mere mention.[9] Arrhenius received the Nobel Prize for Chemistry in 1903 for his new theory on the connection between chemical affinity and electrical conductivity. His name had been put forward for both the physics prize and the chemistry prize, since, while his method was that of a physicist, his results were of prime importance in the domain of chemistry. The Committee found itself in such difficulties that it went so far as to propose that the prize be split, with half awarded in each domain . . . an exemplary case of innovatory interdisciplinarity and of a form of cooperation between two disciplines.[10] In 1907 and 1908 (the year in which P. Ehrlich and E. Metchnikoff shared the Nobel Prize in Physiology or Medicine for their work on immunity), an unusual interplay of reports and counter-reports was observed in the Committee's activities.

In July 1906, K. A. H. Mörner[11] brought all the weight of his authority to bear on the compilation of a sixty-nine-page report on Ehrlich, who had been regularly proposed and retained for consideration since 1901.[12] It was Mörner who, in the following year, revealed the controversy between Arrhenius and Ehrlich in a thirty-two-page document, which gave rise to a nine-page supplementary report (30 August 1907), drawn up by J. E. Johansson, as an addition to the twenty-six-page assessment which C. Sundberg had produced for Ehrlich and Metchnikoff together. In 1908, the year in which the prize was shared by the two immunologists, C. Sundberg drew up two assessments, one devoted to Ehrlich alone (twenty-one pages, 11 August), the other to a group consisting of R. Pfeiffer, J. Bordet, E. Metchnikoff and P. Ehrlich (six pages, 1 September).

The Secretariat of the Committee made a systematic, year-by-year, country-by-country collation of the proposals that had been made since 1901 concerning Ehrlich and Metchnikoff. This was not customary.[13] Following the deliberations of 19 September, the prize was awarded jointly to Ehrlich and

Metchnikoff. A still less customary step was the exchange, after the deliberations, of a series of technical notes between J. E. Johansson – who, from 1908, had ceased to be a member of the Committee, which he rejoined in 1910 – and C. Sundberg, who served on the Committee uninterruptedly from 1901 to 1917, with the Committee itself engaging in the exchange from 15–25 October.[14] The name of Arrhenius was mentioned. The problem in question was not the classical one of priority or contestation, but that of the evaluation of a piece of basic research: What had medicine to do with a unified chemical and stereochemical theory of immunity? Sundberg declared that 'deduktive Forschung in der Medizin' is the precondition for assuring therapeutic effectiveness. Thus, deductive research satisfies the most classical definition of the medical domain, while at the same time recognizing the contiguity of chemistry and medicine and of physics and medicine.

The 'Ehrlich case', reconstructed from the archives of the Nobel Committee for Medicine (which require further translation and discussion), is an exemplary one. Firstly, without these archival documents there would have been no 'Ehrlich case'. For the medical historian, nothing is more impressive than the contribution of the founder of chemical therapy.[15] Within the institution, however, the area of contiguity between chemistry and medicine was considered by some to belong to medicine and by others to chemistry. The Arrhenius-Ehrlich controversy shifted from the Academy of Sciences to the Nobel Committee for Medicine. It was not a 'purely' scientific debate: A Nobel prizewinner in chemistry, whose own position was on the common boundary between chemistry and physics, had interfered – by means of an intermediary, J. E. Johansson – in the affairs of the Nobel Committee for Medicine. . . .

Interdisciplinarity, which is one of the sources of scientific innovation, one of the preconditions for the emergence of new disciplines, is multifaceted by nature. An institution may be innovatory when it agrees to recognize or to embrace a discipline that it has not foreseen;[16] but it may also prove to be a distinct handicap when it transforms a classification of disciplines – provisionally – into canon law, into a full, fixed framework which allows no room for the introduction of a new type of knowledge. In this particular case, the Nobel Committee for Medicine, by granting an award to Ehrlich, had recognized the field of chemical therapy; by setting the seal on the quarrel between Arrhenius and Ehrlich, it had demonstrated its refusal to run the risk of allowing the weight of a controversy within the discipline of chemistry to affect the opening up of a new field in physiology and medicine; and it accorded to the expression 'physiology or medicine' a specific extension of the sense – earlier described as archaic – which it might have had in 1895.

The Dominance of a New Discipline

During the period 1901–1920, seventeen scientists received Nobel prizes for physiology or medicine. Of these seventeen prizes, eight (or forty-seven per-cent) were awarded to 'microbiologists' in the broader sense:

1901 Emile Adolf von Behring, for his work on serum therapy, especially its application against diphtheria
1902 Ronald Ross, for his work on malaria
1905 Robert Koch, for his investigations and discoveries in relation to tuberculosis
1907 Charles Louis Alphonse Laveran, in recognition of his work on the role played by protozoa in causing diseases
1908 Ilya Metchnikoff and Paul Ehrlich, jointly, in recognition of their work on immunity
1913 Charles Robert Richet, in recognition of his work on anaphylaxis

The 1919 prize was awarded in 1920 to Jules Bordet for his discoveries relating to immunity.

This list testifies to the importance of bacteriology, which alone obtained as many prizes as physiology and which represents the posterity of Claude Bernard – as anatomy and cellular physiology represent the posterity of Virchow, and as pure chemistry represents the posterity of von Liebig. In this list one will also find the names of the founders who were still living in 1901 – von Behring, Koch, Laveran – and the names of their successors.

This phenomenon was the result of the gradual emergence of a new discipline, the object of which was the microorganism. From 1860 to 1894, the biological historian discerns different phases, extending from the identification of the object to its role – in the process of fermentation (Pasteur *versus* von Liebig), in the aetiology of certain epizoa (Davaine, Pasteur) and of major human pathologies (Koch, Pasteur, von Behring) – and from the definition of the pathogenic role of the microorganism to its utilization – preventive (anthrax vaccine, 1881) or therapeutic (serum therapy, 1894) – against itself. The beginnings of this late-named discipline[17] were viewed by contemporaries, by part of the scientific community and by the broader public as the promise of a true revolution. This accreditation greatly preceded its actual effects, as is revealed in the quotation from Chekhov which heads this paper. The hygienists – whether doctors or not (civilian, colonial or military) – strongly supported the microbial hypothesis and contributed to its success. Practitioners as a body wisely awaited the statistically significant demonstration in 1894 of the effectiveness of a therapy relating to an extremely common childhood disease with a very high death-rate – diphtheria.[18]

This phenomenon, the establishment of microbiology-bacteriology, should be approached from its international perspective. It can be looked at *geographically*: indeed, exotic pathology, colonial expansion, undertakings such as the digging of the Panama canal or the Simplon tunnel, and commercial imperatives, all have their place in this story, and microbiology makes its appearance as a ubiquitous and influential actor. Here, however, I have looked at the phenomenon *scientifically*: national scientific rivalries, aggravated during the last third of the nineteenth century, now found themselves running up against a structure which, if not neutral,[19] was international in spirit – the Nobel structure.

The quantitative importance attributed to bacteriology from 1901 by the Nobel structure – the Committee and Faculty – through the prize for physiology or medicine was a sign of the international recognition of the place assumed by these new disciplines; but it was only a sign, insofar as the rule of secrecy imposed on the deliberations, past and present alike, moreover allowed for more than suppositions about the mechanisms of proposal, assessment and decision-making, based on isolated and unverifiable indiscretions.[20]

Further to the heavy preponderance (forty-seven percent) of this discipline in the prizes awarded between 1901 and 1920, deriving from the double denomination – physiology (microbial physiology, physiological chemistry?) or medicine (the application of microbiology to medicine, vaccination, immunology and parasitic diseases, or exotic pathology?) – one must also examine the quantitative elements present, in consultations, in the subdivisions into disciplinary groups, and in the assessments, on the basis of the archival material. (See Annex II.)

From the start, two groups out of six covered the field of bacteriology: in 1901, Group V bore the title 'hygiene, aetiology'; in 1902, it became 'bacteriology, hygiene, aetiology', a title which it was to retain. Group VI covered 'immunity'. The number of nominees and nominators in these two groups constantly exceeded the one-third proportion. Only Group II – 'biology, chemistry, physiology, therapeutic means' – encroached upon the pre-eminence of both bacteriology and immunological research, which are anyway involved in work relating to generic terms in biology and chemistry. The Nobel prizes, a new institution, thus moved somewhat more swiftly than the universities towards recognizing the place and the specific nature of these disciplines (particularly Pasteurism), which were founded in the 1880s.

A new element is introduced by examining the lists of personalities from the scientific world who were invited to submit proposals for the prizes in physiology or medicine. The number of 'eminent' scientists approached[21] was indeed large – from 450–900 a year – and there were no serious omissions in the countries included. Japanese science, for instance – and in particular the school of Japanese bacteriologists associated with R. Koch and E. von Behring – was substantially represented: Kitasato, for one, was proposed several times, and H. Noguchi was also considered. American science, occasionally represented during the years 1901–1903, was present in considerable strength as of 1905. Thus, we find ourselves faced with a significant cross-section, although it must be borne in mind that the system lost its meaning as a result of the fact, for instance, that *all* the members of the French Academy of Medicine were invited to put forward proposals (although all did not do so), as were *all* the members of a particular faculty of medicine, even if not on a regular annual basis.

What is immediately striking is the fact that this was a *community* of professors: all the professional details accompanying a name reflected a 'professionalization', either at a university or at one of the large research institutes more or less directly associated with bacteriology and created during the last two decades of the nineteenth century: the Pasteur Institute; The British Institute of Preventive Medicine, London; the Institut für experimental Medizin, St Petersburg; the Institut für Infektionskrankheiten, Berlin; the veterinary schools; the Hygiene Institute of Frankfurt am Main; the Hygiene Institute of Munich; the Rockefeller Institute for Medical Research, New York, etc. Although the medical faculties and the supervisory ministry in France hesitated for a long time before creating a chair of bacteriology or changing the name of a subject of instruction, the records of the Nobel Committee contain an impressive number of professors of bacteriology and hygiene, or of bacteriology alone, from the year 1901.

This was a community of professors which, quite naturally, proposed other professors as possible prizewinners. But this common title – professor – has its roots in a very different reality: on the one hand, there was the tradition of the nineteenth-century clinical school, the axis of which ran from the faculty to the hospital and for which the laboratory was merely an annex. On the other hand, a relatively new phenomenon centred on the laboratory, the object of research – microbe, bacteria, virus, served to bring together people who were not necessarily doctors but who were, to use the terminology of contemporary sociology of science, professionally trained as researchers. In the eyes of the doctors' community, these people were fringe figures,[22] working at the points of intersection between medical, chemical, physical and physiological disciplines. It was no longer the scandal of the experimental physiology of the 1830s that disconcerted some – for the German laboratories had proven themselves – but rather the novelty of the idea that the laboratory was a necessary detour for those aiming at an effective medicine in certain domains. The Nobel Committee and the community of its nominators formed an illuminating microcosm: the institution of the prize was not neutral in substituting the image of the scientist for that of the doctor.

A visible rift now emerged between those who linked teaching, research and discovery – the potential prizewinners – and the professors whose names were put forward purely on account of their teaching reputation. Here one may discern very clearly the beginnings of what has been called, since 1945, biomedical research; here one sees how the successive years in which the prize was awarded, with the cumulative effect of the increased value of the prize itself, the rigour of the selection criteria as seen by the public, and the international acclaim attendant on the awards, have developed in the scientific world a preference for performance over 'beneficial effect', to use the terms of the will.

'Recurrence': the Discoveries and their Rewards

Two questions arise from these remarks, which may have given an impression that the international community of 'doctors' and 'physiologists' of the first twenty years of the Nobel prize has presented an image free from serious conflicts, and that the relations between the Committee and the Karolinska Institute were marked by general consensus (with the exception of the 1901 prize, finally awarded to von Behring, who, as we have seen, had not been selected by the Committee). To what can one attribute this apparent balance, which may seem surprising when compared with the fierce quarrels unleashed by the physics prize (testified to in the archives of the Academy of Sciences, in correspondence and in the publications of the international community of physicists)? Both the theoretical disputes – over the theory of relativity, the quantum theory and their experimental demonstration – and the nationalist conflicts were violent and insoluble and were accompanied by the deferment of prizes. And yet, at times, the international data were identical.

The second question introduces elements of an answer to the first, within a perspective that is not that of the sociology of science – decision-making mechanisms, the influence of national communities, the 'leadership' of certain personalities within the Committee or within the Swedish community as a whole – but that of the history of the disciplines concerned. Why, then, did the 'microbiologists' – in the broader sense – achieve this quantitative pre-eminence?

Their effective pre-eminence was accompanied by reticence, by discussion. After 1902, when the prize was awarded to Ronald Ross, the complexity of the contributions made by some towards a *discovery*, i.e. the solution of a problem – that of malaria, or yellow fever, for example – gave rise to a kind of chain reaction which accounts, among other things, for the awarding of the Nobel prize five years later, in 1907, to C. Laveran.[23]

In their considerations, it was as if the Nobel Committee had to take into account not only the present but also the immediate history of the science: this dual perspective can be perceived in the arguments advanced in support of the proposals, irrespective of the country of origin, in the Committee's discussion of the arguments and in its assessments. Once the bacterial revolution had been triumphantly won, it was both objectivized in its own history and reintegrated into a more general history, that of chemistry (on which it bordered), anatomy and physiology (in particular the central nervous system and the sensory organs).

Here lies the paradox of those first years: the Committee found itself faced with the obligation – one which had been discussed at such length, whatever the discipline, whatever the prizes – of rewarding a discovery made, if not in the very year of the prize, at least extremely recently. At the same time, during the first years of the twentieth century, efficacity became established; recognition was given to the effectiveness of a tremendous revolution of medicine, which, over a period of fifty years, had sprung from the chemical laboratories and moved out, armed with the microscope and the test-tube, backed up by morbidity and mortality statistics, already being used by public health doctors, and over the years had gradually overthrown the notion of aetiology, prevention and therapy in relation to infectious, contagious and parasitic diseases. A one-dimensional medicine, certainly, but one which in the long run could not fail to replace the medical doctrines of the first half of the nineteenth century: Between the cholera epidemic of 1832, when Broussais unsuccessfully treated Casimir Perier by bleeding, and the cholera epidemics of the 1880s, when the teams of Koch and Pasteur were working in the field, in Cairo, and isolated the vibrio, before Haffkine produced a first sort of vaccine in 1891[24] – re-introduced in 1895 by Metchnikoff, Roux and Salimbeni – a certain type of 'medical art' had disappeared. These advances were continued by the medical sciences, with the unqualified sanction of international legislation and national health policies.

Mention should be made of the successive failures of the international health conferences which, from 1852, did not succeed in reaching an agreement; only at the International Health Conference of Vienna in 1897 was a first agreement at last signed, according to which all the countries represented undertook to 'follow the trail of the microbe'. One should note, in France, for instance, the promulgation of the first great public health law (1902), the culmination of lengthy efforts which owed its coherence to the microbial doctrine; or the law of 1905 on alimentary frauds, which was concerned with the bacteriological quality of food in the market system. As a final illustration, and one which reflects in part – but only in part – the terms of Alfred Nobel's will, that the prizes be awarded to those who 'shall have conferred the greatest benefit on mankind', one should recall the health status of the armies engaged in the First World War: antismallpox vaccination, of course, which had been rationally legitimized after a century of use, antityphus and anticholera vaccinations, strict asepsis and antisepsis in field hospitals, etc. No country could have imagined at the time of Solferino that so many millions of men could be kept in action for four years without a major epidemic breaking out.

The pre-eminence of 'microbiology' both tangibly, in the number of prizes awarded, and in the kind of discussions which took place in the Nobel Committee from 1901 to 1920, derived from the fact that the theoretical revolution was over, its period of experimental demonstration (1881–1894) had been concluded, and the resulting transformations of society had become accepted. It was through a movement of recurrence or retrospection that the main body of the international community, from Washington to Tokyo, from Paris to London or Berlin, was able to reach an understanding. The place given in the Committee's discussions to, for instance, the question of malaria,[25] can be understood not only from the history of tropical pathology and parasitology, stemming from the work of Manson and Laveran (Nobel prize 1907), but also from the work of R. Ross (Nobel prize 1902), which was contemporaneous with his prize, and also by the exemplary value of the understanding that an animal vector could transmit a mass-scale disabling disease: malaria and the mosquito, but also plague and the rat (Yersin, 1894), and epidemic typhus and the louse (Nicolle, 1922). Or the importance of immunology, which emerges as an entirely separate discipline within microbiology in the time between the prize of Metchnikoff and Ehrlich in 1908 and the prize attributed to J. Bordet in 1919. The latter prize might equally have been attributed to A. von Wassermann – as far as the Committee's protocol was concerned – had it been possible at the end of the War to recognize the pre-eminence of a German scientist at the same time as that of a Belgian scientist trained in Paris at the Pasteur Institute.

Conclusions

The study of these archives allows the conclusion, provisionally, that the material does not merely reflect the interplay between Swedish, national and international scientific communities; it also makes possible an understanding of the way in which the history of the disciplines was understood, year after year, through an isolated decision which, although it resulted in an award, also weeded out all other potential prizewinners within the discipline, outside the discipline, and in disciplines still being formed. The action of the Nobel Committee for Medicine between 1901 and 1920 can be understood, more than one might have believed, in terms of science at the time of Alfred Nobel, as sanctioning an already accomplished revolution which led to spectacular deepening, implementation and innovation. At the same time as physical theory was introducing uncertainty, microbial theory was bringing in ever-increasing forms of certainty, efficiency, and – it must be said – totalitarianism. It is this period of certainty which seems to me to reflect the pre-eminence of bacteriology in the prize awards: Sir C. Sherrington, proposed from 1902 for his work in neurophysiology, received the prize only in 1932; Sigmund Freud, nominated from 1914 by R. Bárány (Nobel prize 1914, for his work on the physiology and pathology of the vestibular apparatus), was never rewarded; and, during this period, not a single geneticist was considered or even proposed to the Committee.

Notes

1. Cournot, A. A. (1911) *Traité sur l'enchaînement des idées fondamentales dans les sciences et dans l'histoire [1862]*, edited by Levy-Bruhl, p. 235: 'In order to prevent any misunderstanding, it would undoubtedly be best to accept definitively the term *biological sciences*, which is already beginning to gain recognition.'
2. Cf. Canguilhem, G. (1968) *La constitution de la physiologie comme science.* In: *Etudes d'histoire et de philosophie des sciences,* Paris, Vrin, pp. 226–271.
3. Bernard, C. (1856) *Introduction à l'étude de la médecine expérimentale,* Part 3, Chs III and IV.
4. 'We would exhort young scientists not to stray like the sheep of Panurge along a path which would prove fruitless if they were to linger on it indefinitely. Microbes, as a

general system, have seen their day.' *La Lanterne*, 23 July 1887, 'La scie microbienne – les plagiaires de M. Pasteur'.

5. Liljestrand, G. (1962) *The prize in physiology or medicine*. In: Schück, H., ed., *Nobel, the Man and his Prizes*, 2nd ed., pp. 138–139.
6. *Ibid.*, p. 140.
7. Cf. Fruton, this volume.
8. This was the description of Pasteur's laboratory at the Ecole Normale Supérieure in the rue d'Ulm in 1857. As of 1886, E. Duclaux' course at the Faculty of Science of the University of Paris was entitled 'biological chemistry'. Cf. Fruton and Uvnäs, this volume.
9. Liljestrand, G. (note 13), pp. 198–201.
10. Westgren, A. *The Chemistry Prize*, in Schück, H. (note 13), brings out clearly what was at stake from 1903, and what the consequence was of the decision taken by the Academy, following the minority wing of the Chemistry Committee: 'Had the Chemistry Committee had its way and half of each prize been awarded to one scientist, a precedent would have been created which would have rendered the treatment of similar questions later on unnecessarily complicated. *Chemistry also overlaps medicine* . . . and in such cases prizes for chemistry and medicine could scarcely be awarded without joint discussions between the Academy of Sciences and the *Caroline Institute*. . . . It is now generally recognized that the important thing is to decide whether work which can with equal justice be reckoned as chemistry and physics or chemistry and medicine, is in fact worthy of a Nobel prize. If it is, then a prize should be awarded, which prize is of *secondary importance* . . ., and the committees have in fact always tried to collaborate' [my italics], p. 359. See Crawford and Friedman, this volume.
11. K. A. H. Mörner, Professor of Medical Chemistry, was Rector of the Karolinska Institute. He presided over the Nobel Committee without interruption from 1901–1917.
12. In 1901, Ehrlich received one nomination; in 1902, two; in 1904, seven; in 1905, six; in 1906, fifteen; in 1907, fourteen; and in 1908, twenty nominations. The evaluation of 1903 was drawn up by E. Almquist and C. Sundberg; it was twenty-three pages long and stated that experimental confirmation was required.
13. An attempt at quantitative evaluation? I am aware that there is no stable 'ratio' between the number of nominations and the prizes awarded, either in physics and chemistry or in medicine.
14. Source E_1:4, files 20–24.
15. For the elements of this assessment, cf. Witkop, this volume.
16. Cf. Salomon-Bayet, C. (1978) *L'Institution de la science et l'experience du vivant* (method and experiment at the Royal Academy of Sciences 1666–1793), Paris, Flammarion.
17. In Germany, after Ferdinand Cohn (1828–1898), the discipline became known as 'bacteriology'. In France, the word 'microbe' was proposed by C. Sédillot in 1878 after consultation with Littré; 'microbiology' has, as Pasteur stressed, a wider connotation than 'bacteriology'. In Britain, however, as indeed among the French scientific community, both terms were used, in spite of the Pasteurian decree. Cf. *Comptes-rendus de l'Académie des Sciences*, note on the progress of surgery achieved through the works of M. Pasteur, 1878; cf. Roux, E. (1913) Pour la 25ème anniversaire de l'Institut Pasteur. *Rev. Sci.*
18. In France, it took five months – from September 1894 to February 1895 – for the *Concours Médical*, the practitioner's review, to shift from its wait-and-see policy towards the antidiphtheria serotherapy proposed by Roux at the International Hygiene Congress of Budapest, to acceptance of the compulsory use by the practitioner both of biological sampling and of serotherapy in diagnosis. Cf. *La pastorisation de la médecine*, an essay on the first biomedical revolution, in a forthcoming collective work.
19. Cf. Crawford, E. (1980) *Swedish Physics and Chemistry in International Perspective: Consequences for the Discussion and Selection of Nobel Prize Winners, 1901–1914*, Oslo, May.
20. Cf. Annex I.
21. The term 'eminent' is used in the sense attributed to it by Alphonse de Candolle in one

of the first texts on the sociology of science: the recognition, within a specific institutional network (academy, university, learned society of the highest level) of the quality either of innovators (those who make science) or of initiators (those who transmit science): de Candolle, A. (1873) *Histoire des sciences et des savants depuis deux siècles*, Geneva, Basel.

22. 'Monsieur Pasteur, who is not even a doctor!' protested Professor Peter to the Academy of Medicine, during the controversy over the treatment of rabies (1886).
23. The Nobel prize of 1902 was awarded to R. Ross for his work on malaria, and the prize of 1907 to C. Laveran for his work on the role played by protozoa in causing diseases. The difference in designation masks but does not conceal the fact that the same scourge – malaria – was being treated from two different approaches: the causal agent (Laveran) and the vector (Ross).
24. Cf. Piquemal, J. (1959) Le choléra de 1832 en France et la pensée médicale. In: *Thalès*, Paris, Presses Universitaires de France, pp. 27–73.
25. The question of malaria, although not a specifically Swedish problem, was of active concern to the Committee between 1901 and 1907, as Mörner pointed out in 1902 in his address at the award-giving ceremony (*Les Prix Nobel en 1902*, Stockholm, Norstedts, 1905). Between those two dates, 113 pages of assessment were written on the problem and on the scientists who had given it their attention. The details are as follows:

1901:	Evaluation of the question of malaria (Laveran, Ross, Manson, Golgi), by E. Almquist	– 12 pp.
1902:	The question of malaria, by E. Almquist and C. Sundberg	– 7 pp.
	Ross and the question of malaria, by E. Almquist and C. Sundberg	– 15 pp.
1903:	–	
1904:	–	
1905:	Evaluation of Laveran, Castellani and Koch on trypanosomiases, by C. Sundberg. (Laveran, 12 pp.; Castellani, 4 pp.; Koch, 4 pp.)	– 22 pp.
1906:	On Laveran and the protozoa, by C. Sundberg	– 24 pp. + 1 table
1907:	Laveran *et al.*, by J. G. Edgren	– 23 pp.
	On Laveran, by C. Sundberg	– 8 pp.

By way of comparison, the report by Almquist on serum therapy – von Behring, Kitasato, Koch, Roux – covers three pages; C. Sundberg, in one-half a page, presents the mortality statistics for cases of diphtheria treated by serum therapy in different hospitals (5 pp). No evaluation is given by von Behring, who was named on 30 October 1901 as winner of the prize.

Annex I

There remains a great deal of work to be done to restitute the real history of microbiology, so long dominated in France by the figure of Pasteur.[1] Pasteur is most often dealt with in a hagiographic manner – the axis of light, Lister-Pasteur, having for symmetry the counter-axis of darkness, Pasteur-Koch. One must, however, proceed with caution.

The history of Pasteurism, of the quarrels over priority of discovery and over borrowing, of the disagreements between Parisian and provincial laboratories, mentions the prizeworthy quality of a certain number of Pasteur's scientific rivals. 'Prizeworthy' may be taken to mean *proposed* to the Nobel Committee and, possibly, *assessed*. Recognition by the international community, at the same time, of the exceptional scientific quality of a thinker is, then, an argument of some weight for opening or reopening a file, revising a chronology, or modifying attributions and discoveries.

Thus, two figures in the early stages of microbiology, one from Lyon, P. V. Galtier (1846–1908), the other from Montpellier and later Lille, P. Béchamp (1806–1908), have attracted the attention of historians of science,[2] who credit them with having been proposed for the Nobel prize of 1908 – a proposal without effect, as both died that very year. From 1879, Galtier was working on rabies, and it was after a visit that Pasteur made to the laboratory of Chauveau in Lyon that he embarked on his own work; Béchamp, who had been Pasteur's locum in Strasbourg – and was, like him, a pupil of J. B. Dumas[3] – may have recognized more quickly than Pasteur the parasitic nature of a disease of silkworms and the preventive role of cocoon selection in silkworm farms. History, as it is written classically, eliminates a whole part of the real history; this is to be found not only in the *Compte-rendus* of the Academy of Sciences, which are monopolized by a few contributors, but much more in regional publications such as the *Bulletin de la Société centrale d'agriculture de l'Hérault*.

This desire to restitute the real history, to the detriment of Pasteur, has the advantage of bringing into the open the size of a scientific community that is working on identical problems by similar methods. Whenever hagiographies focus their interest on the lone figure, history replies in terms of continuity, of convergent and complementary work, and of collective work.

There was, then, a double reason for attempting to gain access to the archives of the Nobel Committee for Physiology and Medicine: the first, more restricted, being to verify the statements about Galtier and Béchamp; the second, more systematic, to gain an impression of how they were actually viewed at the time by the international scientific community through the hierarchies established between the disciplines and between individuals.

Two indispensable sources were used in researching the first question: (1) the files, established year by year, containing the originals of the proposals made by the invited nominators, as well as proposals that had not been considered, either because they were submitted by a person who had not been officially invited or because they had been made in favour of the nominator himself; and (2) the bound volumes containing the minutes of the Nobel meetings, the list of nominations, the letters of proposal, the Committee's short list for evaluation, and the final evaluations and deliberations. From these documents it emerges that Béchamp was never proposed to the Nobel Committee. As for Galtier, no mention was to be found in the minutes of the meetings of the Nobel Committee for the years 1901–1908 in any of the groups under scrutiny, not even in Group VII, which included late proposals (reaching Stockholm after the deadline of 1 February of the year of the award) to be reserved for the following year. Nevertheless, in Group VIII (file IX), which included all the submissions without proposals (*Försändelser utan förslag*), under number 28, there is a record of a note by Galtier, submitted by M. Bouley, sent by the Paris Academy of Sciences (meeting of 1 August 1881), which was received on 18 November 1907, to the effect that 'injections of rabies virus into the circulatory system do not cause rabies and seem to provide immunity. Rabies may be transmitted by the ingestion of rabid matter.'

A check on the folder of originals (E1:4) made it possible to trace the file, classified under the heading 'submission of no value'. The file was fairly extensive, including an original extract from Galtier's experimental notes, 1879–1880, and six printed documents, including (a) three from the reports of the Academy of Sciences (1879, 1881 and 1888) justifying the use of rabbits for research into the toxicity of rabid liquid and into a means of prevention 'one or two days after the bite'; (b) one from the volume published by Galtier in 1886, 2nd edition, *La Rage envisagée chez les animaux et chez l'homme au point de vue de ses caractères et de sa prophylaxie*; and (c) an article from the *Progrès de Lyon* of 3 December 1906, under the title 'Un précurseur lyonnais de Pasteur: la rage guérit la rage. Dans le doute, inocule-toi. Omelettes antirabiques' (A precursor of Pasteur from Lyon: rabies cures rabies. If in doubt, get inoculated. Antirabies omelettes).[4] The article recorded the homage rendered to Galtier by the Commission on Rabies of the Academy of Sciences in 1887.[5]

This was not therefore a proposal but an old quarrel over an old file, revived from time to time in Lyon and submitted to the Nobel Committee, which was considered to be an acceptable jury. For the historian, this represents the revelation of two fertile 'ideas' – the use of the rabbit as experimental material, and immunity after the bite – which were put forward by the bacteriologist from Lyon but were not theoretically and technically complete. Here again one is faced with the classic confusion between a 'prognostication' – which, to be true, must presuppose an identity between theoretical context and technical approach – and the 'chain' which ends in real discovery and which – over a period sometimes lengthy, sometimes very short – serves to link different laboratories and different scientists, whether competitors or collaborators. The question asked has therefore been answered.

Notes to Annex I

1. In this attempt at restitution, mention should be made of Dubos' fundamental work: Dubos, R. (1950) *Pasteur, Freelance of Science,* Boston; French edition (1955) *Pasteur, franc-tireur de la science,* Paris, Presses Universitaires de France; and Dagognet, F. (1957) *Méthodes et doctrine dans l'oeuvre de Pasteur,* Paris, Presses Universitaires de France.
2. Théodorides, J. (1971) In: *P. V. Galtier, un précurseur méconnu de Pasteur. Congrès International d'Histoire des Sciences, Moscow, Arch. Int. Claude Bernard,* 2. Decourt, P. (1972) *Sur une histoire peu connue, la découverte des maladies microbiennes. Ibid.* 1.
3. The Archives of the Academy of Sciences of Paris contain correspondence between P. Béchamp and J. B. Dumas from 1863–1871 (in the files of J. B. Dumas), which has been studied but remains ambiguous.

4. Empirical recipes of the peasants of the Dauphiné region.
5. Presided over by C. Bouchard, the Commission comprised Marey, Richet, Charcot, Brown-Séquard and Verneuil.

Annex II

The following tables were drawn up on the basis of data from the archives of the Nobel Committee for Physiology and Medicine, Karolinska Institute, Stockholm.

Data used were as follows:

(1) Registers put together each year by the secretariat of the Committee, using the names of nominees, in alphabetical order and by group, the names of the nominators, those candidates singled out for evaluation, the reviewers and the texts of the evaluation, and, finally, the Committee's recommendation to the faculty. These registers contain both handwritten and typewritten sections. Spot checks were made on the originals to verify that the transcription was faithful. Archives nos E_1:1,2,3, etc.

(2) The files of original letters of nomination.

(3) The minutes of the Nobel Committee, 1900–1917 – a bound, handwritten document. Archive no A_3:1.

(4) Bundles, put together by year in a hard cover, for the years 1900–1920 – minutes of the meetings of the professorial staff of the Karolinska Institute held for the awarding of the prizes.

From 1901, the nominations are divided into six groups. Groups V and VI cover the field of our investigation, having as titles 'Hygiene, Bacteriology, Aetiology' and 'Immunity', respectively; the title of Group V was the same from 1903 when 'Bacteriology' was added to 'Hygiene and Aetiology'. In order to limit the size of these tables, I have not listed the names in the category 'Nominations Other Groups' (I–IV) and have simply enumerated them; however, I have mentioned those scientists in Groups I–IV who were evaluated. The names underlined are those of Nobel prizewinners, in the relevant year, in preceding years and in future years; for example, C. S. Sherrington, nominated as of 1902, was evaluated for the first time in 1910 and received the prize in 1932; and R. Koch, proposed in 1906, the year after he received the prize: their names are underlined, no matter what the year.

Prizewinner	Nominations Groups V–VI	Nominations Other Groups	Evaluations Groups V–VI	Evaluations Other Groups
		1901		
Emil von BEHRING	16 nominees	26 nominees	7 nominees	8 nominees
Prizewinners suggested by Committee: R. Ross N. R. Finsen	**Group V** G. Bastianelli A. Bignami C. Golgi G.-B. Grassi R. Koch A. Laveran P. Manson R. Ross **Group VI** E. von Behring P. Ehrlich J. von Fodor A. Högyes S. Kitasato E. Metchnikoff E. Roux A. Yersin		C. Golgi S. Kitasato R. Koch A. Laveran P. Manson R. Ross E. Roux	N. Finsen C. Golgi E. Lang J. Langley J. Loeb E. Müller J. Pavlov S. Ramon y Cajal
	Number of nominators: 37	Number of nominators 74		
		1902		
Ronald ROSS	16 nominees	28 nominees	8 nominees	6 nominees
Prizewinner suggested by Committee: R. Ross	**Group V** V. Balthazard A. Desgrez G.-B. Grassi F. Hüppe R. Koch A. Laveran E. Löffler R. Ross E. Roux **Group VI** J. Bordet P. Ehrlich M. Grüber E. Metchnikoff R. Pfeiffer C. Phisalix F. Widal		J. Bordet P. Ehrlich M. Grüber E. Metchnikoff R. Pfeiffer C. Phisalix R. Ross F. Widal	E. Buchner (chemistry) N. Finsen C. Golgi O. Hertwig O. Overton J. Pavlov
	Number of nominators: 31	Number of nominators: 32		

Prizewinner	Nominations Groups V–VI	Nominations Other Groups	Evaluations Groups V–VI	Evaluations Other Groups
		1903		
Niels R. FINSEN	14 nominees	18 nominees	3 nominees	8 nominees
Prizewinner suggested by Committee: N. R. Finsen	**Group V** A. Agramonte B. Bang J. Carrol F. Durham G.-B. Grassi M. Grüber G. Hansen R. Koch A. Laveran W. Reed **Group VI** P. Ehrlich E. Metchnikoff E. Roux L. Willems		P. Ehrlich G. Hansen R. Koch	F. Dellweiler N. Finsen C. Golgi E. Heman A. Kossel E. Marey J. Pavlov M. Verworn
	Number of nominators: 26	Number of nominators: 33		
		1904		
Ivan Petrovic PAVLOV	13 nominees	27 nominees	4 nominees	7 nominees
Prizewinner suggested by Committee: I. P. Pavlov	**Group V** A. Celli C. Golgi G.-B. Grassi R. Koch A. Laveran P. Manson R. Ross E. Roux A. Yersin **Group VI** P. Ehrlich J. Héricourt E. Metchnikoff C. Richet E. Roux		P. Ehrlich R. Koch E. Metchnikoff A. Yersin	S. Apathy C. Bouchard E. Buchner (chemistry) C. Golgi A. Kossel I. Pavlov S. Ramon y Cajal
	Number of nominators: 49	Number of nominators: 53		

Prizewinner	Nominations Groups V–VI	Nominations Other Groups	Evaluations Groups V–VI	Evaluations Other Groups
		1905		
Robert KOCH	9 nominees	22 nominees	5 nominees	6 nominees
	Group V		H. Carter	C. Achard
	H. Carter		A. Castellani	C. Golgi
	A. Castellani		C. Finlay	J. Langley
Prizewinner suggested by Committee: R. Koch	C. Finlay		R. Koch	S. Ramon y Cajal
	R. Koch		A. Laveran	J. Vassale
	A. Laveran			F. Widal
	E. Roux			
	Group VI			
	P. Ehrlich			
	E. Metchnikoff			
	W. Reed			
	Number of nominators: 48	Number of nominators: 49		
		1906		
Camillo GOLGI and Santiago RAMON Y CAJAL	13 nominees	26 nominees	4 nominees	5 nominees
	Group V		H. Carter	A. Bier
	J. Carrol		P. Ehrlich	C. Golgi
	H. Carter		C. Finlay	J. Loeb
	C. Finlay		A. Laveran	O. Overton
	R. Koch			S. Ramon y Cajal
	A. Laveran			
	J. Lazzar			
Prizewinners suggested by Committee: C. Golgi and S. Ramon y Cajal	E. Metchnikoff			
	A. Neisser			
	W. Reed			
	E. Roux			
	F. Schaudin			
	Group VI			
	P. Ehrlich			
	E. Metchnikoff			
	Number of nominators: 28	Number of nominators: 69		

Prizewinner	Nominations Groups V–VI	Nominations Other Groups	Evaluations Groups V–VI	Evaluations Other Groups
		1907		
Alphonse LAVERAN Prizewinner suggested by Committee: A. Laveran	20 nominees **Group V** Sir D. Bruce J. Carrol H. Carter A. Castellani A. Celli C. Finlay G.-B. Grassi M. Grüber E. Hoffmann A. Laveran E. Metchnikoff L. Rogers E. Roux **Group VI** A. Calmette P. Ehrlich W. Haffkine E. Metchnikoff S. Ramon y Cajal L. Willems Sir A. Wright Number of nominators: 43	21 nominees Number of nominators: 24	8 nominees Sir D. Bruce H. Carter A. Castellani P. Ehrlich C. Finlay A. Laveran E. Metchnikoff L. Rogers	3 nominees S. Arrhenius (chemistry) C. Bohr H. de Vries
		1908		
Elie METCHNIKOFF and Paul EHRLICH Prizewinner suggested by Committee: E. T. Kocher	14 nominees **Group V** A. Castellani G. Hansen A. Laveran A. Neisser T. Smith A. Yersin **Group VI** J. Bordet Sir D. Bruce P. Ehrlich W. Haffkine E. Metchnikoff R. Pfeiffer E. Roux K. Schleich Number of nominators: 56	20 nominees Number of nominators: 39	7 nominees J. Bordet Sir D. Bruce A. Castellani P. Ehrlich E. Metchnikoff R. Pfeiffer T. Smith	5 nominees C. Bohr A. Chauveau E. Fischer (chemistry) J. Langley E. Pflüger

Prizewinner	Nominations Groups V–VI	Nominations Other Groups	Evaluations Groups V–VI	Evaluations Other Groups
		1909		
Emil Theodor KOCHER Prizewinner suggested by Committee: E. T. Kocher	16 nominees **Group V** E. Bosc C. Bouchard Sir D. Bruce A. Castellani A. Chauveau C. Flügge W. Gorgas E. Löffler E. Metchnikoff E. Roux F. Schaudin T. Smith **Group VI** J. Bordet P. Ehrlich E. Metchnikoff R. Pfeiffer Number of nominators: 42	33 nominees Number of nominators: 53	3 nominees J. Bordet R. Pfeiffer T. Smith	6 nominees A. Bier E. Fischer (chemistry) Sir V. Horsley E. Kocher J. Loeb H. Quincke
		1910		
Albrecht KOSSEL Prizewinner suggested by Committee: A. Kossel	15 nominees **Group V** A. Castellani E. Eberth M. Grüber E. Hoffmann P. Manson T. Smith **Group VI** J. Bordet S. Flexner O. Gengou A. Neisser R. Pfeiffer E. Roux P. Ulhenhut A. von Wassermann Sir A. Wright Number of nominators: 47	37 nominees Number of nominators: 95	2 nominees T. Smith A. von Wassermann	6 nominees E. Fischer (chemistry Sir V. Horsley A. Kossel G. Retzius M. Rubner Sir C. Sherrington

Prizewinner	Nominations Groups V–VI	Nominations Other Groups	Evaluations Groups V–VI	Evaluations Other Groups
		1911		
Allvar GULLSTRAND	15 nominees	26 nominees	3 nominees	2 nominees
Prizewinner suggested by Committee: A. Gullstrand	**Group V** S. Arloing Sir D. Bruce A. Castellani G.-B. Grassi G. Hansen Sir P. Manson S. Winogradsky **Group VI** J. Bordet A. Calmette J. Ferran y Clua A. Neisser R. Pfeiffer E. Roux P. Ulhenhut A. von Wassermann		J. Bordet A. Calmette A. von Wassermann	A. Gullstrand Sir V. Horsley
	Number of nominators: 35	Number of nominators: 49		
		1912		
Alexis CARREL	15 nominees	39 nominees	1 nominee	7 nominees
Prizewinner suggested by Committee: A. Carrel	**Group V** A. Agramonte Sir D. Bruce A. Castellani A. Celli C. Finlay G.-B. Grassi Sir P. Manson T. Smith **Group VI** S. Arrhenius (chemistry) J. Bordet J. Ferran y Clua R. Pfeiffer C. Richet E. Roux Sir A. Wright		C. Richet	W. Bechterev A. Carrel J. Langley O. Loew C. Menakow J. Murphy Sir C. Sherrington
	Number of nominators: 42	Number of nominators: 64		

Prizewinner	Nominations Groups V–VI	Nominations Other Groups	Evaluations Groups V–VI	Evaluations Other Groups
		1913		
Charles Robert RICHET	16 nominees	49 nominees	1 nominee	10 nominees
	Group V		C. Richet	E. Abderhalden
	A. Agramonte			R. Bárány
	L. de Beurmann			W. Bayliss
Prizewinner	Sir D. Bruce			W. Einthoven
suggested by	A. Castellani			C. Forlanini
Committee:	C. Chagas			Sir V. Horsley
C. R. Richet	C. Finlay			K. Koller
	G.-B. Grassi			E. Schäfer
	F. Löffler			K. Schleich
	R. Pfeiffer			E. Starling
	T. Smith			
	Group VI			
	J. Bordet			
	J. Ferran y Clua			
	C. Richet			
	E. Roux			
	H. Vincent			
	A. von Wassermann			
	Number of nominators: 32	Number of nominators: 87		
		1914		
Robert BÁRÁNY	20 nominees	43 nominees	2 nominees	9 nominees
	Group V		S. Flexner	E. Abderhalden
	A. Agramonte		H. Noguchi	R. Bárány
	Sir D. Bruce			W. Bayliss
Prizewinners	A. Castellani			W. Einthoven
suggested by	C. Finlay			C. Forlanini
Committee:	S. Flexner			R. Harrison
A. Abderhalden	C. Flügge			J. Langley
and	W. Gorgas			Sir C. Sherrington
J. N. Langley	G.-B. Grassi			E. Starling
	H. Noguchi			
	E. Perroncito			
	T. Smith			
	Group VI			
	S. Arrhenius (chemistry)			
	J. Bordet			
	E. Maragliano			
	R. Pfeiffer			
	C. Pirquet			
	E. Roux			
	P. Uhlenhut			
	A. von Wassermann			
	Sir A. Wright			
	Number of nominators: 43	Number of nominators: 82		

Prizewinner	Nominations Groups V–VI	Nominations Other Groups	Evaluations Groups V–VI	Evaluations Other Groups
		1915		
Reserved	16 nominees	15 nominees	4 nominees	5 nominees
	Group V		J. Bordet	E. Abderhalden
	A. Agramonte		S. Flexner	L. Bolk
	B. Ashford		H. Noguchi	E. Fischer
	Sir D. Bruce		R. Pfeiffer	(chemistry)
	C. Finlay			Sir C. Sherrington
	S. Flexner			S. Sørensen
	W. Gorgas			
	J. Löffler			
	H. Noguchi			
	Sir A. Wright			
	Group VI			
	A. Abderhalden			
	J. Bordet			
	J. Ferran y Clua			
	R. Pfeiffer			
	F. Russel			
	P. Uhlenhut			
	A. von Wassermann			
	Number of nominators: 28	Number of nominators: 22		
		1916		
Reserved	9 nominees	18 nominees	1 nominee	6 nominees
	Group V		S. Kartulis	H. Hamburger
	J. Goldberger			G. Retzius
	S. Kartulis			W. Roux
	A. Neisser			W. Wundt
	T. Smith			G. Zander (twice)
	W. St Clair Symmers			
	Group VI			
	J. Bordet			
	R. Pfeiffer			
	E. Roux			
	A. von Wassermann			
	Number of nominators: 34	Number of nominators: 14		
		1917		
Reserved	6 nominees	27 nominees	1 nominee	6 nominees
	Group V		J. Bordet	C. Eijkman
	A. Agramonte			W. Einthoven
	F. Billings			H. Gutzman
	Sir D. Bruce			R. Harrison
	S. Flexner			B. Krönig
	T. Smith			J. Loeb
	Group VI			
	J. Bordet			
	Number of nominators: 36	Number of nominators: 27		

Prizewinner	Nominations Groups V–VI	Nominations Other Groups	Evaluations Groups V–VI	Evaluations Other Groups
		1918		
Reserved	5 nominees **Group V** Sir D. Bruce H. Vincent F. Widal Sir A. Wright **Group VI** J. Bordet Number of nominators: 10	19 nominees Number of nominators: 30	2 nominees J. Bordet F. Widal	4 nominees H. Quincke W. Roux Sir C. Sherrington (twice)
		1919		
Jules BORDET (1920) Prizewinner suggested by Committee: A. Krogh	13 nominees **Group V** Sir D. Bruce C. Eijkman W. Gorgas Y. Ido R. Inada C. Nicolle R. Pfeiffer E. Roux A. Trillat Sir A. Wright **Group VI** J. Bordet A. Calmette A. von Wassermann Number of nominators: 24	33 nominees Number of nominators: 42	1 nominee C. Eijkman	8 nominees J. Bancroft H. Head S. Henschen A. Krogh C. Neuberg F. Sauerbruch S. Sørensen H. Zwaardemakers
		1920		
Schack August Steenberger KROGH Prizewinner suggested by Committee: C. Neuberg	15 nominees **Group V** A. Besredka A. Castellani R. Cole C. Eijkman S. Flexner C. Flügge R. MacHarrison C. Nicolle H. Noguchi H. Vincent F. Widal **Group VI** J. Bordet A. Calmette J. Ferran y Clua A. von Wassermann Number of nominators: 29	43 nominees Number of nominators: 45	3 nominees J. Bordet R. MacHarrison C. Nicolle	9 nominees J. Haldane H. Hamburger H. Head S. Henschen W. Johannsen A. Krogh J. Mackenzie E. Müller C. Neuberg

DISCUSSION

ARMIN HERMANN

Lehrstuhl für Geschichte der Naturwissenschaften und Technik
Universität Stuttgart, Stuttgart, Federal Republic of Germany

In these comments I have tried to combine the four papers. They really fit nicely together; what may be missing from one can be found in another. These papers have convinced me that from now on history of science and biographies of scientists can no longer be written without making extensive use of the archival material at Stockholm.

I will start with some general remarks. The Nobel prize rapidly reached its position as the most prestigious award in science; its winners gain a high reputation within the scientific community as well as with the public. Elisabeth Crawford spoke at length about the prizes for the field of physical chemistry; Hiebert told us how important the prizes were for the growth of this discipline.

When Adolf Harnack wrote his famous memoire for Emperor William II in 1909, which led to the foundation of the Kaiser-Wilhelm-Gesellschaft, he called Philipp Lenard the 'most competent scholar' (*den berufensten Gelehrten*) in the field of physics. At that time, there were only two prizewinners in physics of German nationality, and the other, Röntgen, was much older, very shy and had withdrawn altogether from public life. The status of the Nobel prize made it possible for Harnack to refer to only one of the group of about fifty chairholders in physics and to regard Lenard's proposals and ideas as justified, without finding it necessary to ask the others. This example illustrates that the internal reputation of the prize developed within a few years; and, undoubtedly, public opinion followed immediately. Nevertheless, it would be of interest to know more about the development of this reputation in different countries. I suspect that the counting of Nobel prizes developed for reasons of national prestige; it was regarded as a symbol of national superiority, like having the most battleships, the fastest cars, the most Olympic gold medals.

The reason for the high internal reputation of the prize is to be found in the fact that the prizewinners are indeed regarded within the community as the most successful scientists and discoverers. If, today, some historian of science tried to make a list of the names of the best physicists and chemists of our century, that list would correspond almost exactly to that of Nobel prizewinners in those fields, in contrast to the situation with the prizes for literature. The list of prizewinners in physics omits only a few names, for example, that of Arnold Sommerfeld; and only in a very few cases would there be reluctance today in giving them the award.

The decisions of the Academy are thus generally well accepted within the scientific community. It would be very interesting to collect comments from that community on the decisions: This might be done by consulting large collections of letters, for example, at the Niels Bohr Institute of Copenhagen, at the American Institute of Physics in New York or in the Office for the History of Science and Technology at Berkeley. Such studies could supplement the work being done with the Nobel archives.

Since the prize has such a high reputation, the decisions of the Royal Swedish Academy of Sciences played, and still play, an important role in the sciences and in science policy throughout the world. As already mentioned, the prize-winners are in general regarded by governmental institutions and by the scientific community as the leading scientists. An embarrassing case is that of Lenard and Stark, who could exercise their evil and destructive influence as deadly enemies of relativity and quantum theory, which they called 'Jewish physics', only because of their prestige gained by winning the Nobel prize.

Some general remarks on the nomination process: A definite procedure was worked out by the Academy, as we learned from these papers, to find the laureates of the year. The first half of this procedure was the nomination process. There were two groups of nominators, those selected annually and the permanent ones. How did the annual selection work? The question is important as regards national affiliation and the different fields. Chemistry is separated into organic, inorganic and physical chemistry, and physics into experimental and theoretical physics. In physics, most chairholders were experimentalists, and the theoretical physicists usually held only associate professorships. This is true at least for the period before the First World War and for the German-speaking countries, where this field was most developed. How many theoretical physicists were there in the group of nominators? I would expect a very low representation before 1920, and then a rapid increase.

Küppers *et al.* discussed at some length the degree of consensus in the nomination of candidates. There are some important conditions to be satisfied before one can really speak of a measurement of consensus, i.e. (1) the absolute numbers must be sufficiently high, and (2) the nominators must really be free in their nominations. On the one hand, there may be previously existing agreements among subgroups (especially national ones); this would lead to a greater 'consensus' than really was the case. On the other hand, a (bad) practice may have developed of nominating candidates for reasons of courtesy, since even nomination represents an honour. This would lead to a lower degree of 'consensus' than really existed.

The awards to Planck and Einstein, which have been studied in some detail, are apparently among the most interesting cases. The awarding of the Nobel prize to these two scientists is not a reflection of the quality of their work, but a test of the nominators, of the Nobel committee and of the Academy itself.

The first problem facing those scientific bodies was the revolutionary character of Einstein's and Planck's work; the second was that their work was on theoretical physics. It was very interesting to learn from the Crawford-Friedman paper about the influence of theoretical physicists on the Committee. Generally, theoretical physics had been neglected by the experimentalists, until this new discipline made its way and assumed a leading role in science.

Planck was nominated first in 1907, which is quite early, and we learned from Bengt Nagel that in the following year he came quite close to winning the prize: It is very, very interesting to learn that Planck nearly made it but failed at the last minute. He nearly made it because the revolutionary character of his work on black-body radiation was *not* known. In 1908, only Einstein and a few others understood what was really behind Planck's theory. As this became more and more generally known, Planck's chances went down. They rose again only after a few years, when the quantum idea became accepted by an increasing number of physicists. So, as Bengt Nagel states in his paper, for the history of quantum theory the Nobel material is of the greatest value.

Turning to Einstein, I want to thank Crawford and Friedman for the valuable information in their paper. It is remarkable that the first nomination was made for 1910. The special theory of relativity had been recognized in 1908, when Minkowski gave his famous lecture in Cologne, and it was known by a small but leading group of theoretical physicists and mathematicians, including Planck, Sommerfeld and Felix Klein. It is interesting that Ostwald, the physical chemist, was the first to nominate Einstein. Ostwald was one of the strongest supporters of positivism, and I am sure that he considered that the special theory of relativity was based on positivistic principles.

Physicists have always regarded it as peculiar that it was Einstein's paper in which he predicted the photo-electric law that was especially chosen by the Academy as the basis for awarding the prize. The relation between the frequency of the light and the energy of electrons, predicted by Einstein in 1905, was confirmed by Millikan's measurements; and experimental proof of an idea or a theory was always — or at least for a long time — essential for the Academy.

I will conclude my remarks with two quotations from Pauli, commenting on the status of theoretical physics in Sweden and the decisions of the Academy. In a letter written in 1930, in which he congratulates Oskar Klein for having obtained a chair of theoretical physics in Stockholm, Pauli wrote: 'Until now there was in Sweden almost no theoretical physics — an injustice especially to experimental physics, which is represented so brilliantly by Siegbahn and Hulthén.' When Heisenberg got the prize in 1933 for his ideas which led to so-called matrix mechanics, Pauli congratulates him: 'Comparison with previous lines of argument (especially Einstein's) and perusal of the statutes of the Nobel Foundation lead me to conclude that you have won the prize for your celebrated, and to date unrefuted, doctoral dissertation on hydromechanics. That does, after all, involve the most direct links with physical experiments on which the Nobel Foundation sets such great value.'

APPENDIX I
LIST OF NOBEL LAUREATES, 1901-1930

Physics

1901 – W. C. Röntgen
1902 – H. A. Lorentz and P. Zeeman
1903 – H. Becquerel, P. Curie and M. Curie
1904 – Lord Rayleigh
1905 – P. Lenard
1906 – J. J. Thomson
1907 – A. A. Michelson
1908 – G. Lippmann
1909 – G. Marconi and C. F. Braun
1910 – J. D. van der Waals
1911 – W. Wien
1912 – N. G. Dalén
1913 – H. Kamerlingh-Onnes
1914 – Reserved
1915 – Prize for 1914: M. von Laue
– Prize for 1915: Sir W. H. Bragg and Sir W. L. Bragg
1916 – Reserved
1917 – Reserved
1918 – Prize for 1917: C. G. Barkla
– Prize for 1918: Reserved
1919 – Prize for 1918: M. Planck
– Prize for 1919: J. Stark
1920 – C. E. Guillaume
1921 – Reserved
1922 – Prize for 1921: A. Einstein
– Prize for 1922: N. Bohr
1923 – R. A. Millikan
1924 – Reserved
1925 – Prize for 1924: M. Siegbahn
– Prize for 1925: Reserved
1926 – Prize for 1925: J. Franck and G. Hertz
– Prize for 1926: J. B. Perrin
1927 – A. H. Compton and C. T. R. Rees
1928 – Reserved
1929 – Prize for 1928: O. W. Richardson
– Prize for 1929: Prince L.-V. de Broglie
1930 – Sir C. V. Raman

Chemistry

1901 – J. H. van't Hoff
1902 – E. Fischer
1903 – S. A. Arrhenius
1904 – Sir W. Ramsay
1905 – A. von Baeyer
1906 – H. Moissan
1907 – E. Buchner
1908 – E. Rutherford
1909 – W. Ostwald
1910 – O. Wallach
1911 – M. Curie
1912 – V. Grignard and P. Sabatier
1913 – A. Werner
1914 – Reserved
1915 – Prize for 1914: T. W. Richards
– Prize for 1915: R. M. Willstätter
1916 – Reserved
1917 – Reserved
1918 – Reserved
1919 – Prize for 1918: F. Haber
– Prize for 1919: Reserved
1920 – Reserved
1921 – Prize for 1920: W. Nernst
– Prize for 1921: Reserved
1922 – Prize for 1921: F. Soddy
– Prize for 1922: F. W. Aston
1923 – F. Pregl
1924 – Reserved
1925 – Reserved
1926 – Prize for 1925: R. A. Zsigmondy
– Prize for 1926: T. Svedberg
1927 – Reserved
1928 – Prize for 1927: H. O. Wieland
– Prize for 1928: A. O. R. Windaus
1929 – A. Harden and H. von Euler-Chelpin
1930 – H. Fischer

Physiology or Medicine

1901 – E. von Behring
1902 – Sir R. Ross
1903 – N. R. Finsen
1904 – I. P. Pavlov
1905 – R. Koch
1906 – C. Golgi and S. Ramon y Cajal
1907 – A. Laveran

1908 – I. I. Mečnikov and P. Ehrlich
1909 – E. T. Kocher
1910 – A. Kossel
1911 – A. Gullstrand
1912 – A. Carrel
1913 – C. Richet
1914 – R. Bárány
1915 – Reserved
1916 – Reserved
1917 – Reserved
1918 – Reserved
1919 – Reserved
1920 – Prize for 1919: J. Bordet
– Prize for 1920: A. Krogh
1921 – Reserved
1922 – Reserved
1923 – Prize for 1922: A. V. Hill and O. Meyerhof
1924 – W. Einthoven
1925 – Reserved
1926 – Reserved
1927 – Prize for 1926: J. A. G. Fibiger
– Prize for 1927: J. Wagner-Jauregg
1928 – C. Nicolle
1929 – C. Eijkman and Sir F. G. Hopkins
1930 – K. Landsteiner

APPENDIX II
MEMBERS OF THE NOBEL COMMITTEES FOR PHYSICS, CHEMISTRY AND PHYSIOLOGY OR MEDICINE ELECTED 1900-1930

Nobel Committee for Physics

K. Ångström 1900–1910	Professor of Physics at Uppsala University
S. Arrhenius 1900–1927	Professor of Physics at the Stockholm Högskola; subsequently, Director, Nobel Institute of Physical Chemistry
V. Carlheim-Gyllensköld 1910–1934	Professor of Physics at the Stockholm Högskola
P. G. D. Granqvist 1904–1922	Professor of Physics at Uppsala University
A. Gullstrand 1911–1929	Professor of Physiology and Physical Optics at Uppsala University
B. Hasselberg 1900–1922	Physicist at the Royal Academy of Sciences
H. H. Hildebrandsson 1900–1910	Professor of Meteorology at Uppsala University
E. Hulthén 1929–1962	Professor of Physics at the Stockholm Högskola
C. W. Oseen 1923–1944	Professor of Mechanics and Mathematical Physics at Uppsala University; subsequently, Director, Nobel Institute of Physics
H. Pleijel 1928–1947	Professor of Electro-technology at the Royal Institute of Technology
M. Siegbahn 1923–1962	Professor of Physics at Uppsala University
R. Thalén 1900–1903	Professor of Physics at Uppsala University

Nobel Committee for Chemistry

P. T. Cleve 1900–1905	Professor of Chemistry at Uppsala University
Å. G. Ekstrand 1913–1924	Engineer, civil servant
H. von Euler 1929–1946	Professor of Chemistry at the Stockholm Högskola
O. Hammarsten 1905–1926	Professor of Physiological Chemistry at Uppsala University

P. Klason 1900–1925	Professor of Chemistry and Chemical Technology at the Royal Institute of Technology
W. Palmaer 1926–1942	Professor of Chemistry at the Royal Institute of Technology
O. Pettersson 1900–1912	Professor of Chemistry at the Stockholm Högskola
L. Ramberg 1927–1940	Professor of Chemistry at Uppsala University
H. G. Söderbaum 1900–1933	Professor of Agricultural Chemistry at the Experimental Agricultural Station (Academy of Agriculture)
T. Svedberg 1925–1964	Professor of Physical Chemistry at Uppsala University
O. Widman 1900–1928	Professor of Chemistry at Uppsala University

Nobel Committee for Physiology or Medicine[1]

J. H. Åkerman 1909, 1911–1913, 1916–1921	Professor of Surgery
E. B. Almquist 1900–1905, 1907–1909	Professor of Public Health
J. V. Berg 1906	Professor of Surgery
J. A. Dalén 1911	Professor of Ophthalmology
J. G. Edgren 1905–1907, 1913	Professor of Medicine
B. E. Gadelius 1912, 1914–1916, 1925	Professor of Psychiatry
H. V. Gertz 1927–1932	Associate Professor of Physiology
G. P. E. Häggqvist 1924	Professor of Histology
E. Hammarsten 1929–1931	Professor of Chemistry and Pharmacology
G. Hedrén 1915, 1918, 1920–1931	Professor of Pathological Anatomy

[1] The Nobel Committee for Physiology or Medicine was composed of three members elected for a period of three years. At the end of each year's nominating period, the professorial staff at the Karolinska Institute appointed two additional members whose terms ran until the end of the year. As shown in the list above, many members served on the Committee in both capacities. With one exception (indicated on the list), the Committee members were all members of the staff of the Karolinska.

E. Holmgren 1902, 1903, 1905, 1906, 1910	Professor of Histology
I. F. Holmgren 1918–1921	Professor of Medicine
H. C. Jacobaeus 1925–1933	Professor of Medicine
J. E. Johansson 1904, 1907, 1910, 1912–1915, 1917–1926	Professor of Physiology
K. G. F. Lennmalm 1908–1910, 1917, 1918, 1920	P. H. Malmsten Professorship in Neuropathology
K. O. Medin 1902–1904	Professor of Pediatrics
K. A. H. Mörner 1900–1917	Professor of Chemistry and Pharmacy
E. Müller 1916, 1917, 1922	Professor of Anatomy
A. Pettersson 1908	Associate Professor of Bacteriology
J. A. Sjöqvist 1919–1922, 1924, 1925	Professor of Chemistry and Pharmacy
C. J. G. Sundberg 1902–1918	Professor of Pathological Anatomy
E. W. Welander 1911	Professor of Syphilology
F. J. E. Westermark 1914, 1917–1919	Professor, Director of the Stockholm Maternity Hospital

ORGANIZING COMMITTEE

President:	Professor Carl-Gustaf Bernhard
Vice-president:	Professor Sven Tägil
Secretary:	Dr Elisabeth Crawford
Members:	Professor Jan Lindsten
	Professor Bengt Nagel
	Dr Stig Ramel
	Dr Per Sörbom

WORKING GROUP FOR PROGRAMME OF THE SYMPOSIUM

President:	Professor Sten Lindroth
Secretary:	Dr Elisabeth Crawford
Members:	Professor Gunnar Eriksson
	Professor Tor Gerholm
	Professor Bo Holmstedt
	Professor Lamek Hulthén
	Professor Sven-Erik Liedman
	Professor Bo Malmström
	Dr Sigvard Strandh
	Dr Jakob Sverdrup
	Professor Sven Tägil
	Professor Börje Uvnäs

LIST OF PARTICIPANTS

Professor Sune Bergström,
Chairman of the Board of Directors,
Nobel Foundation,
Sturegatan 14,
Stockholm 114 36, Sweden

Professor Carl Gustaf Bernhard,
Royal Swedish Academy of Sciences,
Box 50005, 104 05 Stockholm, Sweden

Professor Gunnar Brandell,
Department for the History of Literature,
Uppsala University,
Box 513, 751 20 Uppsala, Sweden

Dr Tore Browaldh,
Svenska Handelsbanken,
103 28 Stockholm, Sweden

Professor L. J. Bruce-Chwatt,
Wellcome Museum of Medical Science,
P.O. Box 129, Wellcome Building,
183 Euston Road,
London NW1 2BP, UK

Dr Elisabeth Crawford,
Groupe d'Etudes et de Recherches sur la Science,
EHESS-CNRS,
10, rue Monsieur le Prince,
75006 Paris, France

Professor Gunnar Eriksson,
Department for the History of Science and Ideas,
Uppsala University,
Box 256, 751 05 Uppsala, Sweden

Dr Robert Marc Friedman,
Institute for Studies in Research and Higher Education,
Norwegian Research Council for Science and the Humanities,
Wergelandsveien 15,
Oslo 1, Norway

Professor Joseph S. Fruton,
350, Kline Biology Tower,
Yale University,
New Haven, Conn. 06520,
USA

Dr L. F. Haber,
Department of Economics,
University of Surrey,
Guildford, Surrey GU2 5XH, UK

Professor John Heilbron,
Office for History of Science and Technology,
470, Stephens Hall,
University of California,
Berkeley, Calif. 94720, USA

Professor Armin Hermann,
Historisches Institut,
Abteilung für Geschichte der Naturwissenschaften und Technik,
Universität Stuttgart,
Friedrichstrasse 10/IV,
Postfach 560, 7000 Stuttgart 1, FRG

Elisabeth Heseltine,
159, boulevard de la Croix Rousse,
69004 Lyon, France

Professor Erwin Hiebert,
Department of History of Science,
Harvard University,
Science Center 235,
Cambridge, Mass. 02138, USA

Professor Torsten Hägerstrand,
Institute for Cultural and Economic Geography,
University of Lund,
Sölvegatan 13,
223 62 Lund, Sweden

Dr A. G. Keller,
Department of History of Science,
University of Leicester,
University Road,
Leicester LE1 7RH, UK

Professor Melvin Kranzberg,
School of Social Sciences,
Georgia Institute of Technology,
Atlanta, Ga. 30332, USA

Dr Günther Küppers,
Forschungsschwerpunkt Wissenschaftsforschung,
Universität Bielefeld,
Postfach 8640,
4800 Bielefeld 1, FRG

Professor Maurice Lévy-Leboyer,
Département d'Histoire,
Université de Paris X Nanterre,
2, rue de Rouen,
92001 Nanterre, France

Professor C. Lichtenthaeler,
Institut für Geschichte der Medizin der Universität Hamburg,
Martinistrasse 52,
2 Hamburg 20, FRG

Civ. Ing. Svante Lindqvist,
Institute for History of Technology,
Royal Institute of Technology Library,
100 44 Stockholm, Sweden

Professor Jan Lindsten,
Secretary, Nobel Committee for Physiology or Medicine,
Karolinska Institutet,
104 01 Stockholm, Sweden

Professor Bengt Nagel,
Secretary, Nobel Committee for Physics,
Royal Academy of Sciences,
Sturegatan 14,
114 36 Stockholm, Sweden

Dr Wilhelm Odelberg,
Chief Librarian, Stockholm University Library,
Universitetsvägen 10,
106 91 Stockholm, Sweden

Dr Nils Oleinikoff,
Hässelviksvägen,
139 00 Värmdö, Sweden

Dr Eugeniusz Olszewski,
Al. Niepodleglosci 222-13a,
00 663 Warsaw 1, Poland

Dr Stig Ramel,
Director, Nobel Foundation,
Sturegatan 14,
114 36 Stockholm, Sweden

Professor Stanley J. Reiser,
Francis A. Countway Library of Medicine,
Harvard Medical School,
Boston, Mass. 02115, USA

Professor Nathan Rosenberg,
Department of Economics,
Stanford University,
Stanford, Calif. 94305, USA

Dr Claire Salomon-Bayet,
Institut d'Histoire des Sciences,
Paris I-Centre National de la Recherche Scientifique,
13, rue du Four,
75006 Paris, France

Professor Brigitte Schroeder-Gudehus,
Institut d'Histoire et de Sociopolitique des Sciences,
Université de Montréal,
Case Postale 6128, Succ. A. Montréal,
Québec H3C 3J7, Canada

Dr Per Sörbom,
Department for the History of Science and Ideas,
Uppsala University, Box 256,
751 05 Uppsala, Sweden

Dr Sigvard Strandh,
Research Scientist, National Museum of Technology,
115 27 Stockholm, Sweden

Dr Nils-Eric Svensson,
Executive Director, Bank of Sweden Tercentenary Foundation,
Box 1649, 111 86 Stockholm, Sweden

Professor Sven Tägil,
Department of History,
University of Lund,
223 62 Lund, Sweden

Professor Rolf Torstendahl,
Department of History,
Uppsala University,
S:t Larsgatan 2,
752 20 Uppsala, Sweden

Professor Börje Uvnäs,
Department of Pharmacology,
Karolinska Institutet,
Fack, 104 01 Stockholm 60, Sweden

Professor Peter Weingart,
Forschungsschwerpunkt Wissenschaftsforschung,
Universität Bielefeld,
Postfach 8640,
4800 Bielefeld 1, FRG

Dr Bernhard Witkop,
Laboratory of Chemistry,
National Institute of Arthritis, Diabetes, and Digestive and Kidney Diseases
National Institutes of Health,
Bethesda, Md. 20205, USA

Professor J. M. Ziman,
H. H. Wills Physics Laboratory,
University of Bristol,
Royal Fort, Tyndall Avenue,
Bristol BS8 1TL, UK

NOTES ON CONTRIBUTORS

Leonard Jan Bruce-Chwatt is Emeritus Professor of Tropical Hygiene, University of London and ex-Director of the Ross Institute. He spent twenty years in Nigeria as a member of the Colonial Medical Service and ten years as a staff member of the World Health Organization. He is a specialist in malaria research and a member of the WHO Expert Panel on Malaria. An author or co-author of four books, the most recent being *Essential Malariology* and *The Rise and Fall of Malaria in Europe*, he has also published many articles on tropical medicine and on the history of medicine.

Elisabeth Crawford is a Senior Research Fellow at the Centre National de la Recherche Scientifique in Paris. She has edited books and written articles on the history and sociology of science, including *Social Scientists and International Affairs* (1969) and *Demands for Social Knowledge* (1976). More recently, she has published articles on the history of scientific prizes – both the Nobel prizes as well as the prizes awarded by the French Academy of Sciences.

Robert Marc Friedman is a Research Fellow at the Norwegian Research Council for Science and the Humanities. He wrote his Johns Hopkins PhD dissertation (1978) on Vilhelm Bjerknes and the Bergen School of Meteorology, 1918–1923. His publications include 'The creation of a new science: Joseph Fourier's analytical theory of heat', *Historical Studies in the Physical Sciences* (1977) and 'Nobel physics prize in perspective' (*Nature*, 1981).

Joseph S. Fruton is Eugene Higgins Professor of Biochemistry at Yale University. His interest in the history of biochemistry is of long standing. His major work in this area is *Molecules and Life: Historical Essays on the Interplay of Chemistry and Biology* (1972).

Ludwig F. Haber is Reader in Economics at the University of Surrey. His field is economic history, particularly the history of the chemical industry; this is the subject of his two major works: *The Chemical Industry during the 19th Century: A Study of the Economical Aspects of Applied Chemistry in Europe and North America* (1958) and *The Chemical Industry 1900–1930: International Growth and Technological Change* (1971).

J. L. Heilbron is Professor of History at the University of California at Berkeley and Director of the Office for History of Science and Technology there. Among his writings are *H. G. J. Moseley: The Life and Letters of an English Physicist* (1974); *Electricity in the 17th and 18th Century: A Study of Early Modern Physics* (1979); *Historical Studies in the Theory of Atomic Structure* (1981); *Literature on the History of Physics in the 20th Century* (1981), with B. R. Wheaton; and *Elements of Early Modern Physics* (1982). He has edited *Historical Studies in the Physical Sciences* since 1979.

Erwin Hiebert is Chairman of the Department of History of Science at Harvard University and President of the Division of the History of Science of the International Union for the History and Philosophy of Science. He has a particular interest in the history of thermodynamics and physical chemistry. He is the author of *Impact of Atomic Energy* (1961); *Historical Roots of the Principle of Conservation of Energy* (1962, reprinted 1981); and *The Conception of Thermodynamics in the Scientific Thought of Mach and Planck* (1966). He has also written articles concerning Ernst Mach, Ludwig Boltzmann, Walther Nernst and Wilhelm Ostwald.

Melvin Kranzberg is Professor of the History of Technology at the Georgia Institute of Technology. He has edited and written many books and articles concerning the history of technology and its impact on society, including *Technology and Culture: An Anthology* (1972); *Technological Innovation* (1978); *Technology in Western Civilization* (1967); *By the Sweat of Thy Brow: Work in the Western World* (1975); *Energy and the Way We Live* (1980); and *Ethics in an Age of Pervasive Technology* (1980).

Günther Küppers is a physicist working in the Science Studies Unit at the University of Bielefeld. Among his works in the area of science policy and the social studies of science are: 'On the relation between technology and science – goals of knowledge and the dynamics of theories. The example of combustion technology, thermodynamics and fluid mechanics' (1978); 'Fusionforschung – zur Zielorienterung in Bereich der Grundlagenforschung' (1979); and, with P. Lundgren and P. Weingart, *Umweltforschung – Die Gesteverte Wissenschaft? Suhrkamp Taschenbuch Wissenschaft* (1978).

Maurice Lévy-Leboyer is a professor at the University of Paris X (Nanterre) and an economic historian. He is the author of a number of books and articles about French industrial enterprises and their managerial structure in the late nineteenth and early twentieth centuries. Among these are: 'Innovations and business strategies in 19th and 20th century France' (1976); *Le Patronat de la seconde industrialisation* (1979); 'Hierarchical structures: the early managerial experience of Saint-Gobain' (1979); 'The large corporation in modern France' (1980).

Charles Lichtenthaeler is Professor of Medical History at the Universities of Hamburg and Lausanne. He is the author of a comprehensive history of medicine, *Geschichte der Medizin,* (2 vols, 1974; 3rd edition, 1982). He has also published studies on Hippocrates and Thucydides; on historical and scientific methodology; and on ethics. He is presently working on a book on the Hippocratic Oath and the case histories of the Hippocratic *Epidemics III and I.*

Bengt Nagel is Professor of Theoretical Physics at the Royal Institute of Technology in Stockholm and Secretary of the Nobel Committee for Physics. He has published articles on nuclear physics, relativistic quantum theory, and general mathematical physics.

Stanley Reiser is Associate Professor of Medical History at the Harvard Medical School and Co-director of the Inter-faculty Program in Medical Ethics there. His book, *Medicine and the Reign of Technology*, appeared in 1978.

Nathan Rosenberg is Professor of Economic History at Stanford University. His major work is *Perspectives on Technology* (1976).

Claire Salomon-Bayet is a Senior Research Fellow at the Centre National de la Recherche Scientifique in Paris and a historian of science. She is the author of *L'Institution de la science et l'expérience du vivant* (1978) and has written a number of articles on the history and philosophy of science. She has recently directed a study of the influence of Pasteurism on French medicine.

Brigitte Schroeder-Gudehus is a professor at the University of Montreal and was until very recently the Director of its Institute for the History of Science and Science Policy. One of her research interests – international scientific relations – is also the subject of her book, *Les Scientifiques et la paix: La communauté scientifique internationale au cours des années 20* (1978). Recent articles are 'Science, technology and foreign policy' (1977) and 'The international scientific community and the advancement of science in Third World countries' (1980).

Rolf Torstendahl is Professor of History of Uppsala University. He has carried out a major study of the training and careers of Swedish engineers, 1860–1940, which resulted in his book, *Dispersion of Engineers in a Transitional Society* (1975), as well as in a number of articles. In another book, in Swedish, he has investigated the administrative and political attitudes in technological education in Sweden, 1810–1870.

Norbert Ulitzka is a sociologist. He is a research fellow in the Science Studies Unit at the University of Bielefeld.

Börje Uvnäs is Professor of Pharmacology at the Karolinska Institute in Stockholm and former chairman of the Nobel Committee for Physiology or Medicine.

Peter Weingart is Professor of Sociology at the University of Bielefeld, where he is also a member of the Science Studies Unit. He has written and edited books and many articles in the fields of sociology of science and science policy, including 'Die amerikanische Wissenschafts lobby' (1970); 'Wissenschaftsoziologie I and II' (1973/4), 'Wissensproduktion und soziale Struktur' (1976); 'Die geplante Forschung' (with W. van den Doele and W. Krohn; 1978); editor with E. Mendelsohn and R. Whitley, *The Social Production of Scientific Knowledge* (1977); editor with W. Krohn and E. Layton, *The Dynamics of Science and Technology* (1978).

Bernhard Witkop directs the Laboratory of Chemistry at the National Institute of Arthritis, Diabetes, and Digestive and Kidney Diseases (National Institutes of Health) in Bethesda, Md. Among his publications are: 'Probing ion channels with natural and synthetic heterocycles; 'The chemist's magic bullets: selectivity as a guiding principle in biomedical research'; 'Natural products, receptors and ligands. Natural products as medicinal agents'; 'Science and progress'; 'Heinrich Wieland: his lifework and his legacy'; 'The venoms of the South American frogs'; 'Progress in protein chemistry through selective chemical cleavage'; and 'Progress in neurochemistry through novel toxins'.

John Ziman is Professor of Physics at the University of Bristol. Among his books on the philosophy and sociology of science are: *Public Knowledge* (1968), *Reliable Knowledge: An Exploration of the Grounds for Belief in Science* (1978), and *Teaching and Learning about Science and Technology* (1980).

INDEX OF NAMES CITED*

*Active prior to 1930